A Dictionary of
ELECTRONICS

A Dictionary of
ELECTRONICS

Er. V. K. Jain
Er. Amitabh Bajaj

CBS PUBLISHERS & DISTRIBUTORS PVT. LTD.
New Delhi • Bengaluru • Chennai • Kochi • Kolkata • Mumbai • Pune

ISBN: 81-239-0359-6

First Edition: 1995
Reprint: 1999, 2001, 2002, 2003, 2006,
         2009, 2010, 2012, 2013, 2016

Copyright © Publisher

Published by:
Satish Kumar Jain for CBS Publishers & Distributors Pvt. Ltd.,
4819/XI Prahlad Street, 24 Ansari Road, Daryaganj, New Delhi - 110002
delhi@cbspd.com, cbspubs@airtelmail.in • www.cbspd.com
Ph.: 23289259, 23266861, 23266867 • Fax: 011-23243014

Corporate Office: 204 FIE, Industrial Area, Patparganj, Delhi - 110 092
Ph: 49344934 • Fax: 011-49344935
E-mail: publishing@cbspd.com • publicity@cbspd.com

Branches:
• *Bengaluru:* 2975, 17th Cross, K.R. Road, Bansankari 2nd Stage,
  Bengaluru - 70 • Ph: +91-80-26771678/79 • Fax: +91-80-26771680
  E-mail: cbsbng@gmail.com, bangalore@cbspd.com
• *Chennai:* No. 7, Subbaraya Street, Shenoy Nagar, Chennai - 600030
  Ph: +91-44-26681266, 26680620 • Fax: +91-44-42032115
  E-mail: chennai@cbspd.com
• *Kochi:* Ashana House, 39/1904, A.M. Thomas Road, Valanjambalam,
  Ernakulum, Kochi • Ph: +91-484-4059061-65
  Fax: +91-484-4059065 • E-mail: cochin@cbspd.com
• *Kolkata:* 6-B, Ground Floor, Rameshwar Shaw Road, Kolkata - 700014
  Ph: +91-33-22891126/7/8 • E-mail: kolkata@cbspd.com
• *Mumbai:* 83-C, Dr. E. Moses Road, Worli, Mumbai - 400018
  Ph: +91-9833017933, 022-24902340/41 • E-mail: mumbai@cbspd.com
• *Pune:* Bhuruk Prestige, Sr. No. 52/12/2+1+3/2,
  Narhe, Haveli (Near Katraj-Dehu Road Bypass), Pune - 411041
  Ph: +91-20-64704058/59, 32342277 • E-mail: pune@cbspd.com

Representatives:

• Hyderabad: 0-9885175004   • Nagpur: 0-9021734563
• Patna: 0-9334159340   • Vijayawada: 0-9000660880

Printed at:
Neekunj Print Process, Delhi

Preface

Electronics had been the buzzword of this country. Rapid progress made by humanity in developing low-current miniature equipment had braught thousands of new equipment and gadgets and millions of new terms. A dictionary of electronics complied in 1980's is no more valid as daily hundreds of new terms are getting added. This dictionary had been therefore compiled to include not all, but most modern terminology in electronics and related field.

Every effects have been made for including the latest terminology from all attributes of electronics and allied science like computer science, micro-electronics, biomics, robutic, artificial intelligence, semantics, logic, satellite and space engineering, avoimics etc.

The dictionary has been compiled keeping in view the needs of students in electronics. A considerable terminology have been desired by the authors from various periodrcals and manals of which individual for ever for all such derivations. The dictionary was compiled as a database on personal computer and authers are very much thankful to miss Jyotsna Bajaj for this purpose.

Er. V.K. Jain
Er. Amitabh Bajaj

1-2-3 : By Lotus Development Corporation Combined financial planning spreadsheet, graphics and rudimentary data management software package that was much talked-about in 1983. Hailed as a new VisiCalc, along with a similar package "Quatropro" it has spawned a whole flock of similar multi-function products that now represent a second generation of spread sheet developments (said products include Lotus's own Symphony package).

3101 : Designation of a simple dumb IBM computer terminal, a display terminal similar to widely-used Teletype in communications terms. The protocol 3101 can be readily imitated by a PC connected to mainframe if the machine's asynchronous communications adaptor is programmed with the right emulation software.

4GL : Clipped from Fourth Generation language represents a family of user's friendly object oriented non procedural programming languages that allows users to simply specify what the output should be without describing all the details of how the data should be manipulated to produce that result Mapper, Linc, Focus Ranils etc are some popular 4GLs.

4TH DIMENSION : 4th Dimension represents one of the best relational databases available for the Macintosh. If you need to use dBASE III Plus files on IBM, FoxBase +MAC is a better choice.

68000 : The microprocessor used in the Macintosh Plus, Macintosh SE, by Motorol. Microprocessor chip that's often cited as a true sixteen-bit processor chip in comparison to the IBM's Personal Computer's 80286.

8088 : Popular for the more powerful class of microcomputer intended for running multiuser systems and/or for the programmers.

68020 : The microprocessor used in the Macintosh II computer.

68030 : The microprocessor used in the Macintosh IIci/IIcx/IIc and the Macintosh SE/30.

68881 : The math-coprocessor that works with the 68000 and 68020 microprocessors.

80286 : The microprocessor used in IBM AT and compatible computers. Used in computers that include "286" in the name, such as the Compaq 286 or the AST Premium 286.

80287 : The math-coprocessor that works with the 80286 microprocessor

80386 : The microprocessor used in computers that include "386" in the name such as Compaq 386 or AST Premium 386. Faster than the 80286 microprocessor.

80387 : The math-coprocessor that works with the 80386 microprocessor, used for processing complex mathemetics at very fast rate.

80486 : The microprocessor used in computers that include "486" in the name, such as Compaq 486 or AST Premium 486. Faster than the 80386 microprocessor.

8080 : By Intel, the first widely-used eight-bit microprocessor chip.

8085 : By Intel, a hotted-up but still eight-bit development of the 8080. Used in the IBM System/23 small business computer.

8086 : By Intel, a relative of the 8080 processor chip, said to be a true 16-bit processor. Used in IBM's highly-successful word processing computer, Displaywriter.

8087 : By Intel, co-processor chip that can be used in the empty socket next to the Personal Computer's 8088 for extra mathmetic power. Needs special software to take advantage of it.

8-BIT CODE : See **ASCII**

A : An optimal algorithm that uses a cost function and an underestimating heuristic evaluation function to find optimal solutions to state space search problems by expanding the partial path that minimizes their sum; when the heuristic is always 0.

ABA : American Bankers Association.

ABACUS : A counting device consisting of beads arranged on threads in several rows, used for counting in primitive days, 3000 BC or even older.

ABAMPERE, ABCOULOMB, ABFARAD, ABOHM, ABVOLT, ETC. : Proposed names for the absolute C.G.S. electromagnetic units of current, quantity, capacitance, resistance, e.m.f., etc.

ABC TELEGRAPH (WHEATSTONE) : A step-by-step dial system of telegraphy, in which a separate key is used for each letter, and the needle of the receiving dial stops at the letter to be indicated, after having been fed forward by the correct number of current impulses.

ABDUCTION : A rule of logical inference that states if we know that A implies B, and if we find B to be true, we can say A is true. Although this is an unsound form of reasoning, it enables diagnostic reasoning with 'educated guesses' about causation from symptom.

ABEND : An acronym for **abnormal end** of task. An abend in the termination of computer processing on a job or task prior to its completion because of an error condition that cannot be resolved by programmed recovery procedures.

ABNORMALLY POLARISED WAVES : Electric waves (as used in radio-communication), in which the magnetic-component is not at right angles to the plane of propagation, e.g. the waves (present more particularly at night) reflected or refracted from the Lonosphere. Cf. NORMALLY POLARISED WAVES.

ABRUPT JUNCTION : A semiconductor junction in which the impurity concentration changes abruptly from acceptors to donors. In practice such a junction may be approximately realized when one side of the junction is much more highly doped than the other, i.e. a *p+-n* or *n+-p* junction. Also known as ONE-SIDED ABRUPT JUNCTION.

ABSOLUTE DETERMINATION : A determination of a physical quantity, made directly from the fundamental standards of *length, mass,* and *time,* without comparison with any previously calibrated standard of its own kind.

ABSOLUTE UNITS : Units derived from the fundamental units of *length, mass,* and *time.* In the case of electrical units, two such systems of units have been derived from the *centimetre,* the *gramme,* and the *second,* depending respectively upon electrostatic and electromagnetic relations. The practical units, the Ampere, the Volt, etc., are convenient multiples or submultiples of the latter. See C.G.S. UNITS.

ABSORBER (in Radio-Communication) : A device for absorbing the energy normally put into the aerial of a transmitter in a non-radiating circuit for testing purposes.

ABSORBER VALVE (in Radio-Communication) : A Thermionic Valve in a circuit controlled by the signalling key, which absorbs a part of the energy of the system during spacing periods.

ABSORPTION (of Electric Waves) : Causes of reduction in amplitude of electric waves during transmission involving dissipation of enrgy. Attenuation of a radiowave due to dissipation of its energy, as by the production of heat.

ABSORPTION COEFFICIENT : (1) Of light. See ABSORPTION FACTOR. (2) Of X-rays. The ratio of the rate of change (with depth) of the intensity of X-rays at a point in the medium, to the intensity at that point.

ABSORPTION COIL : An iron-cored inductor between the two parts of a six phase circuit feeding a mercury vapour rectifier, rendering it equivalent to two three-phase circuits in parallel.

ABSORPTION FACTOR : The ratio of the Luminous Flux (lumens) absorbed by a body to that incident upon it.

ABSORPTION LOSS : The magnitude of the absorption of a radiowave, usually expressed in *nepers* or *decibels.* See also UNABSORBED FIELD STRENGTH.

ABSORPTION MODULATOR : A Modulation apparatus for radio-transmitting stations in which the variable resistance of the microphone produces a variation of the radiation resistance of the aerial, either directly or through an intermediate valve circuit.

ABSORPTION WAVEMETER : An instrument for measuring wavelength of received waves by absorption in an oscillating circuit tuned to resonate with them. The face of resonance can be indicated in a variety of ways.

AC : The abbreviation for Alternating Current.

ACCELERATING ELECTRODE : Any electrode that accelerates electrons in the electron beam of an electron tube. See also ELECTRON GUN.

ACCELERATING GRID (in a Cathode Ray Oscillograph.) : A grid, charged at a moderate potential, placed in front of the screen to accelerate the electron stream and to increase its energy.

ACCELERATOR : An Anode in a Cathode Ray Tube, serving principally to accelerate the electrons forming the beam.

ACCELERATOR BOARD : An expansion board that plugs in a computer to make the computer work faster.

ACCELER METER : An instrument for measuring acceleration. A direct reading electrical accelerometer may be made by a voltmeter coupled through a transformer to a magneto generator, so that it will indicate the rate of change of the voltage produced and consequently of the speed at which the magneto is driven.

ACCEPTOR CIRCUIT : A circuit consisting of a combination of inductance and capacitance in series, permitting the passage of currents or oscillations of one frequency only. Used for avoiding interference in radio-reception. (Cf. REJECOTOR CIRCUIT).

ACCEPTOR LEVEL : See SEMICONDUCTOR.

ACCESS : The process of seeking, reading, or writing data on a storage unit.

ACCESS METHODS : Techniques and programs used to move data between main   memory and input/output devices.

ACCESS TIME : The amount of time, a hard disk drive needs to find data. Often measured in milliseconds, abbreviated as *ms*, such as "28*ms*"

ACCOUSTIC DELAY LINE : See delay line.

ACCUMULATOR : (1) In computers a (2) In electrical engineering single word register that accumulates the partial results of arithmetic or logical operations. An (electrical) accumulator, Storage Battery or Secondary Battery is a group of Secondary Cells, in which the energy of a current can be stored chemically during charging and given out again electrically during discharge. (1) See register. (2) Syn. for secondary cell. See cell.

ACCUMULATOR BOX : A vessel, usually of rectangular form, for containing the plates and electrolyte of an accumulator cell or several cells in separate compartments; made of glass, lead-lined wood, or celluloid, etc.

ACCUMULATOR CAR : A railway, tramway, or road vehicle propelled electrically and deriving its enrgy from an Accumulator Battery carried thereon.

ACCURACY : The degree of freedom from error. Accuracy is often confused with precision, which refers to the degree of preciseness of a measurement.

ACE : Automatic Computing Engine.

ACORD CIRCUIT : A subscriber's operator's Cord Circuit in a telephone exchange.

ACOUSTIC WAVE : Also known as SOUND WAVE. A wave that is transmitted through a *solid, liquid,* or *gaseous* material as a result of the mechanical vibrations of the particles forming the material. The normal mode of propagation is longitudinal, *i.e.* the direction of motion of the particles is parallel to the direction of propagation of the wave, and the wave therefore consists of compressions and rarefactions of the material.

ACOUSTIC WAVE DEVICE : A device used in a signal-processing system in which acoustic waves are transmitted on a miniature substrate in order to perform a wide range of function. Active and passive signal-processing devices formed on a single semiconductor chip have been produced including delay lines, attenuators, phase shifter, filters, amplifiers, oscillators, mixers, and limiters.

ACRONYM : A word formed by the initial letters of words or by initial letters plus parts of several words. Acronyms are widely used in computer technology. For example, COBOL is an acronym for Common Business Oriented Language.

ACTINO THERAPY : The application of ultra-violet rays to curative puposes.

ACTION : A consequent element in the right-hand side of production rule that is performed when the rule fires; it can be one or more facts that become true, procedures that are executed, or messages that are printed, for example.

ACTIVE : Denoting any device, component, or circuit that introduces gain or has a directional function. In practice any item except pure resistance, capacitance, inductance, or a combination of these three is active. Compare PASSIVE.

ACTIVE CURRENT : The component of an alternating current which is in phase with the voltage, and represents power in the circuit.

ACTIVE DIRECTORY : The directory that the computer performs commands on unless instructed otherwise.

ACTIVE DRIVE : The disk drive that the computer performs commands on unless instructed otherwise.

ACTIVE ELECTRODE (in Electrical Precipitation) : The electrode kept at a high potential. Also called the Discharge Electrode. Cf. PASSIVE ELECTRODE.

ACTIVE MATERIAL : The lead oxides or other substances in the plates of accumulator cells, the chemical changes of which effect the storage and recovery of the energy of the charge.

ACTIVE PARTITION : A section of the computer's memory that houses the operating system being used.

ACTIVE VOLTAGE : The component of an alternating voltage which is in phase with the current.

ACTIVE VOLT-AMPERES : The product of the active volts and the current or of the volts and the active amperes, i.e. the watts.

ACTIVE WINDOW : The displayed window that the computer performs comands on.

ACTOR : (1) An entity or individual performing a primitive action in conceptual dependency theory, or (2) a new object-oriented programming language, similar to Small Talk, and used for AI applications.

AD : Analog to Digital.

ADA : Programming language designed by the Department of Defence. Similar to Pascal. See also Modula-2 and Pascal.

ADC : Analog to Digital Converter the, basic builiding block of digital voltmeters.

ADCOCK DIRECTION FINDER : A type of direction finder employing spaced aerials, which may be of various forms to minimise polarisation error.

ADDER : A device that adds two binary numbers, giving a *sum* and a *carry*.

ADDRESS : A *name, label,* or *number* identifying a register, location or unit where information is stored. The location of a character or word in computer memeory. Also the location of a track or record on a random access device.

ADJACENCY : A square of a Karnaugh map that may be combined with another square.

ADMISSIBILITY : A constraint on the heuristic evaluation function used in the "A" state-space search algorithm, whose satisfaction ensures that the procedure will always terminate with the optimal solution to the problem, it requires that all estimates of remaining distance to the goal state be under-estimates of the true distance.

ADMITTANCE : The reciprocal of IMPEDANCE. Symbol: Y.

ADP : Automatic Data Processing.

AEOLIGHT : A cold-catode glow-discharge lamp (see gas-discharge tube) that is filled with a mixture of permanent gases. The intensity of illumination produced varies with the applied signal and it is used as a modulating light for sound recording.

AERIAL : Another name for antenna (mainly U.S.). The part of a radio system that radiates energy into space (transmitting aerial) or receives energy from space (receiving aerial). An aerial together with its feeders and all its supports is known as an aerial system.The system of conductors, supported on high masts or otherwise, used for radiating and receiving energy in the form of electric waves in radio transmitting and receiving stations. Usually connected to earth through the trnsmitter or receiver.

AERIAL ARRAY : An arrangement of radiating or receiving elements so spaced and connected to produce directional effects. Very great directivity and consequently large aerial gain can be produced by suitable design.

AERIAL CABLE : Cable, usually uninsulated or lightly insulated, for aerial lines.

AERIAL CIRCUIT : All the apparatus and connections connected between the aerial and earth, including the aerial itself.

AERIAL CONDUCTOR : Any wire or cable supported clear of the ground for the transmission of current.

AERIAL CURRENT : The R.M.S. value of the current in an aerial, at an antinode. The root-mean square value of the current measured at a specified point in an aerial, usually either at the feed point or at the current maximum.

AERIAL DISCHARGE : A path to earth for high-tension surges in a radio aerial which does not interfere with the signals; sometimes a gass-filled tube.

AERIAL EARTHING SWITCH : A switch provided for connecting the aerial directly to earth and disconnecting it from the receiving set when not in use.

AERIAL EFFECT : Non-directional disturbance in a directional receiving apparatus due to unsymmetrical disposition of stray capacitance.

AERIAL EFFICIENCY : Syn RADIATION EFFICIENCY The ratio of the power radiated by an aerial, at a specified frequency, to the total power supplied to it.

AERIAL FEED IMPEDANCE : The effective impedance of an Aerial as measured at a point to a point to which a feeder is connected.

AERIAL FEED-POINT IMPEDANCE : The impedance of an aerial at the point at which it is fed. The real part of this impedance is the aerial feed-point at which it is fed. The real part of this impedance is the aerial feed-point resistance and the imaginary part is the aerial feed-point reactance.

AERIAL GAIN : (1) (in transmission) The ratio of the power that must be supplied to a reference aerial compared to the power supplied to the aerial under consideration in order that they produce exactly similar field strengths at the same distance and in the same specified direction (2) (in reception) The ratio of the signal power produced at the receiver input by the given aerial to that produced by a reference aerial under similar receiving conditions and trnasmitted power.

AERIAL INSULATOR : An insulator, usually of porcelain, for insulating an aerial from its supports.

AERIAL RADIATION RESISTANCE : The component of the Aerial Resistance due to the radiation of energy. Symbol: R.

AERIAL RESISTANCE : The total power supplied to an aerial divided by the mean square value of the current at a given specified reference point on the aerial, usually the *feed point* or a *current antinode*. A term used to include all effects having a result similar as regards dissipation of energy to true resistance in an aerial in radio-communication. Including the effect of radiation of energy. Symbol: Ra.

AERIAL SWITCH : A switch for changing over the connection of an aerial from the arrangement required for transmitting to that required for receiving, and vice versa.

AERIAL SYSTEM : An aerial and the whole of its supporting structure.

AERIAL TUNING CONDENSER AND AERIAL TUNING INDUCTANCE : A variable condenser or inductance for tuning an aerial circuit to respond to a certain wave length.

AFIPS : American Federation of Information Processing Societies.

AFTERGLOW (in a Gas Discharge Tube) : The persistence of luminosity in the tube or on the screen of a Cathode Ray Tube after the voltage has been cut off or the electron beam has moved away.

AGEING : A change in the properties of a substance which can be brought about mainly by lapse of time, but can sometimes be accelerated by other means. (1) Thus the so called ageing effect in transformer iron by which the hysteresis loss increase after a long period. (2) The non-reversible, generally gradual change of resonance frequency with time in crystal oscillators. (3) In case of magnets the ageing can be effected in a short time by heating to a suitable temperature; also the processes by which the Sub-permanent magnetism can be quickly got rid of in the manufacture of permanent magnets for instruments is sometimes called an "ageing" process.

AGGLOMERATE LECLANCHE CELL : A low resistance form of Leclanche Cell in which the depolariser is made up into blocks and surrounds the carbon electrode without a porous pot, and the negative electrode is in the form of an external zinc cylinder.

AGITATION OF THE ELECTROLYTE : (in Electrode position). A more or less rapid movement of the electrolyte towards the cathode to increase the permissible rate of deposit. Cf. CIRCULATION.

ALDREY : An alloy of aluminium containing magnesium and silicon of high tensile strength, suitable for transmission lines.

ALEXANDERSON ALTERNATOR : A high frequency alternator for radio communication, of the highspeed inductor type, which can give frequencies up to 1,000,000 cycles per second.

ALEXANDERSON SYSTEM OF PHOTOTELEGRAPHY : A system

of phototelegraphy characterised by the production of four grades of strength of the dots of which the reproduced piture consists, attained by superposing the effect a varying number of times during the four revolutions of the drum.

ALGEBRAIC LANGUAGE : A language whose statements are structured to resemble the structure of algebraic expression. FORTRAN is a good example of an algebraic language.

ALGOL : Algorithmic Language.

ALGORITHM : A set of well-defined rules or a procedures to be followed in order to obtain the solution of a problem in a finite number of steps. An algorithm can involve arithmetic, algebraic, logical and other types of procedures and instructions. An algorithm can be simple or complex. However, all algorithms must produce a solution within a finite number of steps. Algorithms are fundamental when using a computer to solve problems, because the computer must be supplied with a specific set of instructions that yields a solution in a reasonable length of time.

ALIGNED-GRID VALVE : A type of thermionic valve in which the power-handling capacity of the valve is increased by arranging for only a very small fraction of the total space current to be intercepted by the grid.

ALL·ELECTRIC SIGNALLING : Automatic or other railway signalling in which the signals are actuated as well as controlled electrically. Cf. ELECTRO-PNEUMATIC SIGNALLING.

ALL MAINS RECEIVER : A radio-receiving set in which all the necessary supplies of voltage and current are drawn from the mains, inclduing high and low tension, and grid bias; employing somoothing circuits and, in the case of A.C. supply, some form of Rectifier.

ALLOYED JUNCTION : A semiconductor junction formed by bonding metal contacts on to a wafer of semiconductor material and then heating to produce an alloy. It is a method commonly used for germanium diodes and transistors and in early silicon devices, the devices being termed alloyed (or alloy) transistors or diodes.

ALMALEC : An aluminium alloy of high tensile strength suitable for transmission lines.

ALPHABETIC : Data representation by alphabetical characters in contrast to numerical; the letters of the alphabet.

ALPHA CURRENT FACTOR : Syn. for common-base forward-current transfer ratio. See TRANSISTOR.

ALPHA CUTOFF FREQUENCY : The frequency at which the common-base forward-current transfer ratio, (α), of a bipolar junction transistor has fallen to $1/\sqrt{2}$ (i.e. 0.707) of the low-frequency value.

ALPHA-NUMERIC : A contraction of the words **alphabetic** and **numeric**; a set of characters including letters, numerals, and special symbols.

ALPHA PARTICLES : Positively charged particles, consisting each of four protons and two electrons. Cf. BETA (β) PARTICLES and see ALPHA (α) RAYS.

ALPHA RAYS : The least penetrating of the three principal kinds of rays given off by radioactive bodies, consisting of projected positively charged particles travelling at 10 to 20 thousand kilometres per second. They are slightly deflected by a magnetic field and are powerful ionisers. The bombardment of certain atoms by these rays can produce their break up and cause transmutation. Cf. BETA and GAMMA rays.

ALTERNATING : Electric and magnetic quantities, such as current, voltage, flux, etc., are said to be "alternating" (as opposed to "direct") when their magnitude successively increase to a maximum in one direction, decreases to zero, increases in the opposite direction to maximum of equal value and decrease to zero again, following the same cycle over and over again in a regular manner.

ALU : Arithmetic and Logic Unit.

ALUDUR : An aluminium alloy, containing magnesium, of high tensile strength, suitable for transmission lines.

ALUMINIUM : This metal is frequently used instead of copper for overhead lines, switch-board connections and other purposes, on account of its weight for equal conductivity being about half that of copper. It cannot however be jointed quite so easily, and is liable to corrosion near the sea, owing to the constant presence of a thin film of oxide on its surface, which has an appreciable insulating property.

ALUMINIUM GATE CIRCUIT : See MOS INTEGRATED CIRCUIT.

ALUMINIZED SCREEN : See CATHODE RAY TUBE.

AM : Amplitude Modulation.

AMALGAMATION : In preparing zinc electrodes for primary cells, they are treated with mercury, which dissolves or "amalgamates" with the surface of the zinc, causing iron and other impurities to float off so that Local Action is avoided.

AMERICAN WIRE GAUGE : See BROWN AND SHARP WIRE GAUGE.

AM/FM RECEIVER : A radio receiver that detects both amplitude-modulated and frequency modulated signals.

AMMETER : Shortened and more usual form of the word "Amperemeter." An instrument for measuring current, with a scale graduated to read directly in amperes. See ELECTROMAGNETIC MOVING COIL, INDUCTION, and HOT-WIRE INSTRUMENTS.

AMMETER SHUNT : See SHUNT.

AMP : Abbreviation for Ampere. The practical unit of Current (named after the French physicist Andre Marie Ampere, 1775-1836). The true ampere is one tenth of the C.G.S. electromagnetic unit of current. The ampere can also be defined as the current produced by one volt applied to a conductor, the resistance of which is one Ohm. See also INTERNATIONAL AMPERE.

AMPERAGE : An expression occasionally used for the current which a circuit is carrying or is designed to carry, expressed in amperes. Cf. VOLTAGE.

AMPERE'S PRINCIPLE : The mutual attraction and repulsion of parallel wires carrying currents in the same or opposite directions respectively.

AMPERE'S RULE : The following rule to rememeber the direction in which a magnetic needle is deflected by current if a man is swimming with the current and is facing the needle, the North seeking pole will be deflected towards his left hand. See also "SNOW RULE."

AMPERE'S THEORY OF MAGNETISATION : The assumption that magnetic property is due to circulatory current within the molecules of the magnet.

AMPERE-TURNS : The product of the number of turns (convolutions) composing a coil and the number of amperes flowing through it. It forms a convenient practical unit of Magnetomotive Force.

AMPERE-WIRES : An expression sometimes used in dynamo design for the product of the number of conductors, per slot, per pole, etc., and the current in amperes carried by each.

AMPLIFICATION CONSTANT (or Amplification Factor) : (1) The ratio of a small change in anode current in a thermionic valve to the change in grid voltage which produces it. (2) Of an amplifier in general, the ratio between the changes of current, voltage or power on the input and output sides.

AMPLIFICATION OF MODULATED HIGH FREQUENCY : A method of radio-transmission, particularly for large broadcasting stations, in which Choke Control is not employed directly upon the main oscillating circuit but upon a small scale. The output is then amplified by several high frequency stages until the requisite power for the aerial is reached.

AMPLIFIER : An apparatus capable of delivering a variable current derived from an independent source of energy, in which the scale of the variations is much greater than that of the variations in the e.m.f. supplied to it; and thus capable of being used to reinforce or "amplify" the effect of weak electrical oscillations; employed for this purpose extensively in radio-communication and line telephony. Sometimes called a Magnifier.

The nomenclature of amplifiers depends on their application, construction, and method of operation according to context. The simple amplifier shown in Fig. *(a)* can be described as a common-emitter, or linear, or class A, or audiofrequency amplifier. An amplifier that produces an increased e.m.f. operating into a high impedance is a *voltage amplifier*. A transistor in common-base connection (Fig. b) acts as a simple voltage amplifier operating with low input impedance and high output impedance. One that produces an appreciable current flow

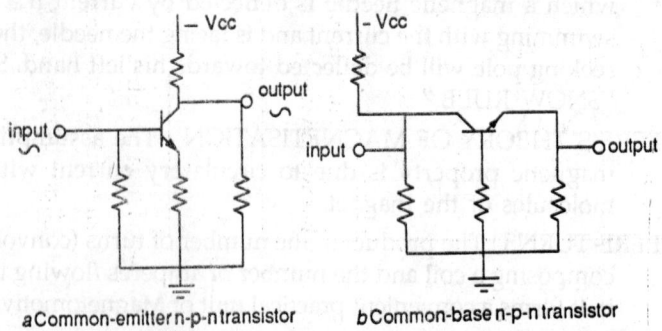

a Common-emitter n-p-n transistor   b Common-base n-p-n transistor
             amplifier

into a relatively low impedance or a large increase in output power is a *power amplifier*.

Alternating-current amplifiers are described either by the range of frequencies amplified, i.e. they are either wideband or tuned amplifiers, or by the region of the electromagnetic spectrum in which they operate, as with audiofrequency or radio frequency amplifiers.

AMPLITUDE : An expression usually employed for the peak value attained in either direction by a quantity executing periodic oscillations, such as an alternating current, and sometimes for the sum of the positive and negative values, especially when equal. The latter is preferably called Double Amplitude.

AMPLITUDE DISTORTION : Alternation of the quality of reproduced sound in radio-teleponydue to reproduction of different amplitudes at the same Frequency in different ratios. Some writers use the term to include Frequency Distortion and to distinguish from Phase Distortion.

AMPLITUDE FADING : A phenomenon that occurs when all transmitted frequencies are attenuated approximately equally, resulting in a smaller received signal. Selective fadding occurs when some frequencies are more attenuated than others, resulting in a distorted received signal. See FADDING also.

AMPLITUDE MODULATION : Modulation by variation of the amplitude of the oscillations. Cf. FREQUENCY MODULATION.

ANALOG : Information expressed in measurable quantities, such as resistance, current, or voltage.

ANALOG TO DIGITAL CONVERTER : A device that converts an analog input to a binary equivalent number. ANALOGY is a method of learing and reasoning, in which a new problem is solved by matching its structure with similar, previously solved problems.

ANAPHORESIS : Electrophoresis in which the particles in question pass towards the anode. Cf. CATAPHORESIS.

AND CONNECTIVE : A condition requiring all inputs to be true in order for the outout to be true.

ANDERSON'S BRIDGE : A Bridge method of measuring inductance in which the inductance to be measured is included in one bridge arm with a condenser between the opposite battery terminal and a point between the galvanometer and a resistance in series with it.

AND GATE : A binary circuit having two or more inputs and a single output in which the output is on (1) only if all inputs are on (1) together, and is off (0) if any one of the inputs is off (0).

ANGLE OF LAG : The angle by which the current in an inductive A.C. circuit lags behind the voltage producing it. See PHASE DIFFERENCE, POWER FACTOR, etc.

ANGSTROM UNIT : A unit of length sometimes used for wave lengths, etc., equal to 10^{-8} centimetre. Abbreviation: A°

ANIONS : The dissociated parts of molecule, or Ions carrying negative charges which during electrolysis move towards the Anode. Cf. CATIONS.

ANISOTROPIC CONDUCTIVITY : Conducttivity which is different in different directions in a body. Cf. UNILATERAL CONDUCTIVITY.

ANODE : The pole or electrode through which, according to the old convention, the current enters an electrolytic cell, vacuum tube or other piece of apparatus, i.e. the Positive Pole, where the electrons leave the medium. In the case of the primary cell, the electrode through which the current enters the electrolyte, i.e. the zinc or other negative pole is sometimes called the "Anode". Derived from Greek Words meaning literally "the way up." See also **Plate**. cf. CATHODE

ANODE A.C. CONDUCTANCE : The receiprocal of Anode A.C. Resistance.

ANODE A.C. RESISTANCE (of a Thermionic valve) : The ratio of a small change in anode voltage to that of anode current, all other quantities being constant.

ANODE BEND RECTIFICATION : Anode rectification by a thermionic valve in which the grid potential is so adjusted that the valve works in the region of the lower bend of the anode-current characteristic.

ANODE CIRCUIT : The circuit connected to the Anode of a thermionic valve.

ANODE CIRCUIT, TUNED : See TUNED ANODE CIRCUIT.

ANODE CONDUCTANCE : The receiprocal of Anode Resistance.

ANODE CONVERTER : A small Converter for obtaining a high direct-current voltage for the anode circuit of a thermionic valve.

ANODE COUPLING,TUNED : See TUNED ANODE COUPLING.

ANODE CURRENT : The current flowing from the anode to the

cathode of a device such as a solid-state rectifier.The total ionic and/or electron current flowing between the anode and the other terminals in a thermionic valve. See also THERMIONIC CURRENT.

ANODE CURRENT CHARACTERISTIC (of a Thermionic Valve) : A curve connecting simultaneous steady values of (1) anode-current and grid. Voltage and (2) anode-current and anode-voltage, all other variables remaining constant. Cf. GRID CURRENT CHARACTERISTIC.

ANODE CURRENT SURFACE (of a Thermionic Valve) : A surface connecting coordinates representing simultaneous values of anode current, anode-voltage and grid voltage. Cf. GRID CURRENT SURFACE.

ANODE DISSIPATION : The ratio of the steady anode voltage and anode current, all other quantities being constant.

ANODE EFFECT (in Electrolysis) : A sudden drop of the current due to the formation of a film of gas on the anode.

ANODE FEED : See FEED CURRENT.

ANODE GLOW : A slight glow seen on the anode of a vacuum tube at high degrees of exhaustion but disappearing when the gas pressure is below 1/10,000 of an atmosphere.

ANODE IMPEDANCE : A term for what is better called Anode A.C. resistance.

ANODE MODULATOR : A modulator in a radio-transmitting apparatus which acts by applying variations of the microphone current to the anode circuit of the oscillating valve through a modulating valve circuit. See CHOKE MODULATOR.

ANODE PICKLING : Removal of surface impurities from metal objects by making them act as the anodes in an electrolytic cell.

ANODE RAYS : See POITIVE RAYS.

ANODE RECTIFICATION : Rectification of a thermionic valve in which the oscillations to be rectified are applied to the Anode Circuit. Cf. GRID RECTIFICATION.

ANODE RESISTANCE (of a Thermionic Valve) : See ANODE A.C. RESISTANCE AND ANODE D. C. RESISTANCE.

ANODE RESISTOR : See RESISTANCE-CAPACITANCE COUPLING.

ANODE SUPPLEMENTARY AND ZERO POINT : See SUPPLEMENTARY ANODE and ZERO-POINT ANODE.

ANODE VOLTAGE (in a Thermionic Valve): The voltage between the anode (plate) and the negative terminal of the filament or the centre of an independently heated cathode.

ANODIC OXIDATION : The electrolytic production of a film of oxide on a metal.

ANOTRON : A vacuum tube rectifier of the cold cathode glow discharge type with a large cathode of sodium or other material and a copper anode.

ANSI : American National Standards Institute.

ANTENNA : See AERIAL.

ANTI-CATHODE : The "target" in an "X" Ray Tube, where the rays are produced by the bombardment of the Cathode Stream.

ANTI-FADING AERIAL : An aerial designed to emit radiation chiefly at low angles to avoid interference with the ground waves of down coming waves reflected from the ionosphere.

ANTI-FADING DEVICE. (in RAdio Receivers) : An arrangement to compensate temporary loss of signal strength due to fading, e.g. Automatic Volume Control.

ANTI-MICROPHONIC VALVE HOLDER : A holder for thermionic valves, supported on springs, etc., to minimise the communication of mechanical vibration to the filament and thus to reduce liability to microphonic noise.

ANTI-NODES : Maximum points in stationary waves or oscillations in a conductor, also called Loops and Internodes.

ANTI-SIDE-TONE CIRCUIT : See SIDE TONE REDUCTION WIRING.

ANTI-SPRAY FILM : A floating film of a special kind of oil on the surface of the electrolyte in accumulator cells to prevent the formation of acid spray by the bursting of bubbles.

ANTI-STATIC AERIAL : An aerial arranged to diminish the effect of load interference and other parasitic waves by special screening and other precautions.

ANTI-SULPHATING ACCUMULATOR : A lead accumulator in which the positive and negative plates are separated by a membrane permeable to ions but not to molecules, thus preventing access of oxygen to the negative plates.

APEC : Automated Procedures for Engineering Consultants.

APERIODIC AERIAL : A term sometimes applied to an aerial circuit with no tuning arrangements therein, inductively coupled to a tuned circuit.

APERIODIC CIRCUIT : A circuit which can have no natural period of oscillation, i.e. where $4L$ is less than R^2C (where L is the inductance, R the resistance and C the capacitance in farads). Cf. OSCILLATING CIRCUIT.

APERIODIC INSTRUMENT : An indicating instrument such as a galvanometer is said to be "aperiodic," when the moving system has no natural period of swing and therefore cannot take up oscillations before coming to rest in a new position. Cf. DEAD BEAT.

APERTURE DISTORTION : A distortion in an image occurs in a scanning system when the scanning spot has finite dimensions rather than infinitely small dimensions.

APL : A Programming Language.

APPARENT EFFICIENCY : The ratio of the output of a piece of alternating current apparatus in true watts to the input in volt amperes.

APPARENT IMPEDANCE : The ratio of voltage to current under such special conditions as those of short circuit of an alternator.

APPARENT INDUCTANCE : The value of the inductance which a coil in an oscillating circuit would have to produce the same oscillation frequency as an actual coil, if it has no Self-Capacitance.

APPARENT POWER : The product of voltage and current irrespective of the power-factor, in an alternating current circuit, expressed in volt-amperes.

APPLETON LAYER : Syn. for F LAYER. See IONOSPHERE.

APPLICATION : The system or problem to which a computer is applied. Reference is often made to an application as being either of the *computational* type, in which arithmetic computation predominate, or of the *data processing* type, in which data handling operations predominate.

APPLICATION PACKAGE : A program or series of programs intended for use by more than one group of users.

APPLICATION PROGRAM : A set of instruction that embody the logic of an application. It should be distinguished from a supervisory program, which controls the operations of the computer.

APPLICATION SOFTWARE : A redundant term used by people to sound impressive when talking about computers, much like telling someone that you own a "car automobile".

APT : Automically Programmed Tools.

ARBITRARY UNITS : Units not dependent upon any relationship with absolute units, such as the (obsolete) Siemens or Jacobi Units of resistance.

ARC : A pointer connécting two nodes is a graph, such as an *operator* in statespace search or a *move* in game playing.

ARC CONVERTER : A device by means of which alternating current is generated from an electric arc.

ARCHITECTURE : The basic underlying structure of a processing system (intelligent or otherwise), representing the level of primitive operations that cannot be changed from the building blocks for all other operations executed by the system; in most computers, the hardware organization is the architectural level.

ARC RECTIFIER : (1) A rectifier in which an arc struck between suitable electrodes is so controlled as to be during the current wave in one direction only, e.g. the Marx rectifier. (2) A style of rectifier similar in principle to the thermionic rectifier except that the cathode is heated by the current itself and not by an external source.

ARC RELAY : A relay depending on the variations in behaviour of a mercury vapour arc according to the potential of an additional electrode.

ARC SYSTEM OF RADIO-COMMUNICATION : A system in which continuous Waves are produced by connecting an Oscillating circuit to an arc of special character so that advantage is taken of certain relations between the current through the arc and the voltage between its electrodes to produce an instability resulting in the continuance of oscillations. Arc systems on the same general principles but differing in detail, are known under the names of Poulsen, Elwell, Telefunken, Federal (U.S. Navy), Colin Jeance, Moretti, Ruhmer and Dubilier.

ARC TRANSMITTER : A radio-transmitting apparatus employs an arc as the source of the oscillations (see above). Cf. SPARK TRANSMITTER and VALVE TRANSMITTER.

ARGON Symbol : Ar. An inert gas, atomic number 18, that is extensively used as the gas in gas-filled tubes.

ARGUMENT : (1) A type of variable whose value is not a direct function of another variable. It can represent the location of a number in a mathematical operation, or the number with which a function works to produce its results. (2) A known reference factor that is required to find a desired item (function) in a table. For example, in the square root function SQR (X), X is the argument. The value of X determines the square root value returned by this function.

ARITHMETIC AND LOGIC UNIT : A logic device capable of performing both arithmetic operations (add, subtract, etc.) and logical operations (AND, OR, etc).

ARITHMETIC REGISTERS : Registers that actually perform arithmetic operations on data, in Central Processing Unit of a digital computer.

ARMOURED CABLE : Cable, generally intended to be laid directly in the ground, covered with ARMOURING.

ARMOURING (of a cable) : An external covering, usually made of steel wire or bands, to give mechanical protection.

ARRAY : (1) An organized collection of data in which the argument is positioned before the function. (2) A group of items or elements in which the position of each item or element is significant. A multiplication table is a good example of an array.

ARTIFICIAL AERIAL : U.S. syn. DUMMY ANTENNA : A device that simulates all the electrical characteristics of an actual aerial except that the energy supplied to it is not radiated. (It is usually dissipated as heat in a resistor). It allows adjustments to be made to a transmitter or receiver before connecting up to the actual aerial.

ARTIFICIAL INTELLIGENCE (AI) : A science that studies primarily techniques for making computers exhibit intelligent behaviour, ways of illuminating human congnitive processes, and the construction of an autonomous, generally intelligent, artificial system.

ARTIFICIAL RADIO-ACTIVITY : The production, without the use of radium, of similar radiations to those produced by its disintegration, by means of such apprratus as the Cyclotron or by bombarding Beryllium or certain other bodies by alpha particles.

ASCII : Acronym for American Standard Code for Information Interchange is a standardized 8-bit code used by most computers for interfacing was developed by the American National

Standards Institute (ANSI) It uses 7 binary bits for information and the 8th bit for parity purposes.

ASCII FILE : A file containing characters that can be used by any program on any computer. Sometimes called a *text* file or an ASCII text file.

ASSEMBLER : A computer program that produces a machine-language program which may then be directly executed by the computer.

ASSEMBLY LANGUAGE : A symbolic language that is *machine*-oriented rather than *problem*-oriented. A program in an assembly lanugae is converted by an assembler to a machine-language program. Symbol representing storage locations are converted to numerical storage locations; symbolic operation codes are converted to numeric operation codes. A language that closely resembles machine language, although mnemonics are substituted for numeric codes in instructions and addreses. Generally, one machine-language statement is produced for each assembly-language statement during the translation process

ASSOCIATIVE MEMORY : See CONTENT-ADDRESSABLE MEMORY

ASYMMETRICAL VOLTAGE : A polyphase voltage system in which the instantaneous sum of the voltages in the various phases is not always zero.

ASYNCHRONOUS : (1) Not having a regular time or clocked relationship. See SYNCHRONOUS (2) A type of computer operation in which a new instruction is initiated when the former instruction is completed. Thus, there is no regular time schedule, or clock, with respect to instruction sequence. The current instruction must be complete before the next is begun, regardless of the length of time the current instruction takes.

ASYNCHRONOUS CIRCUIT : A sequential circuit in which the next event occurs as soon as the present event is completed, without waiting for a clock pulse

ASYNCHRONOUS COMMUNICATION : A way of transmitting data serially from one device to another, in which each transmitted character is preceded by a *start* bit and followed by a *stop* bit. This is also called start/stop transmission.

ASYNCHRONOUS OPERATION : Any operation that occurs out of phase with other operations. For example, in certain CPUs, an instruction look-ahead feature, which fetches instructions before they are needed, operates asynchronously with regular instruction processing.

ATDM : Asynchronous Time Division Multiplexing.

ATLANTIC CABLE : A sumbmarine telegraph cable laid in the bed of the Atlantic Ocean. Although the first Atlantic Cable was laid in 1858, permanent telegraphic communication was not established between Britain and America until 1866. There are now 15 cables across the Atlantic.

ATOM : The unit of matter from which molecules are made up; consisting, acccording to the Rutherford-Bohr theroy, of a number of Electrons or units of negative electricity revolving in elliptical orbits round a nucleus composed of Protons and Neutrons. Other Theories conceive of the outer system as an electronic cloud.

ATOMIC HYDROGEN WELDING : Welding by an intensely hot flame, produced by the recombination of hydrogen atoms which have been dissociated by passing through an electric arc.

ATOMIC NUMBER : The number of electrons circulating in orbits around the nucleus of an atom (not including any which may form part of the nucleus). Cf. MASS NUMBER.

ATTENUATION : The diminution of the amplitude of electric waves at progressive distances from the source including the case of waves propagated along telephone lines or cables of considerable capacitance, when it results in diminished audibility. (Cf. DISTORTION.)

ATTENUATION COMPENSATOR OR EQUALISER : A combination of *resistance, inductance* and *capacitance* to compenstate for the variation of attenuation with frequency in aline.

ATTENUATION CONSTANT : The natural logarithm of the ratio of the amplitudes of waves at unit distance apart along the line of propagation. The attenuation constant of a telephone line is measured by $/ \{^1/_2 \ [\ / \ R^2 + p^2L^2 \ )(S^2 + p^2C^2) + (SR - p^2 \ LC)]\}$ where R = resistance per mile, P=2πn (n=frequency), and S=dielectric conductivity (in mhos per mile).The attenuation constant can also be defined as the reciprocal of the length of the line in which the amplitude is reduced in the ratio of 2.71828 to one.

ATTENUATION DISTORTION : A distortion occurs when the gain or loss of the system depends on frequency.

ATTENUATION FACTOR : The ratio of the received amplitude to the initial amplitude, measured by ε^{ax} where a= the attenuation constant of the circuit and x=the distance.

ATTENUATION LENGTH : The product of the attenuation constant and the length of line in miles. Attenuation due to a line is expressed in Nepres or Bels or in terms of miles of Standard Cable to which the line is equivalent.

ATTENUATOR : A resistor introduced into an oscillating circuit to reduce the amplitude of the oscillations therein.

AU : Arithmtic unit; an ALU of computer incapable of performing the logic functions.

AUDIBILITY : The degree to which a particular signal is audible in a telephone receiver, in radio-communication, measured in practice by ascertaining the resistance of a shunt, S, which must be put across a telephone receiver of resistance R to render the sound only just audible. The audibility then equals $(R+S)/S$.

AUDIBILITY CURRENT : The minimum Audio-current which can give a distinct signal at a particular frequency in a given telephone receiver.

AUDIO-FREQUENCY : A frequency within the range at which telephone currents can produce audible sounds in a receiver, i.e. beetween about 25 and10,000 cycles per second. Cf. VOICE FREQUENCY.

AUDIO-FREQUENCY AMPLIFICATION : The use of one or more amplifying stages in the audio-frequency circuit of a radio-receiver or radio gramophone, sometimes combined in the latter case with Negative Automatic Volume Control.

AUDIO-FREQUENCY AMPLIFIER : An amplifier used in an audio-frequency circuit.

AUDIO-FREQUENCY REDIFFUSION : A form of Wire Broadcasting in which a programme is picked up from a broadcasting station and supplied by wire circuits to subscribers' loudspeakers. CF. CARRIER FREQUENCY WIRE BROADCASTING.

AUDIO-FREQUENCY TELEGRAPH SYSTEM : See VOICE FREQUENCY TELEGRAPH SYSTEM.

AUDIO-RESPONSE : Vocal output produced by a special device that contains prerecorded syllables or synthesizes speech.

AUDION : The name given by L. de Forest to the early form of three elecrode moderate vacuum *Thermionic Valve* used by him as a radio detector and amplifier.

AUDIT TRIAL : A means for tracing data on a source document to an output such as a report or for tracing an output to its source.

AUGMENTED TRANSISTION NETWORK (ATN) : A device for processing and for representing the syntactic structure of complex languages, especially natural languages; It is a recursive transition network with the addition of registers and conditional tests.

AUTO-CAPACITIVE COUPLING : Coupling of two circuits by means of a capacitance common to both.

AUTODIAL : An appliance enabling a subscriber to an automatic telephone exchange to call a limited number of habitual number by single pressure on a key which causes the required train of calling impulses to be sent.

AUTO-DIALER : A feature that lets a modem and communications software dial phone calls automatically.

AUTO-DYNE (RADIO) TRANSMISSION : The arrangement of a Thermionic Valve transmitter with the inductances in the grid and anode circuits forming a common coil or Autotrnasformer. The term autodyne is also sometimes used for Endodyne.

AUTODYNE RECEPTION : See ENDODYNE RECEPTION.

AUTO-HETERODYNE RECEPTION : See ENDODYNE RECEPTION.

AUTO-MANUAL TELEPHONE SYSTEM : See SEMI-AUTOMATIC TELEPHONE.

AUTOMATIC ACCELERATION : Controllers of electric locomotives, cars, etc. are said to be arranged for "automatic acceleration," when the starting resistances are cut out in progressive steps by automatic electromagnetic means, as the speed of the motors, and their back e.m.f.rises, without further movement of the controller handle.

AUTOMATIC DIRECTION FINDER : A development of the Robinson Direction Finder, in which the galvanometer is replaced by a relay system controlling a motor employed to rotate the direction finding coils, until a balance is obtained and thus to keep an indicator on the bearing required.

AUTOMATIC GAIN CONTROL : Control of the amplification factor of repeaters, amplifiers, etc., in line and radio-telephony upon the principle described under Automatic Volume Control.

AUTOMATIC HYPHENATION : A feature that hyphenates words automatically. Often found in word processors and desktop publishing programs related to digital computers.

AUTOMATIC NOISE LIMITER : A device in a radio receiver designed to limit the effect of impulse noise.

AUTOMATIC PAGINATION : A feature that automatically breaks text into pages. Often found in word processors and desktop publishing program.

AUTOMATIC PROGRAMMING : A subfield of AI and software engineering that studies systems that can themselves either write new computer programs or modify existing ones.

AUTOMATIC RADIO CALL DEVICE : An apparatus for actuating a call bell on receipt of a special Radio Call Signal, usually consisting of a relay for starting a selector mechanism which only permits of the bell circuit being closed when the signal impulses coincide with the sequence of closing of the slector contacts.

AUTOMATIC RINGING : The arrangement of telephone exchange connections, whereby a called subscriber's bell rings automatically as soon as his line is connected, without the operator having to manipulate a special key.

AUTOMATIC ROUTINER : An apparatus used in automatic telephone exchanges for carrying out the sequence of operations required for routine tests automatically.

AUTOMATIC SCRUTINEER (OR SCRUTINISER) : An apparatus used in cable and machine telegraphy, etc., where strict synchronism is required to detect inequality in the signal periods.

AUTOMATIC TEST EQUIPMENTS : ATE are versatile and flexible systems which can be easily adaptable to any application. Automation enables more rigorous and comprehensive test performance to be built into the system. Until recently, mainframe computers for their high speed data handling and computational capabilities were in use for this application. Because of matching performance of Personal Computer (PC) and cost/performance benefits better than mini's. ATEs are in general, being designed around PCs these days. The heart of an ATE is an instrument controller which is system integrated to various test and measuring modules over a General Purpose Interface Bus (GPIB). The instrument controller can be realised from a PC by incorporating standard software/hardware package, commercially available. The ATE operates via a test programme, a test fixture provides a hardware interface through which access to the Device Under Test (DUT) is achieved. By system integration and with the controller command, standard, intelligent, bus compatible modulers are made to conduct specific device tests and measurements of parameters. The system developed can test and handle encapsulated devices upto 48 pins configuration, operating at a clock frequency of 25 MHz. ATE works in the BASIC environment and test software is written and compiled in BASIC.

AUTOMATIC TRACKING : A method of holding a radar beam locked on target while the target range is being determined.

AUTOMATIC TRAFFIC SIGNALLING : The lighting up of appropriate traffic signal lamps automatically either to give definite time-intervals of priority to the different streams or traffic or worked in accordance with the approach of vehicles which actuate road contacts. See ELECTROMATIC TRAFFIC SIGNALLING.

AUTOMATIC TUNING CONTROL : A type of automatic frequency control in a radio receiver that adjusts the tuning to the correct setting for a given received signal when the manually operated adjustment has been set only approximately for that signal. It holds the tuning at the correct setting despite any small drift in the components of the receiver.

AUTOPLEX RECEPTION : A system of radio reception employing a single valve for detection and high and low-frequency amplification controlled by variometers in both plate and grid circuits.

AUTO-REDIAL : A feature that lets a modem and communications software repeatedly dial a phone number unit it connects.

AUTO-RELAY : A relay controlling the Auto-Switch in high speed telegraph Repeater Stations.

AUTO-TRANSDUCTOR : A transductor in which the main current and control current are carried in the same windings.

AUXILLARY ELECTRODE (in an Electrolytic Cell) : An interal intermediate electrode for purpose of voltage measurement.

AVALANCHE Syn : TOWNSEND AVALANCHE. A cumulative ionization process in which a single particle or photon produces several ions, each of which in turn gains sufficient energy from an accelerating field to produce more ions, and so on.

A large number of charged particles is thus produced from the initial event. The phenomenon is utilized in the Geiger counter, the avalanche photodiode, and IMPATT diodes.

AVC : See AUTOMATIC VOLUME CONTROL.

AWG : American Wire Gauge.

"A"-WIRE : The wire of a telephone line connected to the "T" wire parallel to the shaft.

AYRTON-MATHER GALVANOMETER : A form of moving-coil permanent magnet galvanometer, with no iron within the moving coil.

B

B : (1) Symbol for Magnetic Flux Density and Susceptance: also for Brightness. (2) β is an iterative search algorithm mainly used for game trees; it is similar to alpha-beta pruning, differing in its maintenance of separate "optimistic" and "pessimistic" evaluations for positions.

BABBAGE, CHARLES : A mathematician who, in 1834, designed the Analytical Engine for doing calculations using special cards. Although it could not be built, the machine had many of the principles of a moderen digital computer.

BACK CONTACT KEY : A signalling key with a contact worked by the back part of the key lever, behind the fulcrum, which keeps the receiving circuit closed when the key is not depressed.

BACK CONTACT SPRING (in Telephony) : A spring which makes contact with a main contact spring in the normal position.

BACKGROUND PROGRAM : In a multiprogramming environment, a program that can be executed whenever the computer is not executing a program having higher priority. Contrast with **foreground program.**

BACKUP : (1) A second copy of data on a diskette or other medium, ensuring recovery from loss or destruction of the original media. (2) One-site or remote equipment available to complete an operation in the event of primary equipment failure.

BACKUP DISK : A duplicate copy of a floppy disk that preserves the data in case the original file is ruined by mistake.

BACKWARD CHAINING : An inference algorithm for production systems that starts with a fact to be proved and finds all the rules with that fact in their action parts; it then tries recursively to satisfy the conditions of those rules, working backwards from a hypothesis through a proof that it is correct.

BAFFLE : A screen surrounding the vibrating surface in a loud speaker of the cone type to prevent air waves from affecting the air pressure at the back, resulting in better production of notes of low pitch.

BAIN'S CHEMICAL TELEGRAPH SYSTEM : An early system of telegrpahy in which the messages are recorded on a strip of paper chemically prepared with potassium iodide and starch, which assumes a bright violet colour when a signal current has passed through it.

BAKELITE : A synthetic insulating material which can be moulded and machined, or used in liquid form for impregnation, made from phenol and formal.

BALANCED AMPLIFIER : Syn. for push-pull amplifier, See PUSH-PULL OPERATION.

BALANCED ARMATURE LOUD SPEAKER : A loud speaker in which a pivoted armature connected to the cone or other diaphragm is surrounded by stationary coils between which it moves.

BALANCED CURRENT (OR VOLTAGE) : A term, used particularly in connection with interference in communication circuits for currents (or voltages), the algebraic sum of which is zero. Cf.RESIDUAL CURRENT (or Voltage).

BALANCED JACK-AND-PLUG : A single conductor jack-and-plug system in which the plug, with a sprung metal tip, somewhat resembles a banana in shape.

BALANCED MODULATOR : An arrangement of two sets of modulating apparatus partly in opposition in a radio-telephone transmitting apparatus so that oscillations of carrier frequency are balanced and eliminated and only the *side bands* of waves are transmitted. See also UPPER SIDE BAND

BALANCED SEA EARTH : A *Sea Earth* in a submarine cable with the earthing core terminating in a length of considerable resistance, such as stranded manganin wire.

BALANCED WIRE CIRCUIT : A circuit having two sides that are symmetrical with respect to earth and other conductors and are electrically alike.

BALANCING AERIAL : An auxiliary aerial, usually at right angles to the receiving aerial of a duplex radio station (i.e. one for sending and receiving simulataneously), adjusted so as to balance the effect of signals transmitted from the same station and thus to prevent them affecting the receiving aerial.

BALANCING BATTERY : A battery connected in parallel with a circuit to help in equalising the load by discharging at times of heavy load and receiving a charge at times of light load.

Sometimes controlled by an Automatic Reversible Battery Booser.

BALANCING NETWORK : An Artificial Line used for balancing purposes. See EQUALISING RINGS.

BALLAST RESISTANCE : (1) A steadying resistance used to limit variations of current in a circuit, such as the resistance used in series with an arclamp to secure stable conditions, or the resistance of iron wire in hydrogen of negative temperature coefficient used in series with the glower of a Nernst Lamp and for other similar purposes. (2) In Track Circuit Signalling, the leakage resistance across the ballast between the two track rails.

BALLISTIC GALVANOMETER : A galvanometer in which the moving system is of considerable moment of inertia and of low damping factor, so that it can be used to measure the total effect of a current impulse due to such a cause as the discharge of a condenser or the sudden change in the flux through a magnetic circuit, by reading the amplitude of the first swing instead of a steady deflection.

BAND : A channel of group of channels that from a loop, as on a magnetic drum.

BAND-PASS FILTER : A Frequency Filter which permits only the passage of oscillations between certain limits of frequency. Sometimes used to improve selectivity of radio-receivers.

BAND PASS RECEPTION : See BAND-PASS TUNING.

BAND-PASS TUNING : A term sometimes used for a system of tuning radio-receivers employing two similar circuits coupled by a reactance arranged so that the coupling is constant over a large range of tuning. Cf. BAND STOP FILTER.

BAND REJECTION OR BAND STOP FILTER : A Frequency Filter which opposes currents within a certain range of frequencies.

BAND'S WIRE GAUGE : The "Brown and Sharp" or "American" Wire gauge used for copper wire sizes in America. The Diameters form a geometrical series in which No. 0000 is 0.46 inch. and No.36 is 0.005 inch. Wires of a particular denomination in B. and S. Gauge are slightly smaller.

BAND SPREAD CONDENSER : A small variable capacitor used in conjunction with the main tuning capacitor in a radio-receiving set, to give an open scale for fine adjustment.

BAND WIDTH : (1) The band of frequencies occupied by a transmitted modulated signal (see modulation) and laying to each side of the carrier-wave frequency. (2) The amount of deviation of frequency that an aerial array is capable of handling without a mismatch. (3) The band of frequencies over which the power amplification in an amplifierfalls within a specified fraction (usually one half) of the maximum value. (4) See receiver. The range of frequencies for signaling; the difference between the highest and lowest frequencies available on a channel.

BANK : A number of similar pieces of apparatus connected in parallel and used like a single piece of apparatus are sometimes said to be "banked," e.g. a "bank" of transformers, or an artificial load formed by a "bank" of lamps. In automatic telephony, a set of rows or Levels, fixed contacts, usually forming Multiples over which the Wipers move in a Selector.

BANK-CABLE *(in Automatic Telephony)* : A multi-conductor cable connecting a Bank on a selector to a terminal rack.

BANKED WINDING : A method of winding inductance coils for radio apparatus in which successive turns do not form a continuous layer but are "banked up" in groups to diminish self-capacitance.

BANKWIRES *(in Automatic Telephony)* : Wires connecting the respective contacts in the Banks in selectors in multiple.

B.A. OHM OR B.A. UNIT : A unit of resistance agreed upon as a standard by the British Association in 1865. equal to 0.9866 International Ohm. Cf. LEGALOHM.

BARRAGE RECEPTION : A method of reception used in telecommunications in which the receiving aerial consists of an array of several directive aerials of different orientations. The received signal is input from selected aerials in the array that are chosen so as to minimize inteference from a particular direction.

BARREL AND PINCUSHION DISTORTION : A distortion seen when the lateral magnification is not constant but depends on image size (see diagram). Barrel distortion occurs when the

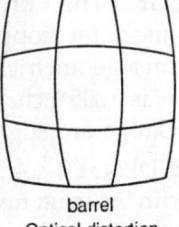

barrel

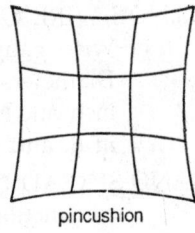

pincushion

Optical distortion

magnification decreases with object size, pincushion distortion when it increases with object size.

BARRETTER : (1) An instrument used in early radio-telegraphy systems, in which small current impulses are detected or measured by their heating effect in a fine wire, by observation of the change in resistance produced. (2) A Ballast Resistance.

BARRIER FILM RECTIFIER : A rectifier of the class in which a film of unilateral conductivity is in contact with metal or other normally conducting plates. The best-known example is the copper-oxide rectifier. Also called ELECTRON RECTIFIER.

BASE : (1) The number of single digits in a number system (2) Short for base region. The region in a bipolar junction transistor between the emitter and collector into which minority carriers are injected. The electrode attached to the base is the base electrode. See also transistor; semiconductor. (3) Short for base electrode.

BASE BAND : See CARRIER WAVE.

BASE ELECTRODE : See BASE.

BASE LEVEL (of a pulse) : See PULSE.

BASE LIMITER Sn : Inverse limiter. See LIMITER.

BASE REGION : See BASE.

BASIC : Acronym for Beginner's All-purpose Symbolic Instruction Code is a computer programming language developed at Dartmouth College as an instructional tool in teaching fundamental programming concepts. This language has since gained wide acceptance as a time-sharing language and is considered one of the easiest programming languages to learn.

BASIC FREQUENCY : In any oscillatory signal composed of several different sinusoidal components, the frequency of the component considered to be the most important.

BATCH : To accumulate to form a group, as of records of sales made during one business day.

BATCH COMPUTER SYSTEM : A computer system characterized by indeterminate turnaround time for output. Data and programs are collected into groups, or batches, and processed sequentially.

BATCH FILE : An ASCII text file containing DOS commands. See also ASCII FILE.

BATCH PROCESSING : A method of operating a computer so that

a single program or set of related programs must be completed before the next type of program is begun. A method of operation used with a large computer system in which a number of previously prepared programs are collected together and input to the computer as a unit. The computer then performs the operations as time becomes available in the system. The programs forming the batch are input from a central input/output device or from a small number of remote job entry location but are non-interactive.

BATTERY : The term "battery" has come to mean a group of a number of connected similar pieces of apparatus; more particularly primary or secondary cells connected together in series, parallel, or otherwise, or a "battery" of condensers. It is however often used incorrectly for a single cell by which a current is produced as the result of chemical source of e.m.f., such as a magneto generator for shot firing. (The term appears to be due to the Analogy of a battery of artillery, although, in this sence, the word really refers to its purpose for bombardment rather than to the multiplicity of the guns).

BATTERY CHARGING BOOSTER : See BATTERY BOOSTER.

BATTERY CUT-OUT : An appliance for automatically taking a battery out of the circuit should the voltage drop below the charging value; sometimes arranged to close the circuit again when the voltage has risen sufficiently.

BATTERY DIALLING : A system of dialling in automatic telephony, employing break impulses in an earth return circuit and a battery at the distant station. Cf. LOOP DIALLING.

BATTERY GAUGE : A portable voltmeter for ascertaining the condition of individual cells in a battery.

BATTERY METER : A meter arranged to integrate the charge and discharge of a battery in ampere-hours, either on two separate dials or with a single mechanism (or two connected together) on one dial, so as to read forward for charging and backward for discharging; with or without a compensating device to allow for the battery losses, so that the proportion of the charge remaining in the battery at any time can be read off.

BATTERY RINGING : The ringing of telephone bells by current obtained from a battery.

BATTERY SPEAR : A spike for making contact between the leads of a battery testing voltmeter and the plates of an accumulator cell.

BAUD : A unit of measurement of data processing speed. The speed is bauds is the number of signal elements per second. Since a signal element can represent more than one bit, baud is not synonymous with bitsper-second. Typical baud rates are 110, 300, 1200, 2400, 4800, and 9600.The rate of sending a signal, usually given in bits per second. It is used as a measure of data flow between computers and/or other requirement.The time interval occupied·by the signal element in telegraphy, variable in number for different leters in the Morse Code but constant in the Baudot and other five-unit codes. The signalling speed is defined as so many "bauds" per second.

BAUDOT MULTIPLEX *(telegraph)* **Sytstem** : A Multiplex printing telegraph system in which several Baudot instruments are worked at the same time over the same line by the use of synchronous revolving contact makers.

BAUDOT PRINTING TELEGRAPH SYSTEM : A type printing telegraph system in which the depression of a distinctive combination of five keys (representing a letter) determines the direction of five successive current impulses in the line. These, by means of a revolving contact maker cause the same combination to be reproduced by the armatures of five instrument, and these changes of position cause the required letter to be printed by mechanical means. Cf. MURRAY, CREED, HUGHES and STELJES PRINTING TELEGRAPH SYSTEMS.

BAYESIAN INFERENCE : A menas by which a reasoning system can use Bayes's Theorem to handle uncertain information using conditions and prior probabilities, either alone or in combination with a rule-based system; it is a more formal technique than the certainty factors used in MYCIN and its derivatives.

BAYONET CAP (B.C.) : The form of incadescent lamp cap with a brass collar, about 7/8 in. diam., which engages in a Bayonet Holder. Other types of holders are Edison-Screw Cap, Central Contact Cap, and Bottom Loop Terminals, etc.

BCD : Binary-coded decimal; 4 bits of binary information can be used to encode one decimal digit. When the successive digits of a decimal number are coded in this way, the number is said to be in the BCD code.

BEACON : A signal station or the signal transmitted by such a station, which acts as a reference point. A beacon that transmits an identifiable signal is a code beacon. A *homing* beacon is one that guides an object, such as an aircraft, to a target, such as an airport. At the airport, signals from the localizer beacon associated with the instrument landing system are picked upon abling the aircraft to be guided to land. A beacon that employs radar signals is a *radar* beacon and one employing radio-frequency waves is a *radio* beacon. The receiver used for detecting the signals from a beacon is a *beacon receiver*.

BEAM *(in a Cathode Ray Tube)* : See ELECTRON JET, and CATHODE RAY OSCILLOGRAPH.

BEAM AERIAL : An aerial for radiating a concentrated beam of nearly parallel waves or receiving such a beam from a particular direction.

BEAM AERIAL SYSTEM : The whole combination of aerials, reflecting and tuning apparatus required to project a conncentrated beam or to receive from a particular direction only.

BEAM ARRAY : See AERIAL ARRAY.

BEAM BENDING : An unwanted effect that occurs in television camera tubes. The electron team used to scan the target area can be deflected from its intended position by the electrostatic charges stored on the target. This can result in misalignment of the picutre image in the receiver with respect to the original optical image.

BEAM COUPLING : The production in a circuit of an alternating current between two electrodes upon passage of an intensity-modulated electron beam. The beam-coupling coefficient is the ratio of the alternating current produced to the beam current.

BEAM CURRENT CHARACTERISTIC : A curve showing the relation between beam current and applied potential in a cathode ray tube.

BEAM LEAD : A connecting lead on a silicon chip that is formed chemically and cantilevered across a void or space on the chip. The lead may be cantilevered either from the chip to the interconnection pattern or vice versa.

BEAM RECEIVING AERIAL : A beam aerial used for receiving waves from a particular direction only.

BEAM TRANSMISSION : Radio-transmission employing short waves, concentrated into a nearly parallel "beam" by specially constructed directional aerials, with or without reflectors;

suitable for long distances and requiring much less power than non-directive transmission.

BEAM TRAP : An arrangement in a cathode ray oscillograph for deflecting the beam off the screen when not actually in use, to avoid fogging the plate.

BEAT : When wave motions, such as sound or electric waves, of nearly but not quite equal frequencies are superposed, a pulsation is caused by the alternate coincidence and opposition of the waves. These pulsations of "beats" have a frequency which becomes lower as the two original frequencies become nearer. See BEAT FREQUENCY, HETERODYNE RECEPTION.

BEAT FREQUENCY : The frequency of the beats caused by the super position of an auxiliary frequency slightly different from the original frequency, as in the Endodyne and Heterodyne systems of radio-reception.

BEAT OSCILLATOR : See HETERODYNE OSCILLATOR.

BEAT RECEPTION Syn : *Heterodyne reception.* A method of radio reception that employs beating. The received radio-frequency (r.f) oscillations are combined with r.f. oscillations generated separately in the receiver by a beat oscillator to produce beats (usually audiofrequency which affects the telephone Audio-Frequency). See ENDODYNE and HETERODYNE RECEPTION.

BEATS : The periodic signal produced by interference when two signals of slightly different frequencies are combined. The amplitude is equal to the sum of original amplitudes; the frequency (the beat frequency) is equal to the difference between the original frequencies.

BECQUEREL RAYS : The name originally given to the rays emitted by uranium salts, etc. (discovered by A.H. Beequerel, 1852-1908). More active sources have since been discovered and the rays have been divided into several classes. See RADIO ACTIVITY, and RAYS, etc.

BEGA : A prefix meaning 10^9 times, i.e. a billion times according to French and American nomenclature, or a thousand million times according to English nomenclature.

BEGOHM : 10^9 ohms, i.e. a thousand megohms.

BEL : A unit used for measurement of power, sound, etc., on a logarithmic scale and in the case of telephone circuits, forming a unit equal to ten times the American "Transmission Unit" (T.U.). Cf. NEPER.

BELIN SYSTEM : *(of Phototelegraphy)*. The earlier systems under this name employed an original prepared in relief form a bichromated film over which passes a stylus connected to an apparatus resembling a microphone in construction which varies its resistance according to the range of movement of the stylus and therefore to the thickness of the film.This variation of resistance controls the line current. In later forms of the system a transparent original and a light sensitive cell are used. A delicate galvanometer in the receiving apparatus causes the deflection of a beam in accordance with the variations of the line current and thus regulates the amount of light falling upon the synchronously moving receiving film.

BELL *(Electric)* : A bell struck by a hammer through the action of an electromagnet attracting an armature to which the hammer is attached; arranged either to give a single stroke, or to continue repeating the strokes, by the automatic action of a *Trembler*, or by the supply of alternating or intermittent currents.

BELLINI-TOSI DIRECTION FINDER : An early form of direction finder in which two aerials in phases at right angles to each other are used, and the relative strength of the signals picked up by them are compared by an instrument called a RADIO-GONIOMETER.

BELL RECEIVER : The original form of electromagnetic telephone receiver invented by Graham Bell in 1876 in which a steel diaphragm is in the field of a permanent magnet, the pull of which on the diaphragm is varied according to the wave form of the sounds to be reproduced by the superposition of a variable field due to the received telephone current which passes through a coil on the pole of the magnet. The same instrument was also used as a transmitter,without any battery, as the movement of the diaphragm induced currents in the winding on the magnet pole.

BELL SYSTEM OF PHOTO-TELEGRAPHY : A system of Phototelegraphy, in use in America, in which a Potassium Cell is placed within a transparent revolving drum carrying the photograph to be transmitted so that the beam of light passing through successive parts of the film falls directly on it. The modulations so produced are applied in the recording apparatus to an electromagnetic light relay.

BELL TELEPHONE : See BELL RECEIVER.

BELL TRANSFORMER : A small transformer for working electric bells from an alternating current.

BERRY TRANSFORMER SWITCH AND SYSTEM : See SERIES TRANSFORMER SYSTEM.

BETA (β) RAYS : The most easily deflected (by a magnetic field) of the three kinds of rays given off by radio-active substances. They affect photographic plates and are more peneterating than rays but less penetrating than rays. They consist of negatively charged Paticles travelling with a speed comparable with but less than that of light, i.e. up to 295,000 Km. per second. Cf. Alpha and Gamma RAYS.

BETA (β) RAY SPECTRUM : A diagram showing by the position of lines the velocities of β rays emitted by a radio-active substance from observations of the relative extent to which they are deflected by a magnetic field.

BETA CIRCUIT : See FEEDBACK.

BETA CURRENT GAIN FACTOR : Syn. common-emitter forward-current transfer ratio.

BETA DECAY : The spontaneous transformation of a nuclide into one of its isobars with the emission of an electron plus *antineutrino* or a *positron* plus *neutrino*.

BETATRON : A cyclic accelerator that employs magnetic induction to produce high energy electrons. A magnetic field is used to deflect the electrons into a circular orbit. The magnetic orbital flux increases with time, giving rise to an induced circumferential electric field that accelerates the electrons.

BETULANDER AUTOMATIC TELEPHONE SYSTEM : An automatic system of telephone exchange working of the successive electromechanical selector switch type. Cf. RELAY AUTOMATIC TELEPHONE SYSTEM.

BIASED AUTOMATIC GAIN CONTROL : Syn. delayed automatic gain control. See AUTOMATIC GAIN CONTROL.

BIASED DIFFERENTIAL PROTECTIVE SYSTEM : A feeder protection system with pilot wires carrying a current derived from current transformers at the two ends of the line and arranged in such a way that inequality between the current supplied by these transformers up-sets the balance of the relays and trips the circuit breakers. (Also known as McColl protective system).

BIASING TRANSFORMER : A current transformer used in a Biased Differential Protective System.

BI-BAND TELEPHONY : A system of carrier current telephoney in which a different carrier frequency is employed for speech in each direction.

BICHROMATE CELL : See BOTTLE BATTERY AND FULLER CELL. A Primary Cell, originated by Poggendorf, having a zinc negative and one or more carbon positive electrodes, with an Electrolyte ususally of dilute sulphuric acid with potassium bichromate as a depolariser; either of the single fluid or double-fluid type.

BIDIRECTIONAL TRANSISTOR : A transistor that has substantially the same electrical characteristics when operated with the *emitter* and *collector* interchanged.

BIFILAR OSCILLOGRAPH : An Electromagnetic Oscillograph in which the moving system consists of two parallel strips.

BIFILAR WINDING : A method of winding consisting of two contiguous insulated conductors connected so that they carry the same current in opposite directions. This

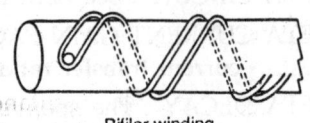

Bifilar winding

results in a negligible magnetic field being produced. The technique is commonly used to wind non-inductive resistors.

BIGRID VALVE : A Thermionic Valve with two grids, e.g. for combining function of two valves into one. Cf. SCREENED VALVE.

BILLI : Prefix meaning a thousand millionth (10^{-9}), i.e. a billionth according to the French and American nomenclature. (In Great Britain, a billion is used to mean a million million (10^{12}).

BILLOHM : 10^{-9} ohms. See also BILLI.

BIMETAL FUSE : A fuse element of two different metals, e.g. a copper wire coated with lead.

BIMETALLIC STRIP : Tripping gear of a circuit-breaker actuated by the deflection of a strip of two metals owing to their unequal expansion when heated by an excess current.

BIMETALLIC WIRE : Wire composed of a central core of one metal, over which is electro-deposited a sheath of another metal, e.g. a copper-coated steel wire for use where good conductivity is required with considerable tensile strength, as in long overhead spans. Sometimes the copper coating is only protective. Such wire is also used for high frequency currents where the current is practically confined to the surface (see SKIN EFFECT), and by combining metals of positive and negative Temperature Co-efficients in proportions to give a temperature coefficient of zero.

BINDING ENERGY : (1) The total energy released when protons and neutrons bind together to form a nucleus. (2) The energy required to strip all the extranuclear electrons from a nucleus is the total electron binding energy. See IONIZATION POTENTIAL.

BIOT : A unit of current in the obsolete CGS electromagnetic system of units. It is equal to 10 amperes.

BIOT AND SAVART'S LAW : See LAPLACE'S LAW.

BIOTRON : A combination of two Thermionic Valves connected to obtain a particularly steep characteristic.

BIPOLAR INTEGRATED CIRCUIT : A type of monilithic integrated circuit based on bipolar transistors. See INTEGRATED CIRCUIT.

BIPOLAR RECEIVER : A telephone receiver with both poles of the magnet close to the diaphragm.

BIRMINGHAM WIRE-GAUGE (B.W.G.) : A gauge used in Great Britain for sheet metals and to a certain extent for wires. The wire diameters differ through most of the range only by a few per cent from the corresponding figures for the British Standard Wire Gauge. (The Birmingham gauge for gold and silver is different).

BISTABLE : Short for BISTABLE MULTIVIBRATOR. A circuit having two stable states. See FLIP-FLOP.

BLACK AND-WHITE TELEVISION : Syn. for *monochrome television*. See TELEVISION.

BLACK BOX : Any self-contained unit or part of an electronic device that may be treated as a single package. Such a circuit may be approached mathematically in terms of its input and output characteristics, irrespective of its internal elements.

BLACK OUT : A temporary loss of sensitivity of any electronic device following the passage of an intense transient signal.

BLAKE TRANSMITTER : An early form of Telephone Transmitter or Microphone in which a platinum bead made a varying contact with a carbon disc.

BLANKING : The rendering of a device or channel ineffective or inoperative for a desired time. For example, blanking eliminates the *return* trace from the screen of a cathode-ray tube. There are usually two blanking components to eliminate the horizontal and vertical components of the *return* trace.

BLATTNERPHONE : A modern form of the Telegraphone of Ponlsen in which sounds are recorded in the form of a change of the magnetisation along a steel strip. Used for recording and reproducing broadcast programmers and for sound film production.

BLIND POWER : An expression proposed for the total Reactive Volt amperes in a three-phase circuit.

BLIND SPOTS : A popular term for localities where radio signals from a certain source are exceptionally weak, although they are received more strongly at a greater distance.

BLOCKING CAPACITOR OR CONDENSER : A small condenser placed in a radio-circuit where it is desired to allow high frequency oscillatory currents to pass but to prevent the transmission of steady currents or oscillatory currents of low frequency.

BLOCKING OSCILLATOR : A form of oscillator used in connection with scanning in television in which a capacitor is charged through one impedance and discharged through another.

BLONDEL OSCILLOGRAPH : The original form of moving conductor reflecting Oscillograph, afterwards perfected by Duddell.

BLOW : A fuse is commonly said to "blow" when it breaks the circuit owing to the action of an excess current, particularly when the current is heavy and the action is violent.

BLUE GLOW (or "AURA") : A blue luminescence due to gas ionisation, which appears in a thermionic valve when the vaccuum is low, but may be present to a certain extent in a soft valve without harm.

BOARD OF TRADE UNIT : The unit by which electrical energy is commonly sold by supply authorities in Great Britain, equivalent to one kilowatt-hour, 3,415 B.Th. U., 2.6552 x 10^6 ft.-lbs. or 3.6 x 10^6 ioules; called by some writers the KELVIN commonly spoken of as a "Unit."

BOBBIN : Properly the reel on which is wound a coil of insulated wire such as in the small electromagnets on electric bells, etc., but now some times used for electromagnet windings themselve, even up to the largest field coils in machines. See SPOOL, and CF. FORMER.

BONE OF HORUS : Name used by the Ancient Egyptians for the lodestone or magnet (after Horus or Apollo).

BOOSTER : An auxiliary generator. usually motor-driven, or a transformer, included in a circuit to give an increase or decrease of voltage, e.g. for battery charging or compensation for feeder drop: controlled automatically or by hand.

BOOSTER TRANSFORMER (or boosting transformer) : A transformer with its secondary in series with the line for injecting a voltage to compensate for line drop on an A.C. feeder or for some other purpose.

BORON RESISTOR Syn : BOROCARON RESISTOR. A resistive film resistor that has a small percentage of boron introduced into the carbon film to add stability.

BOTTLE BATTERY : An old-fashioned name given to a form of single fluid Bichromate Cell in a bottle shaped glass vessel with two fixed carbon plates and a zinc plate between them, which can be drawn up out of the electrolyte into the neck of the bottle, when the cell is not in use, to avoid local action.

BOTTOMING : The operation of an electronic device in such a way that, when in the conducting state, the lowest instantaneous output voltage is determined by the device characteristics rather than the input voltage.

BOUGIE DECIMALE (Decimal Candle) : A French standard of luminous intensity, roughly one-twentieth of the Violle Standard equivalent to the International Candle agreed upon later.

BOURNE CIRCUIT : A circuit for radio reception employing a transformer to couple the aerial to the tuned grid circuit of the first valve, the primary and secondary being connected together at the earthed end.

BRAIDING : An outer plated covering of wire or impregnated fiberous material for portective purposes on cables and wires.

BRAKE HORSE-POWER (B.H.P.) : The output of a motor, engine, etc., as measured on a testing brake on a pulley on its shaft, expressed in Horse-Power.

BREAKDOWN : (1) A sudden catastrophic change in the properties of a device rendering it unfit for its purpose. (2) A sudden disruptive electrical discharge in an insulator or between the electrodes of an electron tube. (3) A sudden change from a high dynamic resistance to a much lower value in a semiconductor device.

BREAKDOWN VOLTAGE *(b.d.v.)* : The voltage under specified conditions at which breakdown occurs.

BREMSSTRAHLUNG : Electromagnetic radiation produced when an electron is suddenly decelerated by a nuclear field.

BRIDGE : An assembly of at least four circuit elements, such as *resistors, capacitors,* etc., together with a curent source and a null point detecting device. Each of the circuit elements· is arranged in one *arm* of the bridge. When the bridge is balanced, i.e. zero response is obtained from the null detector, there is a calculable relationship between the values of elements in the arms given by

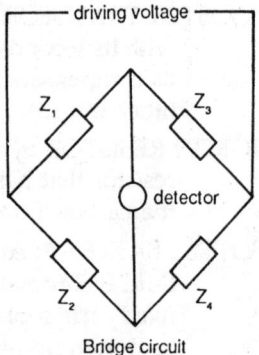

Bridge circuit

$(Z_1/Z_2)=(Z_3/Z_4)$

A measuring equipment or a similar principle to the original. *Wheatstone's Bridge* in the measurement of D.C. resistance,which in later forms has been adapted to A.C. working up to the highest frequencies and to the measurement of capacitance, inductance.

BRIDGE DUPLEX TELEGRAPH SYSTEM : A Duplex telegraph system in which the receiving instrument occupies the position of a galvanometer in a Wheatstone's Bridge, so that the outgoing current does not affect the home insturment although part of the line current flows through the distant receiving.

BRIDGE RECTIFIER : A full-wave rectifier circuit in the form of a bridge, with a rectifier in each arm (see diagram).

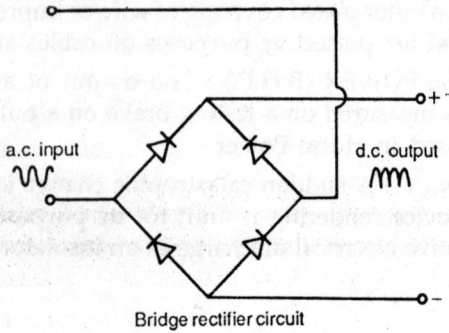

Bridge rectifier circuit

BRIDGING CONDENSER : A condenser of fixed value shunting a piece of apparatus to divert oscillations of the higher frequencies.

BRIDGING TELEPHONE : A telephone receiver placed in parallel with another, or with some other apparatus to divert oscillations of the higher frequencies.

BRIGHT EMITTER VALVE : A Thermionic Valve with a filament which has to be raised to bright incandescence to give sufficient electronic emission. Cf. DULL EMITTER VALVE.

BRIGHTNESS : The luminous intensity emitted by a surface in a particular direction per unit area projected in that direction. See LAMBERT.

BRITANNIA JOINT : A method of jointing bare telegraph wires, etc., in which the ends are laid alongside each other with the tips slightly turnedup (but not twisted together) and bound with fine tinned wire and soldered.

BRITISH STANDARD WIRE GAUGE (S.W.G.) : The legal wire gauge generally used in Great Britain. The sizes run in more regular progression than in the Birmingham or American gauges, but do not differ very greatly. No. 1 is 0.300 in. diam., and No. 50 is 0.001 in. diam. The largest size is called 7/0 and is 0.500 in. diam.

BROADCASTING. The transmission by telephony (radio-or otherwise) of news or entertainment items for reception by anyone provided with suitable receiving apparatus. In addition to the ordinary radio method, systems are in operation ultilising carrier current telephony.

BROWN AND SHARPE WIRE GAUGE (B & S) : Syn. for American wire gauge. See WIRE GAUGE.

BROWN TELEPHONE RELAY : An instrument for producing in a local current a magnified version of the variations of a weak telephone current by varying electromagnettically the length of a gap minute enough to be conducting, between a point and a disc of metal. The adjustment of the gap is maintained automatically by the action of the local current.

BUFFER : (1) An isolating circuit interposed between two circuits to minimize reaction from the output. Usually it has a high input impedance and low output impedance. It may be used to handle a large fan-out or to convert input and output voltage levels. (2) An electronic gadget that stores the output from the digital computer so that its micro processor can take up another work.

BUFFER BATTERY : An Accumulator Battery arranged in parallel with a generating plant to equalise the load by assisting at times of heavy load and taking a charge at times of light load; often controlled automatically by a Reversible Booster.

BUFFER VALVE : A Thermionic Valve used in a radio receiver in an additional coupling stage between a tuned input circuit and high-frequency amplifying valve to obtain stability by compensating the effect of input impedance of the amplifying valve.

BUG : An error or fault in a computer program or in computer equipment. To debug the program or system is to find and correct any errors.

BUILD-UP TIME : The time required for a current to rise to its maximum value in an electronic circuit or device.

BUNCHED CONDUCTORS : A conductor composed of several wires twisted together in the same direction but not in regular layers.

BUNSEN CELL : A Primary Cell of the double fluid type with zinc and carbon electrodes, an electrolyte of dilute sulphuric acid and a depolariser of concentrated nitric acid.

BURGLAR ALARM : A system where by the opening on a door, window, etc., or the treading on a section of the floor, either by closing an electric circuit, or by interrupting a current normally flowing. In the latter case the bell also rings when the wire is cut or broken.

BURIED HEARTH ELECTRODE FURNACE : A Direct Arc Furnce with one electrode embedded in the hearth and covered with crushed carbon, e.g. Tinfos or Lorenzen Furnace.

BURIED LAYER : A layer of high-conductivity semiconductor material diffused into the substrate layer during the manufacture of bipolar integrated circuits and transistors. The buried layer is located below the collector and serves to reduce collector resistance.

BUS : Abbreviation for Bus-Bar.

BUSBAR : The abbreviation commonly used for Omnibus Bar.

BUZZER : An apparatus similar to an electric trembler bell without a gong, to produce a signal by a buzzing sound. A similiar apparatus isused to produce slight oscillations for testing radio-receiving apparatus.

BUZZER WAVEMETER : A wavemeter excited by a buzzer giving short trains of waves so that heterodyning is not necessary for detection.

"B"-WIRE : The wire of a telephone line connected to the "R"-wire inside the exchange. Cf. "A"-WIRE.

BY-PASS CONDENSER OR CAPACITOR : See BRIDGING CONDENSER.

BY-PATH AUTOMATIC TELEPHONE SYSTEM : A development of the Strowger system in which a certain amount of common apparatus is substituted for individual apparatus during conversation for purposes of economy in plant.

BY-PRODUCT CIRCUIT : An unloaded phantom circuit not suitable for telephony, which is available for telegraphy.

❑

CABLE : A conductor for transmitting electric currents composed of several wires or strands laid up together, with or without insulating and protective coverings.

CABLE CONNECTOR : Male/female plug used four connecting cables between a computer and peripherals.

CABLE CORE : That part of a cable consisting of one conductor and its insulation only. Several such cores may be laid up together in one cable.

CABLE DETECTORS, ELECTRODE : See ELECTRODE CABLE DETECTOR.

CABLE RELAY : A sensitiveS Relay used on submarines cable telegraph circuits.

CACHE : Part of memory that makes computer run faster by holding the most recently accessed data from a disk. The next time the computer needs the data, it accesses it from memory rather than from the disk, which would be slower. Sometimes called a RAM cache.

CACHE SIZE : The amount of memory allocated for the cache, measured in kilobytes [K] such as 64K.

CADENCE : A signal in Baudot and other telegraph systems that conditions have been arrived at under which a key may be pressed.

CADMIUM CELL : A standard cell giving an accurately known e.m.f. made up in an H-shaped glass container and having electrodes of mercury and a cadmium mercury amalgam respectively, with a cadmium sulphate electrolyte and a depolariser of mercurous suplphate. See WESTON NORMAL CELL.

CAE : Computer-Assisted Education.

CAGEONS DISCHARGE : The discharge of electricity through gases at low pressure (see VACUUM TUBE). The discharge consists of movement in one direction of negative carriers, consisting (in high vacua) of free electrons and in the other direction of positive carriers of greater.

CAI : Computer-Assisted Instruction

CAL : Computer- Assisted Learning.

CALAMINE, ELECTRICAL : Zinc silicate, so called on account of its Pyroelectric properties.

CALCULAGRAPH : An instrument used in telephone exchanges to record the time of calls.

CALCULATION : A series of numbers and mathematical signs that, when entered into a computers, is executed according to a series of instructions.

CALCULATOR : A device, now usually electronic, by means of which arithmetic operations can be performed on numbers entered from a keyboard. Final solutions and intermediate numbers are generally presented on LCD or LED displays. Present day calculators range from simple calculator having no memory and with only arithmetic operations to those with memory and with verying sets of specialist program for particular fields, engineering, navigation, or business for example may be purchased as accessories to the more expensive calculators, as can small printers.

CALIBRATION : The determinations of the relation between the indicated value on a measuring instrument and the *true of* the quantity to be measured. The *true* value is the values that would be obtained if all sources of error were eliminated.

CALL : (1) To place the necessary initial values of variables in the required addresses and then to transfer control to a subrountine. (2) To transfer controls to a specific closed subroutine. (3) In communications, the action performed by the callings party, or the operations necessary in making a call, or the effective use made of a connection between two stations. Synonymous with cue.

CALLAND CELL : A form of "Gravity" Daniell Cell with a copper cylinder at the bottom, and a zinc cylinder at the top, without any porous diaphragm to separate the coppers sulphate from the zinc sulphate floating on it.

CALLENDER RECORDING BRIDGE : A form of Wheatstone slide wire bridge used for resistance thermometers and pyrometers in which the sliding contact is shifted automatically to the hero position by the current in the "Galvanometer" circuit, so that it can be used to move a recording pen on a resistance chart.

CALL INDICATOR : (1) An apparatus used in manual telephone exchanges linked up to automatic exchanges which translates the calling impulses from the latter into a visible signal. (2) A signalling device in an electric lift car indicating, from which floor a call has been made.

CALLING DEVICE : A device used in automatic telephone systems for sending impulses to make a call, e.g. a Calling Dial.

CALLING DIAL : The apparatus, usually in the form of a rotatable dial attached to the subscribers instrument in an automatic telephone system which actuates a call by being set successively to the digits composing the number required.

CALLING LAMP : A signal lamp on a telephone exchange switchboard, which indicates by its illumination that a call is being made. See LINE LAMP and PILOT LAMP.

CALLING PARTY RELEASE : A system whereby the switches, etc., making a connection in a telephone exchange are released for clearing the line automatically by a caller replacing his receiver.

CALLING PLUG : The plug used by the telephone exchange operator to make connection to the called subscriber's circuit.

CALORIE : The quantity of heat required to raise the temperature of 1 gramme of water one degree Centigrad from fifteen degrees Centigrade: equivalent to 0.00396 British Thermal Units or 4.2 joules, also called *Small Calorie,* or *Gramme Calorie.* One thousand Calories is sometimes narrated as *Great calorie, large calorie* or *kilo calorie.*

CAMAC : A standardized multiplexing intermediate interface. It does not usually connect directly to a processor or a peripheral, but provides a standardized interface to which a number of peripheral interface-adapters and a single computer interface controller can be connected. The peripheral adapters may each have different functions, e.g. digitals to analog converter, level changers, parallel to serial converter, and thus have different interfaces facing outward from the CAMAC. Similarly the controller module connects to the CAMAC interface but the outward facing interface can be chosen to suit the available computer The name CAMAC was chosen to symbolize this characteristic of looking the same in either direction. The adapters are typically a single printed circuit card that plugs into the internal 86-way connector. The outward facing connections are usually mounted on a panel attached to the circuit card or may be made via a second connector mounted above the 86-way CAMAC connection. The interface is widely used for connecting instruments and transducers to computers.

CAMPBELL BRIDGE : An a.c. bridge that is used to measure a mutual inductance, M, by comparison with a standard capacitor, C. The resistances R_1 and R_2 are varied until null deflection is obtained on the indicating instrument I (see diagram).

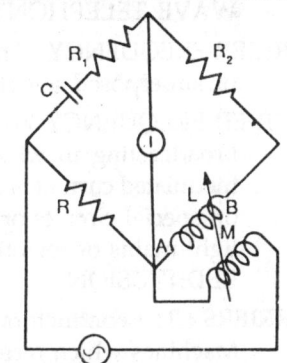

At balance
$L/M = (R + R_1)/R$
$M/C = RR_2$
where L is the self-inductance of the coil AB.

CANCEL : To stop or prevent a command from running. Most often chosen by pressing the Esc key.

CANDIDATE ELIMINATION : A version-space learing procedure for single concept descriptions that removes from the set of possible concepts all candidates that do not cover a new positive example or that do cover a new negative example.

CAPACITANCE : The property of a conducting body or system which determines the magnitude of the charge which is produced therein by unit potential. The Practical Unit of capacitance is the Farad (10 absolute C.G.S. electromagnetic units), but it is usually more convenient to make use of the Microfarad (one million of a farad).

CAPACITANCE BRIDGE : A bridge for comparing capacitances in the various arms; usually employing alternating or intermittent currents and a telephone in place of a galvanometer.

CAPACITIVE COUPLING : See COUPLING.

CAPACITIVE REACTANCE : The component of reactance due to the effect of the capacitance, expressed in ohms by $1/2\pi fc$ where f is the frequency and c the capacitance (in farads). Also called *condensance*.

CAPACITOR : Any system, possessing appreciable capacitance, i.e. in which an appreciable charge is produced by application of an e.m.f. If the e.m.f. is continuous, only a momentary rush of current will be produced (except for such small permanent current that there may be due to leakage), but if the e.m.f. is alternating, a current will surge in and out of the capacitor which will be in advance of the e.m.f. in Phase. Capacitors are therefore used to improve the power-factor of a system. Formerly more usually called condenser or capacity. See also CONDENSER.

CARRIER CURRENT TELEPHONY (and Telegraphy) : See CARRIER WAVE TELEPHONY.

CARRIER FREQUENCY : The frequency of the carrier wave in radio or superposed telephony or telegraphy.

CARRIER FREQUENCY WIRE BROADCASTING. : A form of wire broadcasting in which the current distributed consists of a modulated current at a high carrier frequency, either transmitted on special circuits or superposed on circuits used as electric light mains or for other purposes. Cf. AUDIO-FREQUENCY REDIFFUSION.

CARRIERS : The conductors fixed to the moving plates of Infiuence Machines which receive the induced charges and carry them to the main conductors. Cf. FIELD PLATES.

CARRIER SYSTEM : A transmission system used in telephony and telegraphy in which modulated carrier waves are transmitted along lines or cables. The multichannel carrier system in telephony allows many simulantaneous independent signals to be transmitted on the same circuit.

CARRIER TELEPHONY (and Telegraphy) : See CARRIER WAVE TELEPHONY.

CARRIER TRANSMISSION : The transmission of a signal that is the result of modulation of a carrier wave. Sometimes the carrier wave is not transmitted but only the sidebands resulting from the modulating. This is known as *carrier suppression*. Carrier transmission is used in *telegraphy* and *telephony*. These forms are carrier telegraphy and carrier telephony, respectively.

CARRIER WAVE Syn. *carrier* : The wave that is intended to be modulated in modulation, or, in a modulated wave, the carrier-frequency spectral component. The process of modulation produces spectral components falling into frequency bands at either the upper or lower side continuous high frequency waves propagated eitherin space or along a circuit, upon which the audio-frequency Modulations corresponding to the voice waves are superposed. See SIDEBAND and SUPPRESSED CARRIER WAVE.

CARRIER WAVE TELEPHONY : Simultaneous transmission of several messages each employing a different carrier current of different frequency superposed upon a circuit used for other purposes, and modulated according to the telephone (or telegraph) signals, employing separate receivers for each message tuned to resonate to the individual frequencies of the separate messages. Special arrangements are employed including wave filters, etc., to eliminate the frequencies that are not within the ranges or "bands" actually required. For telegraphy on this system audio frequencies are sometimes used, particularly on the Continent.

CARRY : The admissible mark to be taken to the next higher place and there added when the sum of the marks in the place exceeds or equals the radix.

CARRY DIGIT : The digit generated when numbers are being added together. The carry digit is retained temporarily and is then added to the sum of the next higher-order position.

CARRYING CAPACITY or CURRENT : The current that a conductor or apparatus can carry without exceeding specified limits of heating or voltage drop.

CAR SWITCH (LIFT) CONTROL : A system of electric lift control from a switch or controller in the car. Cf. ROPE CONTROL and PUSH BUTTON CONTROL.

CARTRIDGE : Used on some microcomputers plugged into a socket on the computer. It is actually a ROM containing software in the form of a language or a computer game.

CASCADE AMPLIFIER : An Amplifier consisting of a number of Thermionic Valves in casade connection.

CASCADE CONNECTION : A method of connection of similar pieces of apparatus together so that thhhe output side of each is connected to the input side of next.

CASCADE VOLTAGE SWITCH LOGIC (CVSL) : The basic form of this type of gate is a fully differential network which consists of two complementary n-MOS networks connected to a pair of cross-coupled p-MOS pull up transistors.

CASE (Computer-Aided Software Engineering) : Hardware and software that helps to automate parts of the system life cycle.

CASE SENSITIVE : The ability to distinguish between upper and lowercase letters. Often used when searching for words in a word processor or database.

CASSETTE : Often used for storing programs and files in small computer systems such as micro and minicomputers.

CASSETTE TAPE : A continuous loop of magnetic tape stored in a container referred to as a cassette. The cassette tape used in computer systems is similar to that used in domestic audio systems.

CATCHING DIODE Syn. *clamping diode* : A diode used to limit the voltage at a point in a circuit. Two diodes can be used to keep the voltage within specific limits (see diagram), i.e. within $\pm$ Vd, where Vd is the diode forward voltage. See DIODE FORWARD VOLTAGE.

Catching diodes for voltage limitation

CATEGORY *(OF TABULAR REDUCTION)* : The number of 1's in a binary number.

CATHAUTOGRAPH : A system of transmission of visual writing in which a receiver similar to a cathode ray oscillograph is used with a fluorescent screen on which writing caused by the moving "spot" remains visible for short time. The transmitter contains a stylus regulating two resistances as in the Tel-autograph and thus controls the deflecting fields of the receiver.

CATHETRON : A Grid-Controlled Mercury-Vapour Rectifier with the control electrode external to the tube.

CATHIONS : See CATIONS.

CATHODE : The negative electrode of an electrolytic cell, discharge tube, valve, or solid-state rectifier. The electrode by which electrons enter (and conventional current leaves) a system. The pole or electrode through which according to the old convention the current leaves an electrolytic cell, vacuum tube or other piece of apparatus, i.e. the negative pole, and in a thermionic valve, etc., the electrode from which the electron stream is emitted. In the case of a primary cell, the electrode through which the current leaves the elctrolyte, i.e. the copper, carbon, or other positive pole is sometimes called the cathode. (From Greek words meaning literally the "way down".) Also spelt as KATHODE CF. ANODE.

CATHODE DROP OR FALL (of Potential) : The concentration of

potential gradient which occurs near the cathode in a highly exhausted vacuum tube, due to the higher speed of the negative ions or electrons than that of the positive ions. See GASEOUS DISCHARGE.

CATHODE OSCILLOGRAPH : See CATHODE RAY OSCILLOGRAPH.

CATHODE RAY : Rays consisting of a stream of negatively charged particles or Electrons given off from the cathode in a highly exhausted the cathode in a highly exhausted tube, i.e. when the gas pressure is reduced below 1/100,000 of an atmosphere. They are deflected phorescence of glass, heat and exert mechanical force upon bodies on which they fall, and cause such a target or Anticathode within the tube to emit "X" Rays.

CATHODE RAY DIRECTION FINDER : A radio-receiving apparatus with two frame aerials at right angles to each other, connected through amplifiers to the four plates of a Cathode Ray Oscillograph, the beam of which is so deflected as to show the direction of the received signal.

CATHODE RAY INSTRUMENTS : Measuring instrument, in which the indication is made by the movement of a beam of cathode rays falling on a fluorescent screen or photographic film, e.g. the Cathode Ray Oscillograph.

CATHODE RAY OSCILLOGRAPH : A cathode ray tube arranged with a fluorescent screen at one end upon which a narrow beam of cathode rays falls, and makes a visible spot of fluorescence. The beam is either in the field of two electromagnets, at right angles to one another and excited proportionally to the voltage and current in the circuit under investigation, or, more usually, passes two similarly crossed electrostatic fields. Owing to the deflection of the beam by these fields, the spot is caused to execute energy curves corresponding to the circuit. The instrument can be used up to radio-frequencies as a wattmeter, or for the study of waveforms or phase differences. It can also be arranged for photographic recording. An instrument in which a variety of electrical signals are presented on the screen of a cathode ray tube for examination. The signal under examination is used to deflect the electron beam of the CRT in one direction (usually the vertical) while and another known signal is used in the other direction. The composite signal is shown on the screen. Visualization of the input signal is achieved using the output from a sweep generator, usually called a time-base generator (see time base) and selecting the appropriate sweep speed. The sweep may be generated as a sawtooth waveform or initiated by an external trigger pulse. See also ACCELERATING GRID, FOCUSING ELECTRODE, and BEAM TRAP.

CATHODE RAY TELEVISION TUBE : A cathode ray tube used in a television receiver with a large fluorescent screen opposite the heated cathode, upon which falls the sharply focused electron beam modulated by the simplified incoming signals and deflected by a special set of electrodes in synchronism with the scanning beam in the transmitter. A funnel-shaped electron tube that converts electrical signals into a visible form. All CRTs have an electron gun to produce an electron beam, a grid that varies the electron beam intensity and hence the brightness, and a luminescent screen to produce the display. The electron beam is moved across the screen either by deflection plates or magnets. The *deflection sensitivity* of the tube is the distance moved by the spot on the screen per unit change in the deflecting field. Focusing of the beam may also be done electrostatiscally or electromagnetically or by a combination of methods (*see* diagrams). A greater degree of focusing is required when the electron beam is deflected towards the edges of the screen. The point at which the electron beam comes to a focus is the *crossover area* and the solid angle of the cone of electrons emerging from this area is the *beam angle*. For convenience the deflection and focusing coils are often mounted around the narrow neck of the tube as a single unit, termed a *scanning yoke*. Such an arrangement reduces the overall physical dimensions of the assembly and is particularly important when the tube contains more than one electron beam, as in the *double-beam* CRT or some forms of colour picture tube, and therefore requires more than one set of coils. The screen of the CRT may be coated with aluminimum

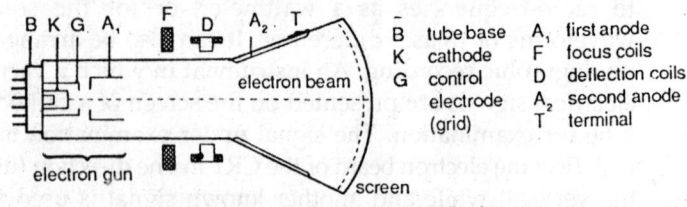

B	tube base	A₁ first anode
K	cathode	F focus coils
G	control	D deflection coils
	electrode	A₂ second anode
	(grid)	T terminal

b Electromagnetic focusing and deflection

on the inside and this coating held at anode potential. Such an *aluminized screen* prevents the accumulation of charge on the phosphor and improves its performance by increasing the visible output and reducing the effects of ion bombardment. In the case of a CRT in which the screen is not aluminized, the maximum potential difference that can be applied between the anode and cathode is limited to the value at which the secondary emission ratio of the screen rises to unity and is known as the *sticking potential*. See also KINESCOPE, ICONOSCOPE and also LABLE SPEED SCANNING.

CATHODE RAY TUBE : A Gas Discharge Tube suitable for the production of Cathode Rays.

CATHODE RAY VOLTMETER : A voltmeter in the form of a cathod-ray tube of known sensitivity of deflection.

CATHODE SPOT : The bright spot on the surface of the cathode of a mercury vapour arc where the action is concentrated, resulting in a very high current density.

CATHODE SPUTTERING : Syn. *cathode disintegration* : The slow disintegration of the cathode in a gas-discharge tube due to bombardment by positive ions. The phenon can be used for the deposition of thin metallic films on a surface. A method of depositing metal on a surface in a special low pressure vessel with parallel plate electrodes, the cathode being of the material to be deposited on the anode by the ionic stream. A voltage of at least 1000 volts is required.

CATHODE STREAM : The stream of ions or electrons, i.e. "Cathode Rays," projected from the cathode in Thermionic Valve, "X" Ray Tube.

CATHODOPHONE : A form of microphone in which the heated cathode causes ionisation of the air between it and a suitably placed anode at a high difference of potential from it. The movement of the diaphragm causes pulsation in the ionic stream resulting in apparatus is used in one system of sound films.

CATIONS (*Also Spelt Cathion, Kathion and Kation*) : The dissociated parts of molecules (or Ions) carrying positive charges which during electrolysis move towards the Cathode. Cf. ANIONS.

CATKIN TYPE VALVE : A thermionic valve in which a metal envelope, also forming the anode, is used.

CATWHISKER : The fine wire which makes contact with the crystal in the most common form of Crystal Detector.

CAUTERY APPARATUS : Apparatus for performing surgical operations by means of an electrically heated wire.

CAVITY RESONATOR Syn. *rhumbatron, resonant cavity* : The space within a closed or substantially closed conductor that will maintain an oscillating electromagnetic field when suitably excited externally. The resonant frequencies are determined by the size and shape of the cavity. The whole device has marked resonant effects and replaces tuned resonant circuits for high-frequency applications for which the latter are impracticable.

CBE : Computer-Based Education.

CDC : Control Data Corporation.

CCD : Charge-Coupled Device.

CCD FILTER : A circuit in which the ability of a charge-coupled device to provide a precise predetermined delay time to an analog signal is utilized in order to produce a desired signal-processing function.

CEASIUM CELL : A Photoelectric Cell with a sensitive surface of the metal ceasium, sensitive to the infrared portion of the spectrum.

CEEFAX : Tradename. See **Teletext.**

CELL : (1) The part of a spreadsheet that contains a number, label or formula. The smallest element of a memory capable of storing a 1 or a 0. (2) The physical location on or in a storage device that is identified by an address. (3) Any arrangement of electrodes between which is a substance subject to a change of electrical properties on a change of other conditions or vice versa, e.g. Selenium Cell, Kerr Cell, etc. A combination of electrodes of dissimilar materials and an electrolyte either capable of giving an e.m.f. owing to chemical action, as in a primary cell, or of producing a certain chemical action when supplied with current, as in an electrolytic cell, or both, as in an accumulator cell. (4) A cubicle or chamber for the accommodation of an item of high voltage switchgear. (5) The insulating tube which lines the slot and projects a little way from the end thereof in high tension armature windings, etc.

CELL CONSTANT : The ratio of the mean distance apart of the electrodes in a cell to the mean cross section of the current path.

CELLULAR AUTOMATA : Advances in microelectronics have made possible design of a wide variety of innovative structure with Very Large Scale Integrated (VLSI) circuits, However, design of complex VLSI circuits within reasonable cost demands two essential prerequisities 'regularity' and 'simplicity'. These two qualities are inherent to the Celluar Automato (CA) structure. A large variety of physical systems have been simulated using this structure.

CELLULOPHONE : A proposed electrical musical instrument in which the sound is produced in a loud speaker by the undulatory current passing through a photoelectric cell under intermittent illumination by a beam interrupted by a rotating perforated disc.

CENTIMETRE : A name sometimes used for the C.G.S. electrostatic unit of capacitance and for the electromagnetic unit of inductance on account of the Dimension of both these quantities being length if K and μ are neglected. This practice is not to be recommended.

CENTIMETRIC WAVES : Electro-magnetic waves of a wave-length between 0.1 and 0.01 metre.

CENTRAL BATTERY SIGNALLING : The provision of current for signalling purposes on a telephone system from a battery at the exchange.

CENTRAL BATTERY TELEPHONE (or telegraph) SYSTEM : A system of working in which current for all purposes is supplied from a battery at the exchange (or central office) instead of employing local batteries for every microphone or instrument. Also called Common Battery System. Cf. LOCAL BATTERY SYSTEM.

CENTRAL OFFICE : A term used in America for a Telephone Exchange, but more commonly limited in Great Britain to a telegraph office connected to a considerable number of lines.

CENTRAL PROCESSOR (CPU) : The heart of the computer system, where data is manipulated and calculations are performed. The CPU contains a control unit to interpret and execute the program and an arithmetic logic unit to perform computations and logical processes. It also routes information, controls input and output, and temporarily stores data.

CENTRE TAP : A connection made to the electrical centre of an electronic device such as a *resistor* or *transformer*.

CENTRE ZERO INSTRUMENTS : Polarised indicating instruments

capable of reading positive and negative values on both sides respectively of a zero in the centre of the scale.

CERAMIC CAPACITOR : A capacitor utilizing a ceramic as the dielectric material. The behaviour of the capacitor is determined by the electrical properties of the ceramic used; these vary widely but most have high permittivity allowing the capacitors to be smaller than most other types.

CERTAINTY FACTOR (CF) : A numeric value, often ranging from - 1.0 to +1.0, with 0.0 indicating no knowledge, that represents the degree of certainty attached to a fact or conclusion; certain inference rules can compute the certainty of inferences based on the certainties of their premises.

C.G.S. UNITS : The system of absolute electrical units adopted by the British Association in 1873, based on the Centimetre, the Gramme and the Second. Two of these systems have been developed depending respectively upon Electrostatic and Electromagnetic relations. The practical units in common use are derived from the latter. See *MKS* and *SI* UNITS.

CHAINING : The use of a pointer in a record to indicate the address of another record logically related to the first.

CHANGE TAPE : A paper tape or magnetic tape carrying information that is to be used to update filed information (the filed information is often on a master tape).

CHANNEL : (1) A computer component with logic capabilities that transfer input and output from main memory to secondary memory or peripherals, and vice versa. A specified frequency band or a particular path used in communications for the reception or transmission of electrical signals. (2) A route along which information many travel or be stored in a data-processing system or a computer.

CHANNEL STOPPER : (1) (in a *p-n-p* junction transistor) A means for limiting channel formation by surrounding the *n*-type base entirely with a ring of highly doped low resistively *p*-type material. (2) *(in an MOS integrated circuit)* A region of highly doped material of the same type as the lightly doped substrate. This increases the filed threshold voltage and inhibits the formation of spurious field-effect transistors caused when interconnection pass between adjacent drain regions.

CHANNEL WORKING *(in Cable Telegraphy)* : The use of one cable circuit as two channels by sending signals from two sources in rapid succession alternately and separating them by synchronous apparatus.

CHARACTER : Any signal letter of the alphabet, numeral, punctuation mark, or other symbol that a computer can read, write, and store. Character is synonymous with the term *byte*. A binary word used to express a number, letter, or punctuation. In general, an admissible mark, more specifically, a letter of the English alphabet, a non-numeric symbol such as H, or a numeric symbol such as 3.

CHARACTERISTIC CURVE : Any curve exhibiting the relation between inderdependent quantities in a piece of apparatus, such as e.m.f. and current.

CHARACTERISTIC RADIATION : The "X" Rays of particular frequency emitted by a substance as Secondary Radiation when bombarded by "X" Rays of sufficiently high frequency; sometimes called Fluorscent "X". It is found that the square roots of the frequencies of the rays given out by the elements form a series corresponding to their atomic numbers.

CHARACTER MODE : Transmission that is serial, a character at a time.

CHARACTERS PER SECOND [CPS] : Unit of measurement for the speed of data transfer. Most often used to describe printing speed, such as 132 cps.

CHARACTER STYLE : The stylistic appearance of text, such as italics, bold, or underline. Also called **typestyle**.

CHARGE : (1) The passage of a current through an accumulator in the direction which causes the chemical changes effecting a storage of energy therein. The number of ampere-hours passed in this way. Used as a verb, to impart a "charge." The quantity of electricity contained by a conductor, i.e. the degree of excess or deficiency of electrons. (See also UNIT CHARGE.) (2) By analogy the distribution of magnetism on a magnet is some times spoken of as a charge of magnetism.

CHARGE-COUPLED DEVICE (C.C.D.) : A charge-transfer device that consists of an array of MOS capacitors suitably designed so that they are coupled and therefore charges can be moved through the semiconductor substrate in a controlled manner. The CCD can perform a wide variety of electronic functions. The device is essentially an analog shift register and can be used for signal processing, it can be used to form analog or digital serial memories (see CCD filter) The device may also be used for imaging, as in the solid-state camera. A semiconductor device that has the structure of a MOSFET with an extremely long channel and many gates, perhaps 1000 closely spaced between the source and drain electrodes.

CHARGE-STORAGE DIODE : Syns : *snap-off diode; snapback diode; step-recovery diode*. A *p-n* junction diode in which carrier storage is the major factor contributing to the operation of the device.

CHARGE-TRANSFER DEVICE (C.T.D.) : A semiconductor device in which discrete packets of charge are transferred from one location to the next. Such devices can be used for the short-term storage of charge in a particular location provided that the storage time is short compared with the recombination time in the material.

CHARGING CURRENT : (1) The proper current strength at which a particular accumulator should be charged. (2) A current in the direction that will charge an accumulator. (3) A current flowing into a condenser.

CHARGING VOLTAGE : The voltage required to pass the proper charging current through an accumulator battery, varying up to nearly 2.5 volt per cell.

CHARIOT : The arm, rotating in synchronism with the type wheel in the Hughes printing telegraph which is tripped by one of a circle of pins at the moment during its rotation corresponding to a particular letter.

CHECK BIT : A word of a fixed-length group of characters to detect errors.

CHECK BOX : A small box associated with an option in a dialog box. A check mark in the check box means the option is on. An empty check box means the option is off. More than one check box can be checked. See also Dialog box and Radio button.

CHECK DIGIT : A number added to a key as a result of some calculation on the key. When data are entered, the computation is performed again and compared with the check digit to ensure correct entry. A redundant digit carried along as part of a field, and which can be calculated from the other digits in the field.

CHECK OUT : To debug or to test the accuracy of something.

CHECK RECEIVER : A receiver in the control room of a Broadcasting Station room by which the quality of the transmission can be gauged for regulating purposes.

CHECK SUM : An error-detecting word in a sequence of binary data which contains the sum of all the data words.

CHEREIX MODULATOR : A system of modulation in which two transmitting valves are used in phase opposition by which very deep modulation can be employed without distortion.

CHILDREN : The successors of a node in a tree, represented in pictures as nodes below their common parent node.

CHINESE ROOM : A metaphor used by John Searle to describe a computer that is programmed to display intelligent behavior; a non-Chinese can learn the Chinese language well enough to fool people without any knowledge of what he is saying, but he is not really understanding the language, just as a computer cannot really understand the world.

CHIP : Small component that contains a large amount of electronic circuitry. Thin silicon wafer on which electronic components are deposited in the form of integrated circuits. Chips are the building blocks of a computer and perform various functions, such as doing arithmetic, serving as the computer's memory, or controlling other chips.

CHOKE : A term used, particularly in radio-communication, for an Inductor or Choking Coil.

CHOKE CAPACITANCE COUPLING : A method of coupling thermionic valve amplifier stages resembling Resistance, Capacitance Coupling, except that an inductor is employed instead of a resistor in the anode circuit. See COUPLING also.

CHOKE MODULATOR : A Modulator in a radio-transmitting station, in which the oscillations from the anode circuit of a control valve are applied to a choking coil in series with the plate circuit of the main oscillating valve, thus varying the amplitude of its oscillations. Also called Heising Modulator.

CHOPPED WAVE : A surge in a transmission line which has been cut short by a flash over.

CHOPPERAMPLIFIER : An amplifier that amplifies direct-current signals by first converting them into alternating current and then using normal a.c. amplifying techniques. The conversion is achieved using a system of relays or a suitable vibrator.

CHRONOGRAPH (ELECTRIC) : Apparatus for measuring and recording, with a high degree of accuracy, intervals of time; usually by closing contacts so as to send current impulses which energise electromagnets causing a recording pen to make indications on a paper strip travelling at a known speed.

CHRONOPHER : A contact maker for sending impulses over a line from an observatory or other standard clock to give standard time signals.

CHRONOTRON : An electronic device that measures the time interval between events. A pulse is initiated by each event and the time interval is determined by the position of the pulses along a transmission line.

CHUNKING : A mechanism for learning and skill acquisition that generally involves abstracting information from previous problem-solving experiences to create new rules that lessen the amount of search required by skipping over intermediate states.

CIM : Computer Imput Micro Film, i.e. the process, or the input itself; it is not widely used. Input devices that have been produced relied on optical character recognition (OCR) to record alphanumeric data on microfilm or have read special micro-film on which the data was recorded as binary code.

CIRCUIT : (1) The combination of number of electrical devices and conductors that, when interconnected to form a conducting path, fulfill some desired function. (2) A physical (electrical) connection used for communications.

CIRCUIT BOARD : A single rigid board of insulating material on which an electrical circuit has been built. If often has an edge connector at one end for making all the connections to other circuits so that the board may be plugged into a piece of equipment. Circuit boards come in a variety of sizes, some of which are standardized. The term circuit card is often used synonymously but is sometimes considered smaller than a circuit board.

CIRCUIT DESIGN: (Logic Design) : The design of circuits and systems whose inputs and outputs are represented as discrete variables. These variables are commonly binary. i.e. two state, in nature. Design at the circuit level is usually done with truth tables and state tables; design at the system level is done with block diagrams or digital design languages.

CLARK CELL : A mercury-zinc standard cell with e.m.f. defined as 1.4345 volts at 15°C. It has been superseded by the Weston standard cell.

CLASS "A" AMPLIFIER : A linear amplifier in which output current flows over the whole of the input current cycle, i.e. the angle of flow equals 2π. These amplifiers have low distortion but low efficiency. Distortion can occur with large-signal operation due to the device transfer characteristics becoming nonlinear.

CLASS "AB" AMPLIFIER : A linear amplifier in which the output current flows for more than half but less than the whole of the input cycle, i.e. the angle of flow is between π and 2π. At low input-signal levels class "AB" amplifiers tend to operate as class "A" and at high input signal levels as class "B" amplifiers.

CLASS "B" : One of two FCC classification for electronic equipment. Class B means that the item can be used in the home or office setting. Class B is a more stringent requirement than class A.

CLASS "B" AMPLIFIER : Amplification by a valve in which the grid bias is sufficient to reduce the anode current to zero when no exciting grid voltage is present. Particularly suitable in Push-Pull systems of amplification. A linear amplifier operated so that the output current is cut off at zero input signal, i.e. the angle of flow equals π, and a half-wave rectified output is produced. Two transistors are required in order to duplicate the input waveform successfully, each one conducting for half of the input cycle (see push-pull operation). Class B amplifiers are highly efficient but suffer from crossover distortion.

CLASS "B" MODULATION : The application of similar principles as those of Class "B" Amplification to the final stage of a Modulation circuit.

CLASS "C" AMPLIFIER : A nonlinear amplifier in which the output current flows for less than half the input cycle, i.e. the angle of flow is less than π. Although more efficient than other types of amplifier, class "C" amplifiers introduced more distortion.

CLASS "D" AMPLIFIER : An amplifier operating by means of pulse-width modulation (see pulse modulation). The input signal produces a square wave modulated with respect to its mark space ratio. Pushpull switches are then operated by the modulated square wave so that one switch operates with a *high* input level and the other with a *low* input level. The resultant output current is proportional to the *mark space ratio* and hence to the input current.

CLEAR : To resert to zero i.e., to charge the contents to zeros in computer science.

CLEARING KEY : A key on a telephone exchange switchboard for actuating a Clearing Signal.

CLEARING LAMP : A signal lamp in a, junction line position on a telephone exchange switchboard which remains alight only when one end of the junction line is disconnected and thus indicates when the other end may be cleared.

CLEARING RELAY : A relay in a telephone exchange which controls a Clearing Lamp, or other clearing signal.

CLEARING SIGNAL : A signal given automatically or otherwise that a telephone line is finished with and that the connections may be cleared by withdrawing the plugs.

CLICK : To move the mouse pointer over an object or icon and press and release the mouse button once.

CLICK METHOD (Of Wave Measurement) : A method depending upon the face that an audible click is produced in a receiving telephone due to sudden commencement of ceasing of oscillation in a tuned circuit coupled to that in which the frequency of the oscillations to be determined when brought into syntony with it.

CLIPBOARD : Temporary storage that holds text and graphics. The *cut* and *copy* commands put text or graphics into the clipboard, erasing the clipboards's previous contents. The *paste* command copies clipboard data to a document. When you turn off the computer, the contents of the clipboard disappear.

CLOCK : A timing device in digital systems in general which produces pulses at a steady rate. In counter/timers the clock is the heart of the time-base circuitry, producing the time reference for frequency and time measurements. A *processor* contains an electronic pulse generator, known as a *clock*, which transmits synchronised streams of pulses to specific parts of the computer for the *interpretation* and *execution* of instructions according to requirements. Everything that happens in a computer is under the strict control of the clock. Various computers have clocks which operate at different speeds, which is one of the features which must be established when selecting a computer for a given task. A faster clock achieves faster profixed repetition rate. A clock cycle is considered to be complete cycle of the clock signal and will always contain one active transition of the processing, e.g. a 8 MHz (megahertz) computer will operate at twice the speed of a 4 MHz machine.

CLOCKED RS FLIP-FLOP : A circuit with two conditioning inputs which control the state to which the flip-flop will go at the arrival of the clock pulse. If the S(set) input is enabled, the flip-flop goes to 1 state when clocked. If the R (reset) input is enabled, the flip-flop gest to the 0 state when clocked. The clock pulse is required to change the state of the flip-flop.

CLOCK SPEED : The speed at which a microprocessor performs calculations. Often measured in megahertz (MHz). Common clock speeds range from 4.77 MHz to 67 MHz.

CLONE : Term used to describe a computer made by a local computer dealer that uses the same parts and program as a more popular computer. Also used to describe a program that mimics a more popular program. Often a derogatory term.

CLOSE : The act of removing a window from the screen or removing an open file.

CLOSE BOX : The small box that appears in the upper-left corner of window. Clicking in this close box with the mouse removes the window from the screen.

CLOSED CIRCUIT : A complete circuit formed entirely of conductive material. Cf. OPEN CIRCUIT.

CLOSED CIRCUIT SYSTEM : A system of telegraphy, signalling, fire alarms, lift-controls etc., in which, under normal conditions, the circuit is closed and a current passes, and the required signals are made by interrupting the current by opening the circuit.

CLUTCH : An *electromechanical* or *mechanical* device for connecting physically moving parts, as the electric motor to the card-moving mechanism in a punched card punch.

CML : Current-Mode Logic, basically equivalent to ECL.

CMOS : Complementary Metal Oxide Semiconductor, an insulated gate field effect digital logic unit using both P and N--MOS devices, a MOST or IC involving both P-channel and N-Channel MOS-FETs.

CMOS COMPLEMENTARY LOGIC : It consists of two full blown

network: a P-net and an N-net. Each of these networks can be obtained from the K-map by first finding the functions f(0) and f(1) in an optimum manner.

COAXIAL CABLE : A form of cable used for high frequency communication circuits, including television, in which the cores consist of a copper tube surrounding a central core, either supported by discs, with the intervening space filled by air or nitrogen or by a special insulating material.

COBOL : COmmon Business-Oriented Language, a computer language suitable for writing complicated business applications programs. It was developed by CODASYL, committee representing the U.S. Department of Defences, certain computer manufacturers, and major users of data processing equipment and is designed to express data manipulations and processing problems in English narrative form, in a precise and standard manner.

CODE : (1) A set of instructions written in a programming language. Also called source code. (2) A symbol combination for representing information, sometimes, an instruction. (3) A rule for transforming a message from one symbolic form (the source alphabet) into another (the target alphabet), usually without loss of information. The process of transformation is called *encoding* and its converse is called *decoding*. These processes are carried out by an encoder and a decoder respectively. (1) To write instructions for a computer system (2) To classify data according to arbitrary tables (3) To use a machine language (4) To program.

CODE CHECK : A testing for the presence of forbiden or illegal codes.

CODE CONVERTER : A device that converts binary inputs from one code system into binary outputs in another code system.

CODED DECIMAL : A way of representing decimal numbers as equivalent numbers in some other number system while retaining in part a 10's radix, as the binary coded decimal system or the excess 3 system.

CODED DECIMAL DIGIT : A decimal digit which is represented by a pattern of four or more bits.

COGNITION : A process of producing intelligent behavior by computing and/or thinking; the basic talent of AI and cognitive science is that all cognitive phenomena arise from some form of computation.

COGNITIVE MODELLING : An approach to creating intelligent behavior by directly simulating processes known or believed to occur in the human brain, as opposed to creating procedures that seem to "do the right thing" regardless of their mental authenticity.

COGNITIVE SCIENCE : A study of human intelligence (one of the goals of AI) using the techniques and knowledge of experimental psychology, computer science, neurobiology, linguistics, philosophy, and anthropology.

COGNITIVE STYLE : The orientation of an individual to approach decision in a particular way, for example, from an *analytic* or a *heuristic* view.

COHERENT RADIATION : Radiation in which the waves are in phase both spatially and temporarilly. A *coherent oscillator* is one that produces very pure well-defined oscillators, as in a laser.

COINCIDENT CURRENT MEMORY : A memory requiring additive currents in order to activate the memory cell.

COLLATE : To combine parts or all of two or more ordered sets (files) of information on any way such that a similar sequence is observed in the combined set.

COLLATION SEQUENCE : The rank order, usually arbitrary, of the character representation. For example, on collation sequence is the following : 0 through 9, H, @, -, =, +, A through I, ;, J through R, blank $, *, S through Z, „ (, and).

COLLECTOR : (1) Short for collector region. The region in a bipolar junction transistor into which carriers flow from the base through the collector junction. The electrode attached to this region is the collector electrode. See also transistor; semiconductor. (2) Short for collector electrode.

COLLECTOR ELECTRODE : See COLLECTOR.

COLLECTOR REGION : See COLLECTOR.

COLOUR CODE : A method of marking electronic parts, such as resistors, with information for the user. The value, tolerance, voltage rating, and any special characteristic of the component may be indicated using coloured bands or dots painted on it.

COLOUR GRAPHICS ADAPTER (CGA) : A color graphics standard that defines the resolution of a program. CGA video boards and monitors display text with noticeable graininess. The resolution of CGA graphics is 640 x 200. Newer computer use

EGA or VGA graphics. See ENCHANCED GRAPHICS ADAPTER and VIRTUAL GRAPHICS ARRAY.

COLOR SEPARATION OVERLAY (CSO) : A technique used in colour television for superimposing part of one scene on another. When a particular colour, such as blue, occurs in one scene viewed by a camera, the output of another camera filming a different scene is automatically switched into repalce the areas of the chosen colour in the original picture.

COLOSSUS-1 : Special electric computer (British) in 1943 to counterfeit ENIGMA. It was first British relay based computer.

COLUMN : A vertical section of printed text on a page, or a vertical row of cells in a spreadsheet.

COM : Means of a COM recorder connected to a processor or, in off-line mode, recorded on magnetic tape first. Whichever method is used, a COM recorder produces imags either on 16 mm roll film or 105 x 148 mm *microfiche*. The *film* or *microfiche* is then processed and used as the negative for the production of copies for distribution. Information is retrieved by a micro film viewer, reader/printer or a microfiche *viewer* and *demand printer*. The facility has been available since the early 1960s an currently most COM devices are run off-line.

COMA : A plumelike distortion of the spot occurring when the focusing elements of the electron gun are misaligned.

COMBINATION LOGIC : A logic system in which the output occurs in direct, immediate response to the input.

COMBINATIONAL CIRCUIT *(Combinatorial Circuit)* : A logic circuit whose outputs at a specified time are a function only of the inputs at that time. In practice, any physically realizable combinational circuit will have a finite transit time, or delay, between the inputs changing and the outputs changing. The intention of the term combinational is to include algebraic elements (AND gates, OR gates, etc.) and preclude memory elements (flip flops, etc.). *Analysis* and *synthesis* of combinational circuits is facilitated by Boolean algebra and Karnaugh maps.

COMBINATORIAL EXPLOSION : A fundamental problem in any search-based system that limits its practical usefulness with large problem spaces, in which the number of alternatives to explore increases manifold as fast as the search progresses; for example, a search tree with branching factor 5 can only be explored one level deeper by a computer 5 times faster.

COMMAND : A *pulse, signal, word,* or *series of letters* that tells a computer to start, stop, or continue an operation in an instruction. Command is often used incorrectly as a synonym for *instruction.* In fact it is a set of signals initiating specific action in an automatic computer, a set of electronics signals to do part of an operation.

COMMON-BASE CONNECTION : This type of connection is commonly used as a voltage amplifier stage. Syn. *grounded base connection.* A method of operating a transistor in which the base is common to both the input and output circuits and is usually earthed. The emitter is used as the input terminal and the collector as the output terminal.

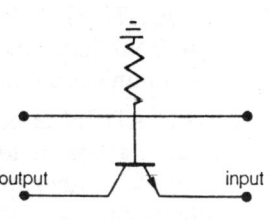

Common-base connection

COMMON-BASE CURRENT GAIN : See common-base forward-current transfer ratio; transistor.

COMMON COLLECTOR CONNECTION: Syn. *Grounded collector connection.* A method of operating transistors in which the collector is common to both the input and output circuit and usually earthed, used in EMITTER FOLLOWER CIRCUITS.

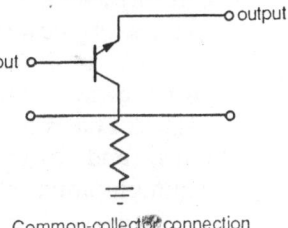

Common-collector connection

COMMON-CONTROL AUTOMATIC TELEPHONE SYSTEM : Telephone system having similar objects as the By path System effected in a different way.

COMMON-EMMITTER CONNECTION : Syn. *grounded-emitter connection.* A method of operating a transistor in which the emitter is common to both the input and output circuits and is usually earthed. The *base* is used as the input terminal and the *collector* as the output terminal. Connection is used for power amplification with a nonsaturated transistor and for switching with the transistor in saturation.

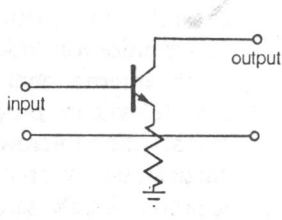

Common-emitter connection

COMMON-FREQUENCY WORKING (Of Broadcasting Stations) : The working of several Relay Stations simultaneously at the same frequency when giving the same programme.

COMMON LANGUAGE : Paper tape, or secondarily punched cards.

COMMON-SENCE KNOWLEDGE : The everyday knowledge about
the world and its contents that underlies most other, domain-
specific knowledge; it is the kind of declarative knowledge
we usually want to represent with semantic networks and
frame systems.

COMMUNICATING PARTITIONS : The communicating partition is
the boundary shared by two partitions through which the
communication between two processes takes place.

COMMUNICATION (Electricals) : The whole subject of the science
and practice of telegraphy and telephony including radio,
and of all other including radio, and of all other methods of
electrical signalling, See also LIGHT CURRENT
ENGINEERING.

COMMUNICATION CHANNEL : A wider term than communication
circuit, including provision for radio or carrier current *telegraphy*
and *telephony*, etc., and all forms of Phantom and Superposed
circuits. Physical means of connecting device at one location
to similar or different device at another location which may
be far away for the purpose of transmitting and receiving
data. Coaxial cables, fiber optics, microwave signals, telephone
lines, and satellite communications are the common
communications channels.

COMMUNICATIONS PROTOCOL : A set of communication rules
that provides for error checking between devices of a
communication system for ensuring that transmitted data are
not lost.

COMMUNICATIONS SATELLITE (COMSAT) : An artificial unmanned
satellite in earth orbit that provides high-capacity
communication links between widely separated locations on
earth. International telephone services and the exchange of
live television programmes and news are achieved by
transmitting microwave signals, suitably modulated, from an
earth station to an orbiting satellite and back to another earth
location. Earth satellites placed in different spots in the
geostationary orbit 36000 km (22500 mile) above the equator
that serve as relay stations for communications signals
transmitted from each stations. These satellites orbit Earth
once every 24 hours, giving the impression that they are

"parked" in one spot over the equator. Once in this orbit, a satellite is capable of reaching 43 percent of Earth's surface with a single radio signal. Most communications satellites are launched by NASA, weight several thousand ponds, and are powered by solar panels. A few communications satellites are Comstar, Westar, Intelsat V, Satcom, and Marisat.

COMPANDOR : A system of improving speech to noise ratio in radio-reception by compressing the range of the emitted signal energy variation by a "compressor" and expanding it again at the receiving apparatus by an "expander."

COMPARATOR : A device used in digital circuits to determine whether two numbers of bits of information are equal: in analog circuits, to determine if two voltage levels are equal.

COMPATIBILITY : (1) Property of some electronic devices that allows programs written for one computer to run on another (compatible) computer, even though it is a different model. (2) Ability of different devices, such as a computer and a printer, to work together. (3) Term used to describe a computer that uses the same parts and programs as a more popular computer such as IBM or Macintosh. Compatible computers, such as Epson or Compaq, have national recoginition.

COMPENSATOR : In general, a piece of apparatus which corrects some disturbing action, e.g. in radio reception, a variable condenser placed between the *grid* of one valve and the *plate* of a scucceeding valve to check self oscilation. Also sometimes used as the equivalent of Auto-Transformer, particularly in connection with its applications to balancing and motor starting, and for machines or apparatus used for Power Factor Compensation.

COMPILE : The process of translating a programming language, such as BASIC or Pascal, into machine code. A computer program that translates a program written in a problem-oriented language into a program of instructions similar to, or in, the language of the computer. A program for a computer that generates programs to be understood by the computer on the assembly language level using simple English input into the computer. A program is compiled before any part of it is run. CF ASSEMBLE.

COMPILING ROUTINE, COMPILER : A routine by means of which a computer can translate a *source* program into an *object* program by assembling and copying from other programs stored in a library of routines.

COMPLEMENT : A quantity which is derived from a given quantity by a reference to the radix or the radix minus 1. Number used to represent the negative of a given number, obtained by subtracting each digit of the number from the number representing its base and, in the case of two's complement and ten's complement, adding unity to the last significant digit. Also called radix complement.

COMPLEMENTARY RING COUNTER : A ring counter in which the inverted output is fed into the input; also known as a Johnson counter.

COMPLIER : A translator for high-level languages. Generally, several machine-language statements are generated for each high-level language statement.

COMPTON EFFECT : The change of wave length of "X" Rays when scattered by incidence on certain surfaces.

COMPUSERVE : Major information service net work in U.S.A. used by individuals as well as businesses. Carriers timely news features, stock market reports, electronic mail, educational programs, programming aids, and more. Personal computer owners can reference the Compu-Serve network via the common telephone system.

COMPUTER : A device or system that is capable of carrying out a sequence of operations in a distinctly and explicitly defined manner. The operations are frequently numerical computations or data manipulations but also include input/output; the operations within the sequence may depend on particular data values. The definition of the sequence is called the *program*. A Computer can have either a stored program or wired program. A stored program may exist in an alterable (read-write or RAM) memory or in a non alterable (ROM) memory.

Most computer systems consist of three basic elements: the central processing unit (CPU), the main memory and peripheral devices (see diagram). The central processing unit controls the operation of the system and performs arithmetical and logical operations on the data. The main memory stores the program and the data is units of bytes or words, each of which has a unique address, so that they may be retrieved quickly by the CPU. A buffer memory is emplyed by larger

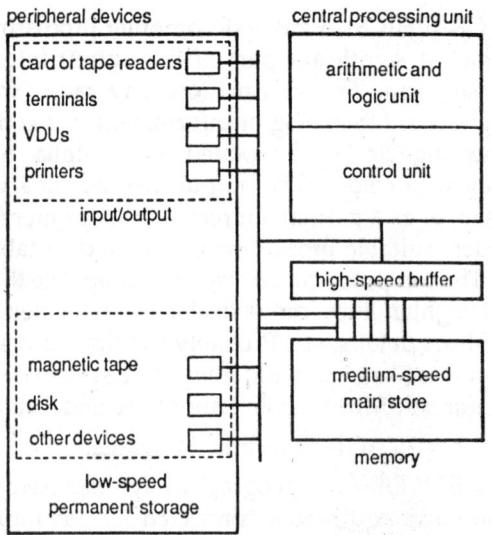

peripheral devices   central processing unit

Elements of computer system

systems: it interacts directly with the CPU and transfers information at extremely high speed. the information currently in active use is held in the buffer. The peripheral devices perform input/output and permanent storage of information. A complete computer system consists of the hardware—the electronic and other devices—and complementary software—the set of programs and data.

COMPUTER-AIDED DESIGN : The use of a computer with sophisticated software for the development of designs of various types using a light pen in conjunction with a video screen is known as *computer-aided design*. The facilities make it possible to modify the shape of an entity and to rotate it to obtain a three-dimensional view in order to assess its features from various perspectives. The technique enables standard shapes to tbe stored on disc and accessed when required for incorporation in other designs. It saves considerable time in the design activity and improves quality as designs can be speedily checked to ensure compatibility with specifications. Erros can be corrected by light pen. The technique is widely used for the design of aircraft, cars and computers as well as a wide range of other products.

COMPUTER-AIDED INSTRUCTION (CAI) : The use of an interactive computer to provide or supplement instruction on some topic.

Also called COMPUTER-BASED TRAINING.

COMPUTER GRAPHICS : A mode of computer processing and output in which a significant part of the output information is in pictorial form. The information may range from a simple histogram or other plog of information to a complex map or engineering design annotated with alpha numerics and displayed in colour. The output may be via a visual display terminal or as a permanent record via a printer/plotter or XY recorder. Suitable input devices inclue data tablets and light pens. The computer can be made to manipulate the information, e.g. straighten lines, move or delete designated area, expand or contract details, etc. Probably the first successful graphics system was Sketchpad and interactive system devised by Ivan Suther and at MIR Lincoln Laboratory and Published in 1963.

COMPUTER LANGUAGE : Machine language.

COMPUTER NETWORK : A geographically dispersed configuration of computer equipment connected by communication lines and capable of load sharing, distributive processing, and automatic communication between the computers within the network.

COMPUTER OUTPUT ON MICROFILM (COM) : This is an alternative form of output from a computer other than printed output. It is approximately ten times faster to produce and reduce stationery costs and storage space requirements. Computer output may be either recorded directly on microfilm by a program. See COM also.

COMPUTER SCIENCE : The study of computers, their underlying principles and use. It comprises topics such as: programming, information structures; software engineering; programming language; cmpilers and operating systems, hardware design and testing; computer system architecture; computer networks and interfacing systems analysis and design; therories of information systems, and computation; applied mathematics and electronics; computing techniques (e.g. graphics, simulatin, artificial intelligence techniques); social, economic organizational, political, legal and historical aspects of computing. It is not a science in the strict sense of being a discipline employing scientific method to explain phenomena in nature or society (though it has connections with physics, psychology, and behavioral science) but rather in the looser sense of being a systematic body of knowledge with a foundation of theory,.

Since however it is ultimately concerned with practical problems concerning the design and construction of useful systems, within constraints of cost and acceptability, it is far more a branch of engineering than it is a science. Note that other branches of engineering have their theoretical foundations in separatly named disciplines such as physics or chemistry; computer science, on the other hand, embraces both theoretical foundations and practical engineering.

COMPUTER VISION : See IMAGE UNDERSTANDING.

COM-RECORDER : Device that records computer output on photosensitive film in microscopic form.

CONCATENATE : To join together data sets, such as files, in a series to form one data set, such as one new file. The term concentrate literally means to "to link together". A concentrated data set is a collection of logically connected data sets.

CONCAVE CATHODE : A cathode in an "X" Ray or other Cathode Ray Tube, pressenting a concave surface towards the anode or anti-cathode.

CONCENTRATOR : A device with some local storage that accepts data from several low-speed lines and transmits them over a single high-speed line to a computer installatior .

CONCEPTUAL DEPENDENCY (CD) : A theory of semantics developed for representing all knowledge in natural language understanding applications that is based on a system of eleven primitive acts into which all others can be decomposed. (See also scripts.)

CONDENSANCE : Another name for Capacitance Reactance; the component of the impedance due to the effect of capacitance in the circuit, corresponding to Inductive Reactance which is the component due to electromagnetic causes. See REACTANCE.

CONDENSER (or capacitor) : A system consisting of two conductors of considerable surface separated by a comparatively thin dielectric, and thus possessing an appreciable Capacitance. A common type consists of a pack of a alternate impregnated with paraffin wax or petroleum jelly, the odd sheets of tinfoil connected to one terminal and the even sheets to the other. Used for a variety of purposes in electrical apparatus on account of their properties of absorbing considerable charges when "charged" and delivering them up when "discharged," of allowing alternating but not direct currents to pass, advancing the phase of an to pass, advacing the phase of an alternating current and balancing the effect of inductance. On account of other meanings of the word, the term Capacitor is now preferred for any system of appreciable capacitance. The term condenser has also been used for other classes of apparatus having a phase advancing effect. See CAPACITOR ROTARY and SYNCHRONOUS CONDENERS.

CONDENSER RECEIVER : An electrostatic telephone receiver consisting essentially of a condenser with one electrode free to vibrate under the influence of the variable attraction of the other.

CONDENSER TRANSMITTER : An apparatus similar to a Condenser Receiver in which the variations in capacitance produced by the vibrations of the diaphragm produce modulations of the line current.

CONDENSIVE REACTANCE : See CAPACITIVE REACTANCE.

CONDITION : An element in the premise part of a production rule; often,the conditions are conjoined, so that all must be satisfied for the rule to trigger.

CONDITIONAL TRANSFER : An instruction which causes the automatic computer or either to continue with the next instruction in the original sequence or to change control or some other stated instruction, depending upon the result of some logic operation.

CONDUCTANCE : The reciprocal of Resistance or, in the case of alternating current, the energy component in the reciprocal of Impedance, Measured in Mhos. Symbol: G or g.

CONDUCTION : The transmission of electric (or heat) energy through a substance that does not itself move. In electrical conductors, such as metals, it entails the migration of electrons; in gases and solids it results from the migration of ions. The transmission of electricity through a body in the form of a current, by a drift of negative electrons in one direction. See also ELECTROLYTIC CONDUCTION.

CONDUCTIVITY : The power of a material to conduct electric currents; measured either as the reciprocal of Volume Resistivity or as the percentage which the conductivity of one substance is of that of a standard substance (such as pure copper). Symbol: G or g. Cf. CONDUCTANCE etc.

CONDUCTOR : Any material passessing an appreciable proportion of free electrons, which therefore permits a current to pass. Any particular object, such as a wire or group of strands in a cable, made of such a material.

CONE LOUD SPEAKER : A Loud Speaker of the hornless type in which a large conical surface of thin flexible material is thrown into vibration by an electromagnetically actuated system.

CONFIDENCE VALUE : See CERTAINTY FACTOR.

CONFIGURATION : In hardware, a group of interrelated devices that constitute a system. In software, that total of the software modules and their inter relationships. The combination of hardware and software making a complete computer system.

CONFLICT SET : A structure that collects all the rules in a production system that can fire in any given cycle of the system; that is, all productions whose antecedent conditions are satisfied by the current contents of working memory.

CONFLICT RESOLUTION STRATEGY : A predefined method for deciding which rule in a production system should fire in a cycle when more than one are triggered (the conflict set contains more than one rule); examples would include selecting the rule with the highest certainly factor, the one highest in a pre-established order, or the first one to enter the conflict set.

CONFUSION REFLECTOR : A device used to produce false signals with radar. Strips of paper or metal foil may be used: long strips of metal foil are called *rope* or *window*.

CONINCIDENCE CIRCUIT : A circuit with two or more input terminals that produces an output signal only when an input signal is received by each input either *simultaneously* or within a *specified time interval*.

CONJUNCTION AND : A used in symbolic logic.

CONNECTIONISM : A framework and methodology for cognitive modelling that represents information processing at a fundamentally lower level than traditional symbol manipulation, a level whose primitive units resemble neurons and their connections rather than symbols and higher level structures built out of them.

CONNECTION MACHINE : A massively parallel computer with tens of thousands of highly interconnected, tiny 1-bit processors; it is suitable for connectionist simulations and other large-input tasks requiring high speed.

CONNECTIVITY : Related to having different computers and devices able to communicate with each other.

CONNECTOR : (1) Any device by which two conductors may be joined together so that current can pass from one to the other. (2) A name sometimes given to the Lins Selector in an Automatic Telephone Exchange which connects the *calling* line to the *called* line.

CONSOLE : A console is that part of an electronic device which facilitates communication. A part of the control unit used by the human operator to monitor computer performance.

CONSTANT : A never-changing value of data item.

CONSTANTAN : A copper-nickel alloy of high specific resistance and almost negligible temperature coefficient at ordinary temperatures; containing about 45 per cent of nickel.

CONSTANT CURRENT SOURCE : A circuit that ideally has an infinitely high output inpedance so that the output current is independent of voltage. In practice sufficiently high output impedances are only achieved for a limited range of output voltages.

CONSTANT VOLTAGE SOURCE : A source of voltage that produces a substantially constant value of voltage independently of the current supplied by it. An ideal voltage source has an internal impedance of zero.

CONSTRAINT SATISFACTION : A powerful problem-solving method in AI that views the problem space as a set of constraints on the solution that can be applied at intermediates points to drastically curtail search; in connectionist networks, local constraints represented by lateral inhibition can be satisfied by a relaxation process. (See also Waltz's algorithm.)

CONSTRAINT SUSPENSION : A new algorithm for candidate generation that represents the system being diagnosed as a network of constraints and then selectively suspends those constraints one-by-one; a component whose behavior beomces globally consistent when the rules governing its operation are suspended is a likely candidate.

CONTACT : The bringing together of two conductors so that current may flow. The resitance at the surface of contact is the *contact resistance*. If the conductors are made from two different materials, a difference of potential will arise when they are placed in contact. This *contact potential* results from a difference in the work functions of the two materials and is usually of the order of a few tenths of volt.

CONTEXT FREE GRAMMER (CFG) : A grammer whose constituents are terminal symbols, non-terminal symbols, and rewrite rules have a single non-terminal on their left-hand sides and a string of any symbols on their righthand sides; therefore, the legal rewrintings of a non-terminal symbol do not depend on the other symbols near it.

CONTEXT-SENSITIVE : Ability to perceive the current conditions of an event. Often used to describe help systems as context-sensitive help systems.

CONTEXT SENSITIVE GRAMMER (C.S.G.) A grammer identical to a context-free grammer but whose rewrite rules may contain both terminal and nonterminal symbols of their left-handed sides.

CONTINUOUS FORM : Paper, mailing labels, or cards, designed for computer printers and connected through perforations at defined lengths.

CONTINUOUS RATING : The load which a machine, or apparatus, can deal with continuously for an unlimited time without overheating or other injury.

CONTRAST AMPLIFIER : See NEGATIVE AUTOMATIC VOLUME CONTROL.

CONTRAST PHOTOMETER : The class of photometer in which two surfaces, illuminated by different sources are contrasted and adjustment of the distances of the sources is made until the illuminations appear equal, e.g. the Lummer-Brodhun Photometer.

CONTROL COUNTER : A register that contains the address of the next instruction to be executed (not always present in multi-address computers).

CONTROL DATA CORPORATION : Large manufacturer of computer equipment, including super computers.

CONTROL ELECTRODE : An electrode to which a signal is applied in order to produce changes in the currents of one or more of the other electrodes. In a bipolar transistor with common-emitter connection, the base electrode is the control electrode; the gate electrode is the control electrode of a field-effect transistor; in a thermionic valve it is the control grid; in a cathode-ray tube it is the modulator electrode.

CONTROL GRID : (1) The particular Grid in a Thermionic Valve with more than one grid, used as the control Electrode, sometimes called the Input Grid (2) A grid to control the discharge in any form of vacuum, gas or vapour arc tube. Cf. SCREEN GRID and SPACE CHARGE GRID.

CONTROLLED VARIABLE : When a system produces outputs, these are measured by a sensor which depicts the actual state of the system. The specific outputs being measured are referred to as controlled variables.

CONTROL OF COMPUTER SYSTEMS : Techniques to ensure the intergity and accuracy of computer processing.

CONTROL PANEL : A panel forming the whole or part of a Control Board.

CONTROL REGISTER : See instruction register.

CONTROL ROOM *(in a Broadcasting Stations)* : The room which contains the amplifiers and apparatus controlling the output of the microphones before being passed on to the modulation system of the transmitter, including provision for combining the output of several microphones at suitable relative strengths. See MIXING. See also FADE UNIT.

CONTROL SEQUENCE : The order in which instruction are executed.

CONTROL STRATEGY : An overall method of control the flow or processing in problem solving algorithms; examples include depth-first or breadth-first for search and forward or backward chaining for production systems.

CONTROL SYSTEM : A system in which the outputs are compared with a standard signal and an error signal fed to the input to modify the process.

CONTROL SYSTEM *(In a Transmitter)* : That part of the equipment which controls the amplification and superposition upon the radio-frequency oscillations of the audiofrequency oscillations from the microphone or incoming line.

CONTROLLER : A device on the IEC/IEEE bus which is capable of setting the ATN line to determine the transmission mode sending bus command, and addressing devices on the bus as talken and/or listeners.

CONTROL UNIT : The part of the hardware of an automatic computer which directs the sequence of operations, interprets the instructions, initiates action, and directs the circuit that execute the commands.

CONVERTERS : A device which converts information presented into a form compatible with instrument's circuity in digital counters a converter is mostly used to exted the frequency range of digital-frequency meter. It converts the input frequency down to a frequency who which can be handled by the counter.

CONVOLUTION : See filtering.

COOLED ANODE VALVE : A Thermionic Valve in which the anode forms part of the envelope and is cooled by air, water, or oil.

COOLIDGE : ("X" RAY) TUBE: A form of powerful "X" Ray tube with incandescent tungsten Cathode and water-cooled or air-cooled Anode working with a higher vacuum than eariler tubes, i.e. with gas pressures as low as about 0.003 millionth of an atmosphere, and having a self-rectifying effect which enables alternating current to be used without a rectifier.

COOLIDGE RADIATOR TUBE : A form of Coolidge Tube without water cooling of the anode,which is fitted with radiating fins.

CO-PROCESSOR : A microprocessor device connected to a central microprocessor that performs specialized computations (such as floating-point arithmetic) much more efficiently than the CPU alone.

COPY PROTECTION : A method that makes copying a floppy disk theoretically impossible. Most often found on games to prevent people from illegally duplicating and distributing the disks to others.

CORAL : A high level programming language for real-time applications developed by the Royal Radar Establishment at Malvern, England.

CORD : A flexiable cable fitted with plugs on a telephone exchange switchboard provided to make connection with the Jacks.

CORD CIRCUIT : A circuit connected to the calling and answering plugs at an operator's position or telephone exchange switchboard, including the operator's instrument, keys, relays, signal lamps, etc., as well as the cords and plugs.

CORDLESS BOARD : A telephone exchange switchboard in which the required connections are made entirely by switches and not by plugs and flexible cords. Practically confined to small branch boards.

CORE MEMORY : A memory using small toroids of ferromagnetic material to store binary data.

CORE STORAGE : A medium of computer storage; for most second and third-generation computers, the term is used synonymously with "primary memory."

CORPUSCULAR RADIATION : The emission of those classes of "rays" which consist of the projection of material particles, e.g. alpha, beta, and delta rays.

COSMIC : Computer Software Management and Information Center.

COSMIC RADIATION (*or Cosmic Rays*) : Radiation of even shorter wave length than Gamma Rays and of great penetrating power, which can be detected particularly at high altitudes by the ionization which they cause, also called Ultra Gamma and Penetrating Radiation or Rays. Thought by some to originate in the upper atmosphere and by others to be extra-terrestrial cosmic origin and variously explained as due to disintegration of atoms or to radiation due to the shock of the recreation of matter from radiation.

COST FUNCTION : A procedure, modelled as a mathematical function mapping states onto numbers, that computes the expense of getting to a state (that has already been reached) in a state-space from the start state.

COULOMB (C). : A "practical" unit of quantity of electricity (named after C.A. de Coulomb, 1736-1806), being the quantity which has passed when one ampere has flowed for one second, i.e. 10^{-2} absolute C.G.S. electromagnetic units, at one time called the Weber. For engineering purposes the amperehour is more commonly used. (One *Ampere hour* = 3,600 coulombs). See also FARADAY.

COULOMB'S LAW : The mechanical force between two charged bodies is directly proportional to their charged and inversely proportional to the square of the distance between them.

COUNTER : (1) A device which records the number of pulses it has received at its input. Depending on the circuitry, it can count and store in binary or another code. Abbreviation for digital counter/timer. (2) A device that detects and counts individual particles and photons. The term is applied to the detector and to the instruments itself.

COUNTERPOISE : (1) A substitue for a conductive earth connection to a radio aerial, consisting of a considerable area of metal sheet or net spread over the ground or supported clear of it, and serving as one plate of a large capacitor on which the earth is the other or, in the case of aircraft, of a system of wires on the frame of the machine. (Also called *Balancing Capacitance, Capacitance Earth* and *Lower Capacitance*.) (2) An earthed conductor or group of conductors above or below ground, connected to a transmission tower to lower earth resistance.

COUPLING : (1) The process of inter connecting or integrating two or more related sub-systems for the purpose of increasing administrative efficiency. (2) Any method of joining circuits so that energy can be interchanged between/among them.

COUPLING COEFFICIENT (Of Two Oscillating Circuits) : The ratio of the common impedance by which two circuits are coupled to the geometric mean of the impedances, of like kind, of the two circuits. Symbol : k.

COUPLING ERROR *(In Direction Finders)* : A slight error due to the magnetic coupling of the field and search coils not following an exact cosine law owing to the spread of the windings.

COURSE-INDICATING RADIO BEACON : A *Radio Beacon* giving characteristic signals in certain definite direction only. Cf. OMNI-RADIO BEACON.

COVER : The property of a term to be a 1 upon input of the covered minterm.

CP : Central Processor.

CPC : Card Programmed Calculator.

CPM : Critical Path Method.

CPU (Central Processing Unit) : The part of the computer that controls the interpretation and execution of instructions, the arithmetic functions, and the I/O channels; the CPU contains a number of registers.

CRAM : Card Random Access Memory.

CRASH : A state when any electronic machine stops working unexpectedly.

CRAY RESEARCH : A company formed in 1972 by Symour Cray at which a series of high-speed super computers have been developed, including Cray-1 Cray-2, and Cray's XMP.

CREDIT ASSIGNMENT PROBLEM : in learning is that of determining which procedures in the performance element contributed to a correct solution and which did not, so that the learning element can modify the appropriate portions of the system.

CREED PRINTING TELEGRAPH SYSTEM : A system of type-printing

telegraphy in which the message is first unched in the ordinary Morse code on a paper strip and put through an instrument similar to the Wheat stone transmitter. The received signal currents actuate through relays a punching instrument which reproduces the original strip. This is then passed through the printer, where a mechancial selecting device depending upon the alignment of holes in a pile of strips displaced according to the position of the holes in the punched strip, causes the right letter to be printed by a typewriter mechanism . In some of the later forms, the five unit alphabet replaces the Morse alphabet. cf. MURRAY, HUGHES and STELJES SYSTEMS.

CREED RECEIVING PERFORATOR : The instrument in the Creed printing telegraph which punches the strip at the receiving end.

CRITICAL PATH METHOD (CPM) : A project planning, management, and scheduling method that devides a project into activities, each with a statement of time and resources required, and their precedent dependences. These activities are then connected in a graph that expresses the *dependencies* and the *times*. The critical path is defined as the path through the graph that requires the maximum time. Variations of this method allow for statistically distributed time and resources, time/resource tradeoffs, etc.

CROOKE'S TUBE : A vaccum tube so exhausted so as to create Crookes Dark space to fill the tube and the phenomena of cathode rays to be observed. It works at a pressure below 1/10000 of an atmosphere.

CROSSED-FIELD MICROWAVE TUBE Syn.: M-type microwave tube, See MICROWAVE TUBE.

CROSS MODULATION : The effect of curvature in the amplification characteristic of a radio-frequency amplifying valve, due to secondary emission from the screen-grid in decreasing the selectivity when a frequency for which the circuit is tuned is present.

CROSSOVER AREA : See CATHODE RAY TELEVISION TUBE.

CROSSOVER DISTORTION : A distortion that occurs in push-pull operation when the transistors are not operating in the correct phase with each other.

CROSSOVER FREQUENCY Syn. : *Dividing Network.* A type of filter circuit that divides the frequency range passed between two paths. Frequencies above a specified value pass through one path and those below that value through another. The frequency at which the output passes from one channel to other is the crossover frequency and at that frequency the outputs are equal. Such networks are widely used with loudspeakers to separate the *bass* and *treble* components. See loudspeaker.

CROSSTALK : Inductive interference between separate circuits in a tenency of one telephone circuit to cause cross talk in another by measurement of current induced in the disturbed circuit by a standard vibrator connected to the resultant field. A signal that has leaked or "crossed" from one communication channel to adjacent channel. This interferes with (causes errors on) the second channel. Cross talk is usually associated with physical communication channels, such as an Rs. 232 connection.

CRT : (CATHODE-RAY TUBE) A terminal resembling on ordinary television set that can display a large number of characters rapidly; many also have graphics capabilities.

CRYPTODYNE RECEPTION : A system of radio reception in which Bigrid Valves are used to reduce the number of valves required by combining the functions of two valves in one.

CRYPTOGRAPHY : The protection of a message by either of two encryption methods, code or cipher, that transform *plain* text to cipher text. The first of these involves a *one* for one substitution of a cipher code group for a plain text code group. The list of these substitutions is called a *code book*; the codebook must be kept secret in order to protect the information. A cipher process involves changing the plain text into cipher text by means of a cryptographic transformation, usually in such a way that each bit, character, or word of cipher text. The cryptographic transformation is accomplished by a specific algorithm or device, which is normally in two parts.

CRYSTAL CONTROLLED TRANSMITTER : A transmitter in which the carrier frequency is produced by a piezoelectric oscillator.

CRYSTAL COUNTER : A radiation counter that detects and counts subatomic particles. A particle or photon striking a crystal with a potential difference across it produces electron-ion pairs that increase its conductivity. This results in current pulses that may then be counted.

CRYSTAL CUTTER : A means of cutting gramophone records using a piezoelectric crystal. The electrical signals from the recording system cause mechanical displacements in a piezoelectric crystal used as a cutting stylus.

CRYSTAL DETECTOR : A detector of electric waves used in radio-communication depending upon the uni-directional conductivity of a contact between a crystal and another substance: used in a suitably tuned circuit without a battery in series with a telephone receiver in which the successive rectified oscillations produce a cumulative effect on the diaphragm and become audible.

CRYSTAL DRIVE : A method of Independent Drive in a radio-transmitter employing a Piezoelectric Oscillator to control the frequency.

CRYSTAL FILTER : A filter that uses piezoelectric crystals to provide its resonant or anti-resonant circuits.

CRYSTAL MICROPHONE : See PIEZOELECTRIC MICROPHONE.

CRYSTAL RECEIVER *(Or Crystal Set)* : A simple receiving set for radio-telephony, etc., consisting of a Crystal Detector in series with a telephone in a tuned aerial circuit.

CTL : Complementary Transistor Logic; a logic system using emitter-coupled circuits with a combination of PNP and NPN transistors.

CUMULATIVE GRID RECTIFICATION : Rectification by a thermionic valve depending upon the uni-directional conductivity between the *grid* and the *cathode*. A small condenser is included in the grid circuit which allows a small negative potential to build up on the grid, due to excessive high-frequency oscillations but provided with a Grid Leak to prevent it reaching a value which would prevent grid current from passing. Also called Grid Leak Rectification. Cf. ANODE BEND RECTIFICATION.

CURB TRANSMISSION : Methods of telegraph transmission ensuring a sharp end to a signal current in a long submarine telegraph cable to improve the speed of signalling by sending at the end of each signal a "curb" current in the opposite direction, or by a suitable arrangment of condensers.

CURRENT *(Electric)* : The Passage of electricity through a body by conduction i.e. by a drift of negative electrons through the body in one direction. Usually measured by the practical unit, the Ampere, which is one tenth of the Absolute or C.G.S. Electromagnetic Unit.

CURRENT DIRECTORY : The directory that the computer will use when given commands.

CURRENT DRIVE : The disk drive that the computer uses when given commands.

CURSOR : A cursor is a moving spot on a video screen which indicates the next position for entering data on the screen. Sometimes a blinking cursor is used, which is useful for drawing attention to a specific element of data or to indicate the part of the screen in use when entering data or instructions. The keyboard of a microcomputer has cursor control keys for positioning the cursor at particular points on the screen when editing.

CUT : Removing text or graphics from a document and placing it in the Clipboard. See also CLIPBOARD.

CUT-OFF : Syn. *Black-out Point.* The point at which the current flowing through an electronic device is cut off by the control electrode. In a transistor, the cut-off point is the minimum base current at which the device condutcs; in a valve it is the minimum negative grid voltage (the grid base) required to stop the current. In a cathode ray tube the cut-off bias is the bias voltage that just reduces the electron-beam current to zero. In all cases the values are dependent on the conditions at the other electrodes, which must be specified.

CUT-OFF FREQUENCY : The frequency at which the attenuation of a passive network changes from a small value to a much higher value; this is the theoretical cut-off frequency. The effective cut-off frequency is that frequency where the insertion loss between two specified impedances has risen by a stated amount compared to the value at a reference frequency. An active network has the same cut-off frequency as a passive one with the same inductances and capacitances. The term is also applied to the limiting frequency of a filter. The frequency above or below which the attenuation of a circuit (e.g. a loaded cable or a piece of apparatus, such as a Microphone) rises rapidly.

"C" WIRE (IN TELEPHONY): See "S" WIRE.

CYBERNETIC CONTROL PROCESS : This process is identical to management by exception based on exception reporting, the basic elements of which may be analysed as follows. The use of resources is planned and relevant control parameters, known as *reference inputs*, are established as a basis of control. The results of a system's operations, i.e. its outputs, are measured by a sensor indicating the actual state of the system, i.e. the magnitude of the output signal. the measured output is referred to as the controlled variable, e.g. units produced or sold, level of costs incurred, number of units scrapped, etc. The output signal is then communicated from the *sensor* to the *comparator*. This is known as *feedback*. The comparator compares the output signal (the actual state of the system) with the required state (reference input) and notes the difference. The *difference* in accounting terms is known as a *variance*. In the latter instance the computer is an integral part of control system. Error signals are communicated to the appropriate executive responsible for controlling the variable under consideration. In cybernetic terms the *executive* is known as the *effector* as he is responsible for adjusting the controlled variable by modifying the use of resources to increase or decrease production or sales as the case may be. All this is done to achieve a state of homeostasis, i.e. a state of balance.

CYBERNETICS : A discipline concerned with control and communication in *animal* and *machine*. Cybernetics attempts to build a general theory of such systems, independent of their makeup, which could, for example, be electronics, organic, or clock work.

CYCLE : A group of events that repeat in a regular pattern. The changes completed during a complete period, including two half-waves in opposite directions, in an alternating current,voltage, etc. See FREQUENCY CYCLE

CYCLE TIME : A interval of time in which one set of events or phenomena is completed. It is usually the time required for one cycle of the memory system, the time between successive accesses of a computer, and is sometimes considered to be a measure of computer power. (2) Any set of operation that is repeated regularly and in the same sequence. The operations may be subject to variations on each repetition.

CYCLOGRAM : A record obtained from a Cyclograph. A name sometimes given to an instrument with an optical or electron-jet "pointer" moving in two dimensions under control respectively of different variables (to distinguish it from a true Oscillograph in which the pointer moves in only one dimension under control of a single variable while the time element is taken into account by movement of the screen or otherwise). In the case of regular periodic functions a closed figure or Cyclogram similar to one of Lissajou's figures is produced.

CYCLOTRON : An accelerator travel within two semicircular hollow metal electrodes (known from their shape as dees). A unipolar magnetic field, H, at right angles to the plane of the dees, causes the particles to execute circular orbits within the dees, orbital radius being proportional to *particle velocity* and *magnetic field*. Particles are accelerated by a radio frequency field as they pass between the dees and the orbital radius thus increases.

CYCLIC REDUNDANCY CHECK [CRC] : An error-checking method used when transmitting data. Often used when *sending* and *receiving* data using modems and communications programs.

CYLINDER : A pair of tracks that lie opposite one another on both sides of a hard disk platter. The storage space formed in a multiple-surface disk system by addressing the same track on all surfaces.

CYMOMETER (Fleming's) : An instrument for measuring the frequency of electric waves, in which a capacity formed by one brass tube sliding over another can be varied simultaneously with an inductance consisting of a solenoid of bare wire, and arranged so that the Oscillation constant at which resonance takes place, as indicated by the glowing of a Neon Tube, can be read off on a scale.

CYMOSCOPE : A detector of electric Waves.

CYSTADYNE RECEPTION : A system of radio reception employing a crystal detector polarised by battery which gives it the property of negative resistance, enabling it to be used somewhat like a thermionic valve for amplification, reaction, etc.

DAMPED OSCILLATIONS : Oscillations which rapidly decrease in amplitude and die away.

DAMPED WAVES : Wave-trains produced by Damped Oscillations, as employed in spark systems of radio-telegraphy. Sometimes called Type "β" Waves.

DAMPING : A term used to mean both the cause of the energy loss-friction, eddy currents, etc., and the progressive decrease in amplitude. The amount of damping is termed critical if the sysem just fails to oscillate. Greater or lesser degrees lead to overdamping and underdamping respectively. The damping factor of underdamped oscillations is the ratio of the amplitude of any one of the damped oscillations to that of the following one. The natural logarithm of the damping factor, the logarithmic decrement, is sometimes quoted.

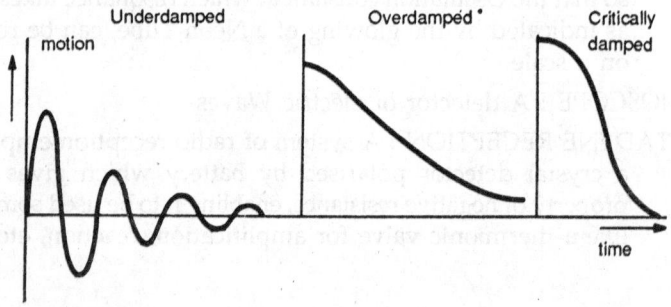

DARLINGTON PAIR : A compound connection of two transistors that operates as if it were a single transistor with an extremely high forward-current transfer ratio.

DARSONVAL GALVANOMETER : A direct-current galvanometer in which the current to be measured passes through a small rectangular coil suspended between the poles of a permanent horse-shoe magnet. The magnetic field produced in the coil reacts with the field of the magnet producing a torque and causing the coil to rotate about the vertical axis in the field.

D'Arsonval movement is used in most forms of galvanometer since it combines a high degree of sensitivity with low resistance and high damping.

DATA : A general term used to denote any or all facts, numbers, letters, and symbols used for processing by digital equipment. In instruments it generally refers to output (i.e., the measuring result, often in BCD code) or input data (the information for remote programming of the instrument).

DATA BASE : A collection of organized information stored in a disk file. Also refers to a program that creates, organizes and sorts information. Examples of a data base include dBASE IV, PARADOX, FOXBASE, SYBASE, FOXPRO, SOFTBASE, DEVBASE and R: Base.

DATA BASE ADMINISTRATOR : The individual in the organization with responsibility for the design and control of databases.

DATA BASE MANAGEMENT SYSTEM : Software that organizes, catalogs, stores, retrieves, and maintains data in a database.

DATA BUS : The eight lines (DIO 1-8) of the IEC bus used to transfer (multiline) remote messages.

DATA BYTES : These multiline device-dependent messages can be sent by an addressed talker as measurement data or received by an addressed listener as display data or program data.

DATA CONVERSION : The process of changing data stored in a particular way to another way.

DATA DEFINITION LANGUAGE : The language used with a database management system to describe the relataionships among data elements.

DATA DICTIONARY : A component of a database managment system that contains names of data elements and information about them.

DATA ELEMENT : The smallest physical data unit.

DATA FILE : A collection of related data records organized in a specific manner. Data files contain computer records which contain information, as opposed to containing data handling information or a program.

DATA PROCESSING : Operations in accordance with definite rules of procedure on a set of data to obtain another set of data which differs in at least one particular field from the original set. For example, the set of data constituting an income statement is obtained from the set of income and expense data by doing data processing.

DATA REDUCTION : The application of mathematical, or statistical techniques to obtain or extract only the needed information from a larger amount of related information, as in the preparation of computer input.

DATA SELECTOR : A device that selects one of several inputs depending to the control leads, and gates this input to the output; a multiplexer.

DATA STRUCTURES : The relationships among different fields of data on secondary storage.

DAY MODULATION : A means of doubling the use of a radio channel by transmitting two carrier waves in quardrature, each separately modulated with different signals.

d-BASE : Popular database program, by Ashton-Tate, that runs on the IBM computers family. Versions include dBASE II, dBASE III, dBASE IIIPlus, and dBASE IV.

DBMS : Data Base Management System.

DCTL : Direct-Couple Transistor Logic; a system of transistor logic in which the collector output of one gate is connected directly to the base input of the next gate.

DCU : Decimal Counting Unit; heart of a digital counter, consisting of a number of decade assemblies in cascade. There are as many decade assemblies as there are digits in the display.

DDA : A Digital Differential Analyzer, a machine for manipulating differential equations.

DDM : Digital multimeter; a multifunction version of the DVM, which also measures ac voltages, currents, resistance, temperature, etc.

DEATH RAY : A popular name for the powerful rays for which great destructive effect is claimed by inventors of secret methods for their production. They are reported to be of wave-lenghts on the border land between ultra-voilet light and "X" Rays, and to be produced in concentrated beams. The claims made have not received satisfactory experimental confirmation.

DEBROGLIE WAVES : A set of waves that are associated with a moving particle and represent its behaviour in certain situations, as when a beam of particles undergoes diffraction.

DEBUG : The process or checking the logic of a computer program to isolate and remove mistake from the program or other software.

DEBUGGER : A special program designed to simplify the process of locating errors or bugs in a program.

DEBUGGING : The process of seeking, finding, correcting, and eliminating errors in a program.

DEBUNCHING : The spreading of electrons in an electron beam or in a velocity-modulated tube that results from their mutual repulsion. The angle of spread of the electron beam is the divergence angle. See also kylstron.

DEC : Digital Equipment Corporation.

DECADE : A group of 10 units. In digital techniques generally a counter which counts and stores up 10 (see counter).

DECADE ASSEMBLY : The basic unit of a DCU, generally consisting of five units: a decade counter, a memory , BCD-IN-decimal decoder, a numerical indicator, and an indicator driver.

DECADE COUNTING UNIT : A device that will output 1 pulse for every 10 received.

DECAMETRIC WAVES : Electro-magnetic waves of a wavelength between 100 and 10 metres.

DECI-Symbol : d. A prefix to a unit, denoting a submultiple of 10^{-1} of the unit.

DECIBEL : On-tenth of a Bel, See also TRANSMISSION UNIT.

DECIBEL METER : An instrument for measuring acoustic response in the form of a thermionic voltmeter in which advantage is taken of the fact that a screen grid valve can be adjusted to possess an approximately exponential anode-grid current characteristic.

DECIMAL NUMBER SYSTEM : A number system composed of 10 Indo-Arabian digits, 0 through 9.

DECISION : In computer operations, the processes of detecting the existence of specified patterns of relationships in the data being handled, and of taking alternative course of action based upon the difference detected.

DECISION LANGUAGE : A programming language using decision tables.

DECISION PROCEDURE : A precise algorithm that takes as input a string of symbols and decides whether or not it is a member of a language; therefore, a particular decision procedure is taken to define a non-finite language.

DECISION SUPPORT SYSTEM : A system designed to support decision maker, generally involving interactive computing and focused on one particular business problem.

DECISION TABLE : An array that lists combinations of conditions and consequences.

DECLARATIVE KNOWLEDGE : A set of ordinary knowledge about concepts and the relationships between them, independent of any procedures to manipulate them.

DECODER : (1) A device used to convert information from one code form into a more useful form (e.g., BCD-TO-decimal decoder). (2) Syn. *demodulator*. A circuit, apparatus, or circuit element that is used in communications to demodulate the received signal i.e. to extract the signal from a carrier with minimum distortion.

DEDICATED FILE SERVER : A computer used exclusively for running a network.

DEDICATED PACKAGE : A software package designed for a specific task, such as accounts receivable or payroll.

DEDUCTION : A sound process of logical inference that uses the following rule: given A implies B, if we know A to be true, we can conclude that B is also true; it is this sound rule, called modus ponens, which underlies virtually all logical arguments and proofs.

DEFAULT : An action or value that the computer automatically assumes, unless a different instruction or value is given. A predefined action or command that the computer chooses unless you specify otherwise.

DEFAULT VALUE : In a frame system is a value that a property of a new concept automatically receives, normally through inheritance, until it gets its own value directly.

DEFECT CONDUCTION : Conduction in a semiconductor due to the presence of holes in the valence band. The presence of the holes is due to an imperfection in the crystal lattice of the semiconductor.

DEFORMATION POTENTIAL : An electric potential caused by mechanical deformation of the crystal lattice of semiconductors and conductors. See PIEZOELECTRIC EFFECT.

DEGENERACY : A condition that arises when an atomic or molecular system with a number of possible quantized states has two or more distinct states of the same energy.

DEIONISATION : The reduction of a mass of gas, which has been made conducting by becoming ionised, to its previous non-conducting state. There are a number of ways, including cooling, diffusion, absorption, an air blast or a magnetic field, by which a spark gap in radio telegraphy can be rapidly deionised, in order to prevent an arc following it.

DEKA : Prefix signifying ten times same as deci.

DEKÁMETRIC WAVES : Electromagnetic waves of a wavelength between 100 and 10 metres. CF. MICROWAVES, ULTRA SHORT WAVES.

DEKATRON : A type of multielectrode cold-cathode scaling tube that has ten sets of electrodes that function in turn. As voltage pulses are received, a glow discharge moves from one set of electrodes to the next. A visual display of counts in the decimal system is provided using the tubes in cascade. The tubes may also be used for switching. Compare DIGITRON.

DELANEY MULTIPLEX (Telegraph) SYSTEM : One of the earlier forms of Multiplex telegraph systems depending upon synchronously revolving commutators at the *sending* and *receiving* ends of the line.

DELAY : (1) The time interval between the propagation of a signal and its reception; (2) The time taken for a pulse to traverse any electronic device or circuit. In a switching transistor, for example, it is the time between the application of a pulse to the *input* and the appearance of a pulse at the *output*.

DELAY DISTORTION : A distortion that is a change in the waveform because of the variation of the delay with frequency.

DELAYED AUTOMATIC VOLUME CONTROL : Automatic Volume Control which does not begin to come into action until a certain signal strength is attained, in order to prevent undue magnification of background noise while the tuning control is passing from one station to another. Also called Quiet Automatic Volume Control or Suppressed Automatic Volume Control.

DELAYED TIME PROCESSING : Processing that operates with historical information that is, information after it can no longer affect its source, as information about the amount of goods shipped, after the shipment has been made.

DELAY EQUALIZER : A network or filter that compensates for the effects of delay distortion and thus maintains the wavefront of a transmitted wave.

DELAY LINE : Any circuit, device, or transmission line that introduces a known delay in the transmission of a signal. Coaxial cable or suitable L-C (inductance-capacitance) networks may be used to provide short delay times but the attenuation is usually too great when longer delay times are required.

DELIMITER : A character that marks the beginning or end of a unit of data on a storage medium. Comman, semi-colons, periods, and spaces are used as delimiters to separate and organize items of data.

DELTA RAYS : A class of rays emitted by radio-active substances consisting of the projection of negatively charged corpuscles at a much slower speed than that in the case of Rays. Cf. ALPHA, BETA and GAMMA RAYS.

DEMODULATION : The action of the receiver in Radio or Carrier Current Telephony, in disentangling and responding to modulations superimposed upon the carrier current or wave.

DEMORGAN'S THEOREMS : A set of Boolean formulas like A.B = A + B and A.B = A + B.

DEMOUNTABLE VALVE : A high power radio transmitting valve in a metal container with procelain insulation, which can be taken to pieces for inspection and renewal of the electrodes and re-exhausted.

DEMULTIPLEXER : A logic circuit that greats its input to one of several outputs, depending on the binary word on its control leads. Performs the opposite of a multiplexer; it separates different signals travelling on a common channel.

DEPLETION LAYER : A space charge region in a semiconductor that has a net charge due to insufficient mobile charge carriers. Depletion layers are inevitably formed at the interface between two dissimilar conductivity types of semiconductor, in the absence of an applied voltage.

DEPLETION MODE : A means of operating field-effect transistors in which increasing the magnitude of the gate bias decreases the current. cf ENHANCEMENT MODE.

DEPOSITION : The application of a material to a base (such as a substrate) by means of vacuum, electrical, chemical, screening, or vapour techniques.

DEPTH-FIRST SEARCH : A search technique that pursues paths in the state-space as far down as possible before backing up and choosing different alternatives; in infinite state spaces it will never terminate, but it often finds solutions quicker than breadth-first search.

DERATING : Reducing the maximum performance ratings of electronic equipment or devices when operated under unusual or extreme conditions. This ensures an adequate safety margin.

DERIVED UNITS : Units which are functions of one or more of the fundamental units. Thus although the units of time and distance are *fundamental*, that of speed is *derived*.

DERRICK : The structure on the roof of a telephone exchange to which over-head wires are brought.

DESAUTY BRIDGE : A four-arm bridge used for the direct comparison of capacitances. The capacitors are charged or discharged using the key if no response is observed from the ballistic galvanometer then $R_1 \times C_1 = R_2 \times C_2$.

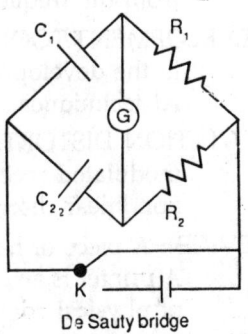
De Sauty bridge

DESAUTY'S METHOD : A method of comparing capacitances in which they are balanced in two arms of a Wheatstone bridge. The other two arms contain resistances which are adjusted till no kick is produced in the galvanometer when the battery key is closed.

DESIGN STUDIO : Design Studio is designed to compete as a high-end desktop publishing program for the professional. Ready, Set, Go! is designed as a powerful, but simpler program, Design Studio's main competitor is Quark Xpress Ready, Set Go! main competitor is Pagemaker.

DESK-TOP : The computer screen displaying a menu bar, windows, and icons. Commonly used to describe the Macintosh and Microsoft Windows.

DESK-TOP PUBLISHING (DTP) : Using a computer to design and print pages for publication.

DESK-TOP SYSTEM : A computer that fits completely on an office.

DESTOP PRESENTATION : Presenting text, video and graphics on the computer for display to others.

DESTOP VIDEO : Using a computer to create and edit video images.

DETAIL FILE : A data file composed of records having similar characteristics, but containing data which is relatively changeable by nature, such as employee weekly payroll data. Compare to master file.

DETECTOR : Any device used to detect the presence of a physical property or phenomenon, such as radiation.

DETECTOR CIRCUIT : The part of the circuit in a radio receiving station which contains the detector.

DETECTOR TUBE OR DETECTOR VALVE : A Thermionic Valve used as detector in radio-reception. See also ANODE BEND RECTIFICATION, and TWO-ELECTRODE RECTIFICATION.

DETUNE : To adjust the frequency of a tuned circuit so that it differs from the frequency of the applied signal.

DEVELOPMENT TOOL : A program designed to assist programmers in the development of software. Intelligent tools incorporate AI techniques.

DEVIATION DISTORTION : A distortion that occurs in frequency-modulated receivers that have an inadequate bandwidth or non-linear discriminator.

DEVICE : A piece of hardware that can perform a specific function. A printer is an example of a device. A functional combination of physical components; an integrated part of the hardware.

D-FLIP-FLOP : A flip-flop whose output will be set to the values of the D input upon receipt of the next clock pulse.

DFT : Discrete Fourier Transform.

D-G FLIP-FLOP : A flip-flop whose output may be changed to that on the D input lead by activating the G control lead.

DIAGNOSTIC PROGRAMS : Special programs used to align equipment or isolate equipment malfunctions.

DIAGNOSTIC ROUTINE : A rountine designed to locate either a malfunction in the computer or a mistake in coding.

DIAL (Automatic Telephony) : (1) Noun; see CALLING DIAL. (2) Verb; see DIALLING.

DIALLING : Making a cell by the operation of a Calling Dial in automatic telephone systems.

DIALOG : Communication between a person and a computer.

DIALOG BOX : A box containing a message and one or more options for the user to choose.

DIAL-TONE : A signal heard in the receiver of an automatic telephone to making a call, consisting of a "purring" sound indicating that the subscriber's line switch has made connection with further selecting apparatus and that all is ready for Dialling to commence.

DIAPHRAGM : (1) A thin flexible plate, such as that in a telephone transmitter or receiver, capable of vibrating in accordance with sound waves. (2) A porous separator in a primary or electrolytic cell, preventing mixture of the anolyte and the catholyte, but allowing the free passage of ions. (3) A porous sheet used as a Separator in an accumulator cell.

DIAPHRAGMLESS MICROPHONE : A telephone transmitter without a mechanical vibrating system, e.g. where the sound waves act upon a flame, arc, glow discharge, or ionised space between a Nernst glower and a cold electrode.

DIASY WHEEL PRINTER : A printer that uses a wheel made up of 'petals" that each contain one character. Printing occurs by spinning the wheel and striking individual petals.

DIBBLE-DOBBLE : The double-up and add method of converting a binary number of decimal.

DICHRONOUS CIRCUITS : Circuits having the same natural period of oscillation.

DICKE'S RADIOMETER : An instrument that measures microwave noise power precisely by comparing it with the noise from a standard source in a wave guide.

DICTAPHONE RECEPTION : A method of reception of high speed radio-telegraphy signals in which the diaphragm of the receiving telephone actuates the stylus of a "dictaphone" or phonograph, obtaining a permanent record on a wax cylinder which can afterwards be run slower for reproduction.

DIE : A single integrated circuit separated from the wafer in which it was made. After separation, leads are bonded to the die and it is mounted in chip-carrier. Also, referred to as a *chip* when it is mounted in a chip-carrier.

DIELECTRIC ISOLATION : A method of isolating individual regions in an integrated circuit, particularly a bipolar interated circuit, by surrounding the region with insulating material (dielectric) rather than with isolating diffusions. There are many different methods used to achieve dielectric phase angle.

DIELECTRIC PHASE ANGLE : The difference between the phase

angle of the alternating (sinusoidal) voltage applied across a dielectric material and the phase angle of the resulting alternating current. The difference between the dielectic phase and angle and 90° is the dielectric loss angle. The cosine of the dielectric phase angle (or the sine of the dielectric loss angle) is the dielectric power factor.

DIFFERENTIAL AMPLIFIER : An amplifier that has two inputs and produces an output signal that is a function of the difference between the inputs. An ideal differential ampifier produces an output signal of zero when the inputs are identical. In practice a small positive or negative signal may occur. The common-mode rejection ratio is a measure of the ability of a differential amplifier to produce a zero output for like inputs.

DIFFERENTIAL ANALYSER : A device for solving differential equations, usually a type of analog computer although a mechanical device is sometimes used.

DIFFERENTIAL CAPACITOR : A variable capacitor having two sets of fixed plates and one set of moving plates. As the moving plates rotate between the fixed plates, the capacitance to one set of plates is increased while that to the other is decreased.

DIFFERENTIAL GALVANOMETER : A type of galvanometer that gives a deflection that is a *function* of the difference between two currents. The currents are passed in opposite directions through two identical coils. The difference between the currents determines the *magnitude* and *direction* of deflection.

DIFFERENTIAL LINEARITY : The variation of the input voltage size of an A/D converter necessary to produce a change of one bit on the output.

DIFFERENTIAL MICROPHONE : A form of microphone in which extra sensitivity is obtained and the effect of non-linear relation between *pressure* and *resistance* of carbon is avoided by utilising two carbon elements, one subjected to increase of pressure on one side of the diaphragm and the second to decrease of pressure on the other side. (Also called Push-Pull Microphone and Double Button Microphone).

DIFFERENTIAL RESISTANCE : The resistance of a device or component part measured under small signal condictions.

DIFFERENTIAL WINDING : Two or more coils or two windings of a single coil arranged so that when carrying a current their magnetomotive forces are in opposition.

DIFFERENTIATOR Syn. *Differentiating circuit*. A circuit that gives an output proportional to the differential with respect to time, of the input. Compare INTEGRATOR.

DIFFUSED JUNCTION : A junction between two different conductivity regions within a semiconductor formed by diffusion of the appropriate.

DIFFUSION CAPACITANCE : The capacitance of a forward-biased semiconductor diode junction caused by unequal doping.

DIGIT : One character in a number. There are 10 digits in the decimal number system. There are 2 digits in the binary number system.

DIGITAL : A method of representing a number by discrete units.

DIGITAL CLOCK : An electronic device, generally a stable oscillator, that generates a repetitive series of pulses, known as clock pulses, whose repetition rate, or frequency, is accurately controlled. The clock rate is the frequency, expressed in hertz, at which active signal occur. The active transition may be from a low to a high voltage level, or vice versa, but will always be followed after a fixed time by an opposite in active transition. The clock signal is thus formed as a series of fixed width pulses having a clock signal and will always contain one active transition of the clock.

DIGITAL COMPUTER : A computer which handles symbols representing information.

DIGITAL TO ANALOG CONVERTOR : A device that converts a binary input to an analog output.

DIGITIZE : Convert an image into a series of dots. Often used to store images on disk for the computer.

DIGITRON : Syn. Nixie tube. : Type of cold-cathode scaling tube that has several cathodes (usually ten) shaped into the form of characters (usually the digits 0 to 9). As voltage pulses are received, the cathode required is selected by a switching connection to one side of the power supply and a glow discharge issuminates the character. These tubes are widely used for display purpose in calculators, counters, etc. Compare DEKATRON.

DIMENSION : A statement which gives the size of an array. For a two-dimensional array this would contain two numbers, one giving the number of rows and the other the number of columns. If plural, it refers to the values obtained when an object is measured.

DIO : Data input/output lines of the IEC bus, the data bus.

DIODE : Any electronic device that has only two electrodes. There are several different types of diode, their voltage characteristics determining their application. Diodes are most commonly used as rectifier; those used for other purpose have special names, as with the Gunn diode. A Thermionic Valve with two electrodes, the cathode and the anode, e.g. the original Fleming oscillation valve. Cf. TRIODE.

DIODE FORWARD VOLTAGE : Syns. diode drop; diode voltage. The voltage across the electrodes of a diode when current flows. The current increases exponentially with voltage and therefore the voltage is substantially constant over the rage of currents in common use: a typically value is about 0-7 volts at 10 milliamps, making diodes very useful as catching diodes. The diode may also function as a voltage reference diode when it is used to provide a reference voltage, equal to the diode forward voltage, across its terminals. See CATCHING DIODE.

DIODE TRANSISTOR LOGIC : (DTL) A family of intergrated logic circuits in which each input signal comes through a diode and the output is taken from the collector of an inverting transistor (see inverter). The basic circuit is a NAND gate. A family of logic using diodes and transistors as logic elements within the integrated circuit.

DIP : Dual-in-Line Package.

DIP SWITCH : A set of small on/off switches mounted on circuit boards used for choosing different options for the circuit board.

DIPOLE AERIAL : An aerial commonly used for frequencies below 30 megahertz. It consists of a centre-fed open aerial excited in such a way that the standing wave of current is symmetrical about the midpoint of the aerial. There are several different types of aerial and fed at the centre of one of the dipoles; a multiple folded dipole consists of more than two parallel half-wave dipoles. Originally an aerial (used for short waves) consisting of two straight bars in line of a length equal to half the wave length to be received, but extended to include any single straight rod aerial about one-half wave-length long. A half wave dipole has a length equal to half the wavelength, l a full-wave diploe has a length of one wavelength; a folded dipole consists of two parallel half-wave dipoles separated by a small fraction of the wavelength, connected at their outer ends.

DIRECT (Telephone) LINE : A telephone line from an exchange to a single subscriber, as opposed to a Party Line serving several subscriber, called in the U.S.A. an "Individual line."

DIRECT CIRCUIT : A telegraph circuit not worked through a Relay.

DIRECT COUPLED AMPLIFIER : D.C. amplifier. An amplifier in which the output of one stage is fed directly to the input of the next or through a resistance. Such an amplifer is used for direct-current amplification and the misnomer direct-current amplifier is sometimes used.

DIRECT COUPLING : Syn. *resistance coupling*. Coupling between electronic circuits or devices, such as amplifier stages, that is not frequency dependent; resistive coupling is a form of direct coupling.

DIRECT CURRENT (D.C.) : A unidirectional current of substantially constant value. See also CURRENT.

DIRECT LOGIC : The method of sequential design in which the output is taken directly off the state flip-flop.

DIRECT PICK-UP : Reception of radio signals by conductors in a receiving circuit other than the aerial itself.

DIRECT PRINTER : A printing telegraph receiver adapted for actuation directly by the received signals without perforation of a tape.

DIRECT RAYS : (In Radio-Transmission) : The part of the radiation, approximately parallel to the ground, which reaches the receiving apparatus without downward reflection or refraction from the Ionosphere.

DIRECTION FINDING Syn. *radio direction finding* : The practice and principle of locating the origin of a radio signal. A discriminating aerial and some form of receiver is required. Automatic direction finding carriers out the process automatically using either a rotating directive aerial or two such aerials at right angles. The rotating aerial is often used in conjunction with a cathode-ray tube (CRT) as an indicator to display strength of signal against direction. The direction of maximum strength is the bearing of the radio source.

DIRECTIONAL RADIO-COMMUNICATION : Radio-communication in which the waves are confined as nearly as possible to the direction between the transmitting and receiving stations. See also BEAM TRANSMISSION.

DIRECTIONAL RECEIVER : A radio receiver having maximum sensitivity to signals coming in from a particular direction and also from a directly opposite direction and minimum sensitivity as regards directions at right angles to this.

DIRECTIONAL RELAY : See DISCRIMINATING RELAY.

DIRECTIVE AERIAL : An aerial arranged to send out waves of maximum amplitude in a particular direction and usually also in the directly opposite direction but nearly zero in directions at right angles to this. Cf. UNIDIRECTIONAL AERIAL.

DIRECTIVE EFFICIENCY *(Of Directive Aerial)* : A measure of the field strength given in one particular direction compared with that in all directions.

DIRECTIVE RADIO-BEACON : A Radio-Beacon with a Directive Aerial. Cf. OMNI-RADIO BEACON and COURSE INDICATING RADIO BEACON.

DIRECTIVE TRANSMITTER : A radio transmitter employing a Directive Aerial.

DIRECTLY HEATED CATHODE : A cathode of a Discharge Tube directly heated by current passing through it. Cf. INDIRECTLY HEATED CATHODE.

DIRECTOR : An apparatus used in certain step-by-step automatic telephone systems, which, after being connected to the subscriber's line by a Line Switch or Preselector, not only receives and stores up the signal impulses forming the call until further vacant selecting a suitable vacant Junction Selector. (Also called a Register Translator.)

DIRECTOR AUTOMATIC TELEPHONE SYSTEM : The automatic telephone system adopted by the British Post Office based on the Strowger step-by-step system with the addition of the Director and other special features.

DIRECTORY : A table that gives the name, location, size, and the creation or last revision date for each file on the storage media. A list of files stored on a disk. Directories within existing directories are called *subdirectories*.

DIRECTOVOTU (Of A Directive Aerial) : A term sometimes used both for Directive Efficiency and Sharpness of Directivity.

DISCHARGE LAMP : A tubular or other lamp in which the light is produced by luminescence of a gas or vanour at a law pressure through which a discharge is passed between electrodes, sometimes with assistance of fluorecent material on the inner surface of the tube.

DISCHARGE TUBE : A closed glass or other vessel containing gas at a very low pressure and fitted with electrodes or otherwise arranged for an electrodes or otherwise arranged for an electric discharge to pass through the gas when a sufficiently high voltage is applied, for the production and investigation of phenomena occurring only at low gas pressures. See GAS DISCHARGE TUBE, VACUUM TUBE, CATHODE RAY TUBE, THERMIONIC VALVE, "X" RAY TUBE, DISCHARGE LAMP, etc.

DISCOURSE ANALYSIS : A level of natural language processing that tries to understand a sentence and resolve its ambiguities by placing it in the framework of the overall conversation; it involves considerations of intention, motivation, prejudice, and so on.

DISCRIMINATING RELAY : A term originally used for directional relays etc but now more commonly limited to apparatus designed to isolate only the faulty section of the network.

DISCRIMINATION NETWORK : A tree-structured device that classifies objects into categories by branching through tests on their features until it reaches a leaf that names the proper class.

DISCRIMINATOR : (1) A circuit that converts a frequency-modulated or phase-modulated signal into an amplitude-modulated signal. (2) A circuit that selects signals with a particular range of amplitude or frequency and rejects all others.

DISJUNCTION : OR gate or logic as used in symbolic logic.

DISK : A round, flat, magnetic storage medium. Floppy disks are made of flexible material and enclosed in 5.25 inch or 3.5 protective cases, Hard disks are rigid.

DISK-DRIVE : A device that reads and writes data to a floppy or hard disk.

DISK-DRIVE HEAD : The part of a disk drive that reads data from a floppy or hard disk.

DISK-OPERATING SYSTEM : A collection of procedures and techniques that enable the computer to operate using a disk drive system for data entry and storage. Disk Operating System is usually abbreviated to DOS.

DISKETTE : A flat, flexible platter coated with magnetic material, enclosed in a protective envelope, and used for storage of software and data. A direct-access storage medium that is flexible; read-write heads of the drive actually touch the surface of the diskette.

DISKLESS WORKSTATION : A computer on a network that has no disk drives. Used primarily for accessing files through the network.

DISPLAY : A device which converts the electrical (decimal) data (e.g., the measuring results) into a visible form. Typical display devices are the NIT, LED or cathode-ray tubes.

DISPLAY DATA : A device-dependent message indicating text or other data to be displayed on or stored in a device.

DISPLAY TIME : The time during which the data are presented on the display(s) A special (hold) switch also makes it possible to set the display time to infinity. When the display time is over, a new measurement starts. See also SAMPLING RATES.

DISRUPTIVE DISCHARGE : A sudden large increase in current in an insulator semicondutor, or gas when the material breaks down under the influence of an applied electric field.

DISSIPATION Syn. *loss* : A loss of power due to the tendency of electronic circuits and components to resist the flow of current. In a resistive circuit the power dissipated is equal to I^2R, where I is the current and R the resistance.

DISTORTED WAVES : An expression sometimes used for waves which differ in wave-form from a true sine wave-form.

DISTORTION : Change of wave-form during transmission of radiated or propagated waves such as that causing impairment in the quality of line or radio-telephone reproduction, produced in long distance telephone lines of considerable inductance and capacitance, owing to the unequal velocities of propagation of the components of different frequency, and due to a variety of cause in radio-telephony. Cf. ATTENUATION.

DISTRIBUTED PROCESSING : The dispersion and use of computers among geographically separated locations; the computers are connected by a communications network.

DIVERSITY SYSTEM : A communication system that has two or more paths or channels. The outputs of these are combined to give a single received signal and thus reduce the effects of fading. The diversity gain is the gain the reception achieved by using a diversity system.

DIVIDER : A device, also called a *scaler*, that divides an input frequency by a certain factor (e.g., a decade scaller divides the input frequency by a factor of ten). See also *prescaler*.

DLT : Decision Logic Translator.

DMA : Direct Memory Access.

D-MOS : MOS circuits or transistors that are fabricated using double diffusion. Regions of different conductivity type are formed by successive diffusion of different impurities through the same opening in the oxide layer. D/MOS devices are short-channels high-performance devices that were originally developed for microwave applications. They have a very precise channel length that is determined by the double diffusion rather than the inherently less precise method of photo lithography.

DMS : Data Management System

DOCUMENT : (1) Data created by a program. Most commonly used to describe a word processor file. See also File. (2) A business form, voucher, or written evidence of a transaction.

DOCUMENTATION : The printed instruction manuals that are supposed to explain how to use equipment or programs. Written descriptions of a system, usually with instruction on how to operate the system.

DOMAIN : An area or field of expertise and knowledge, either of a human being or computer program, that delimits the class of problems to which the system can be applied.

DOMAIN EXPERT : A person with expertise in the domain of the expert system being developed. The domain expert works closely with the knowledge engineer.

DOPING : The addition of a particular type of inpurity to a semiconductor in order to achieve a desired n-conductivity or p-conductive: donor impurities are added to form an n-type semiconductor and acceptor impurities a p-type. The impurity added is the dopant.

DOPPLER EFFECT : The change in the apparent frequency of a source of electromagnetic radiation (or sound) when there is relative motion between the *source* and the *observer*. The effect is utilized in Doppler navigation, which is a navigation system (in a moving object) that operates by ground reflection. Doppler radar employs the Doppler effect to distinguish between fixed and moving targets: the measurement of the change in the frequency of the reflected wave is used to determine the velocity and direction of the moving target.

DOT GENERATOR : A test generator used with a television receiver to adjust the convergence of the picture tube. A pattern of evenly spaced dots or small squares is produced on a dark background and the dynamic focusing (see colour picture tube) is adjusted until a satisfactory image is formed on the screen.

DOT MATRIX PRINT : Print type that uses a series of dots for printing characters. A printer that uses 9-pins, 18-pins, or 24-pins for printing. The higher the number of pins, the sharper the printing looks.

DOTS PER INCH (DPI) : A unit of measurement for the resolution of monitors laser printers, and scanners. Laser printers tend to offer 300 dpi (90,000 dots per square inch).

DOUBLE CLICK : Clicking the mouse button twice in rapid succession. Double clicking usually selects and opens a file.

DOUBLE CURRENT SOUNDER : Telegraph working in which the release of the key sends a reverse "spacing" current instead of simply interrupting the circuit.

DOUBLE DENSITY : A type of diskette that has twice the storage capacity of standard single-density diskettes.

DOUBLE DIFFUSION : A method of forming diffused junctions in which successive diffusions of different impurity types are made into the same well-defined region of semiconductor.

DOUBLE DIODE : The equivalent of two Diodes in a single bulb, one for use as a speech detector and the other to supply a variable bias proportional to the signal strength for Automatic Volume Control.

DOUBLE DIODE PENTODE : A Diode type detector and a Pentode type amplifier with separate electron streams, all contained in one bulb, used for the final stage in Super-heterodyne Receivers.

DOUBLE DIODE TRIODE : A Diode type detector and Triode type amplifier contained in one bulb, with separate sets of electrodes for each.

DOUBLE NEEDLE TELEGRAPH SYSTEM : A system intermediate between Cooke and Wheatstone's original Five Needle, and later Single Needle systems, involving simultaneous deflection at two needles and with two keys, one worked with each hand.

DOUBLE NOTE AMPLIFLER : A *note* amplifler with two valves in Cascade.

DOUBLE PENTODE : A valve for the Quiescent Push-Pull Type of low frequency amplification in radio receiving sets, in which two separate sets of pentode electrodes are contained in one bulb.

DOUBLE PHANTOM CIRCUIT : A circuit each side of which consists of the four conductors of a Phantom Circuit in parallel.

DOUBLE PRECISION : The use of two computer words to represent each number. The technique allows the use of twice as many digits as are normally available and is used when extra precision is needed in calculations.

DOUBLE PURPOSE VALVE : A Thermionic Valve with two grids which can be employed simultaneously as a detector and as an amplifier.

DOUBLE RECEPTION (RADIO) : Simultaneous reception of two messages on the same aerial on different wave-lengths.

DOUBLE RETROACTION : Retroaction in a radio receiving set, applied simultaneously at two different frequencies.

DOUBLE SIDED : A term that refers to a diskette that can contain data on both surfaces of the diskette.

DOWN COUNTER : A counter that decreases its binary value by 1 even time a clock pulse is received.

DOWN LOAD : To copy files from another computer (usually a mainframe or electronic bulletin board) through a modem.

DOWN LOADABLE FONTS : Type faces stored on disk that must be copied into the printer's memory before you can use that particular font.

DOWN TIME : Time when an automatic computer is not operating correctly, or not in a condition to operate correctly, own to component failures. The term contrasts with up time.

DRAFT QUALITY : High-speed but low-resolution printing commonly found with dot matrix printers.

DRAG : To move the mouse pointer on an object, hold down the mouse button, and move the mouse the mouse while keeping the mouse button held down. Often used to select groups of items or to move pictures around the screen.

DRAG QUALITY : High-speed but low-resolution printing commonly found with dot matrix printers.

DRAIN : The electrode of a field-effect transistor through which carriers leave the interelctrode space.

DRAM : Acronym for Dynamic Random Access Memory. See also Dynamic Ram.

DRIFT : The deviation of the operation frequency of a crystal oscillator from its nominal value, generally owing to temperature variations. The variation with time of any electrical property of a circuit or apparatus. Drift often occurs during warm up or when the device is nearing the end of its useful life. In a voltage regulator or reference standard the variation of output voltage with respect to time is the drift rate.

DRIFT MOBILITY Syn. *Carrier Mobility* : The average velocity of excess minority carriers in a semiconductor per unit electric field. In general holes and electrons have different mobilities.

DRIFT TRANSISTOR Syn. *Graded-base Transistor* : A transistor in which the impurity concentration in the base varies smoothly across the base region. The doping level is high at the emitter-base junction dropping to a low doping level (therefore high resistivity) at the base collector junction.

DRIFT VELOCITY : The average velocity of electrons in a conductor or of ions in a gas, in the direction of applied electric field.

DRIVE : A device that holds and manipulates magnetic media so that the CPU can read data from or write data to them.

DRIVE HEAD : The part of a disk drive that moves across a floppy or hard disk surface, reading and writing data on it.

DRIVE OSCILLATOR : An auxiliary oscillatory circuit in which special precautions are taken to obtain constancy of frequency used to control the frequency of a radio transmitter.

DRIVER : (1) A program that lets the computer send or receive information from an external device, such as printer or a mouse. (2) A circuit or device that provides the input for another circuit or controls the operation of that circuit. (3) An expression sometimes used for a source of oscillations in radio-telegraphy, particularly in connection with testing operations.

DRIVER VALVE (In Radio Receiving Sets) : A term sometimes used for the valve in the amplifying stage previous to the final output stage in Class "B" Amplification.

DRIVING-POINT IMPEDANCE : The ratio at the input in terminals of a network of the root-mean square (rms) value of the applied sinusoidal voltage to the rms value of the resulting current between the terminals.

DRO : Destructive Read Out.

DRUM DUMP : See STORAGE DUMP.

DRY BATTERY : A battery composed of Dry Cells.

DRY CELL L : A Primary Cell, usually of the Leclanche type, in which the electrolyte is absorbed by a semisolid mass, for the sake of portability without spilling, as in the small batteries used in Pocket Lamps, Torches, etc., and in radio receiving sets.

DRY ELECTROLYTIC CONDENSER : A special form of Electrolytic Condenser containing only dry materials, used for Smoothing Circuits and other purposes.

DTL : Doide-Transistor Logic : A logic system in which the logic decisions are carried out by a group of diodes and the resulting output coupled through a transistor output stage.

DTM.F.: Dual-tone multifrequency; used in Touch-Tone telephones. Dual tones generated by each digit and * and # for signaling.

DTP : Acronym for Desk Top Publishing.

D-TYPE FLIP-FLOP : A flip-flop that will propagate whatever information is at is D-(data) conditioning input prior to the clock pulse, to the Q output, on recepit of a clock pulse.

DUAL AMPLIFICATION : The use of a single Thermionic Valve in a radio-receiving set, to effect both high frequency and low frequency amplification simultaneously. A circuit arranged in this way is called a *Reflex Circuit*.

DUAL-CHANNEL SOUND : A technique used in television receivers to separate the sound and video signals after the common first detector stage. Separate intermediate-frequency stages are employed for each signal.

DUAL-SLOPE INTEGRATION : A method of analog to digital conversion whereby integrator is chaged when connected to the analog input and then timed as it is discharged through the reference voltage.

DUDDELL OSCILLOGRAPH : An oscillograph in the form of a reflecting galvanometer in which two current carrying strips, in the powerful field of an electromagnet, deflect a mirror. The moving parts have a very smallfree period of vibration and are effectively damped by immersion in an oil bath.

DULL EMITTER VALVE : A Thermionic Valve in which the filament contains special substances which cause it to emit electrons at a much lower temperature than the original tungsten filaments. The valves to which this name applied were run with the filaments at a dull red heat, but later valves are worked at a lower temperature still, giving no visible glow.

DUMB AERIAL : A closed non-radiating oscillating circuit, having the same oscillation frequency as that of the main aerial, which is connected to the oscillator during the spacing periods in signalling tokeep it constantly loaded.

DUMMY : A substitute used to fulfill formal specifications that is replaced by the required thing or data when needed. For example, a dummy address, such as 00000, is often used in writing programs when the required address will be provided by address modification.

DUODYNATRON : A form of Dynatron on which an additional grid oscillating circuit produces a second series of oscillations of different frequency from that of the main oscillations.

DUPLEX : Data transmission in both directions simultaneously using two separate paths.

DUPLEX DIALLING : A system of Dialling employing both conductors of a telephone circuit separately, each with an earth return, with batteries at the opposite ends of the line.

DUPLEXER : A two-cannel multiplexer (see multiplex operation) that uses a transmit-receive (TR) switch so that one aerial may be used for both *reception* and *transmission*. The switch protects the receiver from the high power of the transmission. Duplexers are commonly used in radar, the TR switch operating in the time between transmission of the pulse and reception of the return echo.

DUPLEX (Telegraph) SYSTEM : A telegraph system in which two messages can be sent, one in each direction, simultaneously over the same line.

DUPLEX OPERATION : Simultaneous operation of a communications channel in both directions. Half-duplex operation occurs when the operation is limited to either direction but not both directions at once. Compare *simplex operation.*

DUPLEX RADIO-TELEPHONY : Radio Telephony arranged for simultaneous transmission and reception in both directions by the use of two different wave-lengths or otherwise.

DUST OSCILLOGRAM : A record from a Cathode Ray Oscillograph with a screen of insulating material upon which the moving electron beam has made a trace in the form of a negative charge rendered visible by dusting with an Electroscopic Powder.

DVM : A voltmeter which measures analog voltages and presents the results as a decimal figure.

DVORAK KEYBOARD : An alternate keyboard layout designed to be faster and easier to learn for typing.

DYANA : Dynamics Analyzer.

DYNAMIC CHARACTERISTIC (Of A Thermionic Valve) : A Characteristic relating to working conditions.

DYNAMIC DISPLAY : A display where the presentation of each decimal digit takes place in sequential form (see BCD OUTPUT, serial mode).

DYNAMIC IMPEDANCE Syn. dynamic resistance : The impedance at resonance of parrallel resonant circuit. It is purely resistive by definition (see resonant frequency).

DYNAMIC MEMORY : A memory that will lose its data unless periodically refreshed.

DYNAMIC RAM : Type of memory chip that requires recharging periodically to avoid loss in data. Less expensive but slower memory chip than static RAM or SRAM chips.

DYNAMIC RANGE The range over which an active electronic device can produce a suitable output signal in response to an input signal.

DYNAPHONE : (1) A form of telephone receiver in which the forces or the diaphragm are partly due to the action of eddy currents induced therein. (2) An electrical musical instrument employing Kipp Oscillations.

DYNATRON : A form of Thermionic Valve used as an oscillator having a negative current-voltage characteristics.

DYNATRON OSCILLATIONS : Oscillations produced in thermionic valve circuits in virtue of a negative current, voltage characteristic.

EAM (Electronic Accounting Machine) EQUIPMENT : The first devices used to manipulate punched cards. These devices had wired logic plugboards but no stored program.

EARTH CONDUCTION TELEGRAPH : A system of short distance telegraphy without line wires in which currents are passed through the earth from one earth plate to another comparatively near together at the transmission station and detected by the small potential difference produced between two earth plates in the receiving apparatus.

EARTH WIRE : (1) A wire making an earth connection. (2) A wire, connected to earth, running above the conductors in h.t. transmission lines, to minimise the effects of lightning, to conduct earth currents in case of leakage and in other ways to lessen interference with neighbouring communication circuits.

EBCDIC : Extended binary coded decimal interchange code; and 8-bit code used by one computer manufacture (IBM) to represent characters.

E-CELL : A solid-state timing device consisting essentially of a thimble-shaped electrolytic cell with a central gold electrode, an outer silver-plated electrode, and a paste electrolyte consisting of a suitable silver salt. When a current passes, silver is lost from the outer electrode, which is made the anode, and is deposited at the same uniform rate on the central gold cathode.

ECG : Electro-cardiogram.

ECL : Emitter-coupled logic; a logic system using emitter-coupled transistor circuits.

paste electrolyte

silver-plated electrode

gold electrode

E-cell

ECTL : Emitter-Coupled Transistor Logic.

EDDY CURRENT TRANSMITTER : A telephone transmitter consisting of a disc of aluminium foil vibrating under the influence of the sound waves in a constant magnetic field r ·-

duced by two slab coils. Suitable connections are made to render the eddy currents induced in the disc available for reproduction.

EDGE : A boundary between two surfaces in a visual scene, the detection of which is the major problem for early low-level vision.

EDGE-TRIGGERED FLIP-FLOP : A flip-flop whose output changes are dependent on only one edge of the clock pulse.

EDISON (Loud Speaking) TELEPHONE : A Telephon Receiver in which the movements of the diaphragm were controlled by the variable friction produced by variations in the current between a revolving disc of chalk and a brass band moistened with a chemical solution. (Originally called Electromotorgraph.)

EDIT : To arrange, rearrange, or arbitarily alter information, especially in format.

EDITO : Pertaining to the name of circuits or parts which carry picture signals.

EDITOR : A program that uses ASCII characters and saves files in ASCII format. Cannot change fonts, formats, or styles. Commonly used for writing programs. Also called a TEXT EDITOR.

EDLIN : Acronym for EDIT LINE. A simple editor that comes with MS-DOS and lets you edit one line at a time.

EDP : Electronic Data Processing.

EDSAC : Electronic Delay Storage Automatic Calculator.

EDVAC : Electronic discrete Variable Automatic Computer.

EFFICIENCY : The ratio of the useful output of a piece of apparatus to the total input; usually expressed as a percentage. See SPECIFIC CONSUMPTION.

EISA : Acronym for Extended Industry Standard Architecture, which in an agreed-upon standard for building the bus in an IBM compatible computer.

ELECTRIC HYSTERSIS LOSS : Electrical loss in a dielectric material due to internal forces in the material produced by a varying electric field. The loss usually appears in the form of heat.

ELECTRO-ADHESION TELEPHONE : A telephone reciver in which the pull on the diaphragm is varied by the alteration of the electrostatic adhesion between a rotating cylinder and a brake band.

ELECTRO-CAPILLARY MICROPHONE : A microphone in which the alteration of the depth of immersion of a capillary tube dipping into mercury is varied by the action of the sound waves, causing a varying e.m.f. between the mercury and the electrolyte by what appears to be the converse of the action in an electrocapillary Electrometer.

ELECTRODE : (1) A solid conductor by which a current passes to or from a liquid or gas or to another solid conductor of different material, as in electrolytic apparatus, primary cells, discharge tubes, arc-lamps and electric furnaces. (2) A device that emits, collects, or deflects electric charge carriers. It is usually in the form of solid plate, a wire, or a grid that controls current into and out of an electrolyte, gas, vaccum, dielectric, or semiconductor. Liquid mercury electrodes are also used. The electrode current is the current that flows through a specified electrode, such as a collector, grid, or drain. (3) A musical instrument similar in principle to the ETHEROPHONE, but with the volume control actuated by a foot pedal or otherwise. See ANODE and CATHODE.

ELECTRODE CABLE DETECTOR : An apparatus for locating submarine cable faults by towing a pair of weighted electrodes over the track of the cable while an alternating or interrupted current of audio-frequency is flowing. The electrodes are connected to telephones and amplifiers on board, and a sound can be heard when near a fault on the cable.

ELECTRODE DISSIPATION : The heat dissipated in a particular electrode, usually the anode, as a result of bombardment by electrons, ions, or radiation from other electrodes.

ELECTRODE EFFICIENCY : The ratio of the quantity of a metal electro-deposited to the quantity theoretically deposited according to Faraday's Laws.

ELECTRODE POTENTIAL : The difference of potential between an electrode and the electrolyte in an electrolytic cell, sometimes called Single Potential.

ELECTRO ENDOSMOSIS : See electro-osmosis.

ELECTROGRAPH : (1) A recording electrometer for the investigation of the potential gradient in the atmosphere. (2) An American system of Photo Telegraphy employing an etched original with the etched portionsfilled up with insulating material.

ELECTROGRAPHIC OSCILLOGRAPH : An Oscillograph in which the variation of charge along a moving bar subjected to a varying stream of cathode rays is afterwards rendered visible by an Electroscopic Powder.

ELECTROGRAPHY : (1) A term sometimes used for Electrotyping. (2) A process of picture transmission in which the receiver contains a synchronously moving insulated plate exposed to a modulated electric field producing a distribution of charge which can be made visible by dusting the plate with lycopodium powder.

ELECTROLYTIC CONDUCTION : Conduction in which processions of dissociated portions of atoms or ions carry positive and negative charges, in opposite directions, to the electrodes.

ELECTROLYTIC MAGNIFIER : An apparatus fulfilling a similar function to a Relay in cable telegraphy, except that, instead of containing actual contacts, it has a moving system employed to alter the relative positions of the wires dipping into an electrolyte, thus making changes in resistance which affect the balance of the duplex circuits and actuate the receiving apparatus accordingly.

ELECTROLYTIC SEPARATION : The separation of isotopes by electrolysis, based on the fact that different isotopes of an element are liberated at an electrode at different rates.

ELECTROMAGNETIC LENS : An Electron Lens in which the focusing is effected electromagnetically.

ELECTROMAGNETIC LOUD SPEAKER : Any form of loud speaker utilising electromagnetic forces. See MOVING IRON, MOVING COIL, and MOVING CONDUCTOR LOUD SPEAKERS.

ELECTROMAGNETIC MICROPHONE : Any form of microphone utilising electromagnetic induction.

ELECTROMAGNETIC OSCILLOGRAPH : An oscillograph depending for its action upon electromagnetic reactions, e.g. BIFILAR OSCILLOGRAPH and SOFT IRON OSCILLOGRAPH.

ELECTROMAGNETIC RADIATION : Energy that is radiated by a charged particle undergoing acceleration. Transverse sinusoidal electric and magnetic fields are propagated at right angles to each other and to the direction of motion, the instantaneous values of the fields being related to the charge and current densities by Maxwell's equations. These equations define the field as electromagnetic waves propagated through free space with a constant velocity c, where c is the velocity of light and is equal to 2.988 x 108 metres per second.

ELECTROMAGNETIC RETROACTION : Retroaction in a radio-receiving set in which the coupling between the plate and grid circuits is electromagnetic (inductive).

ELECTROMAGNETIC SPECTRUM : The whole of the known rangs of electric waves, from the longest employed in radio-communication to the shortest Gamma Rays.

ELECTROMAGNETIC UNITS : The system of absolute electrical units upon which the practical units (volt, ampere, etc.) arc based: derived from the fundamental units of length, mass and time (the centimetre, the gramme and the second), magnetic pole which exerts unit mechanical force on an equal pole at unit distance. Cf.

ELECTROMATIC TRAFFIC SIGNALLING : A system of traffic signalling in which the passage of an approaching vehicle over a strip in the road completes a relay circuit which gives it priority over vehicles gives it priority over vehicles approaching from another direction. See AUTOMATIC TRAFFIC SIGNALLING.

ELECTROMETER VALVE : A type of Thermionic Valve for amplification of input to measuring instruments, etc., designed for extremely small grid current.

ELECTROMYOGRAPH : An instrument for recording electrical changes in muscles.

ELECTRON : See NUCLEAR THEORY . The transference of electrons from one atom to another and their independent movement constitute the phenomena of electricity. The charge carried by as electron is equal to 4.8025 x 10 electrostatic units, or 1.592 x 10 coulomb mass of an electron is 9.1073 x 10 gramme, or about 1/1800 that of an atom of hydrogen. (1) The Greek work for amber from which the word Electricity is derived. (2) The smallest negative charge that is believed to beable to exist by itself, and being that carried by, or consisting of, one of the detachable constituents found in all atoms.

ELECTRON CAMERA (BAIRD) : A television camera in which an optical image causes electron emission from a photo-electric screen. This is focused and deflected, so that the "electron image" produced is scanned by a fixed point, the varying potential of which controls the modulat.

ELECTRON DIFFRACTION : The effect similar to diffraction of light
which produces on a photographic plate a series of rings
when a concentrated beam of cathode rays which has passed
through a very thin metallic film falls on it. This effect appears
to indicate that the nature of moving electrons partabkes of
that of wave motion as well as that of discrete particles.

ELECTRON GUN : A name sometimes used for two hole of the
apparatus used for generating and focussing the electron jet
or beam in a cathode ray or similar tube.

ELECTRON THEORY (or ELECTRONIC THEORY) : The theory which
explains all electrical phenomena by the interatomic transfer
or independent movement of electrons.

ELECTRON COUPLED FREQUENCY CHANGER : A valve in which
a single stream of electrons is first made to oscillate with a
locally controlled frequency, and the frequency of the received
signal is superposed upon it, producing a heterodyne intermediate frequency; either of the Heptode or Octode 1 type.

ELECTRONIC CELL : A Primary Cell consisting of two dissimilar
plates, e.g. copper and zince, the space between which is
occupied by radio-active material.

ELECTRONIC CLOUDS : See HEAVISIDE LAYER.

ELECTRONIC DEVICE : A device that utilizes the properties of electrons
for ions moving in a vaccum, gas, or semiconductor.

ELECTRONIC DISCHARGE : A discharge in a very highly exhausted
tube consisting entirely of electrons without gaseous ions,
e.g. in a Coolidge Tube.

ELECTRONIC MAIL : A system in which computers users have an
electronic mailbox and send messages using terminals, communications occur at the convenience of the user without
interruptions.

ELECTRONIC OSCILLATIONS : Oscillations in a circuit connected
to a thermionic valve of high frequency owing to conditions
of inertia of the electrons of the system; the period of oscillation being dependent upon the time of travel between the
electrodes. Cf. REACTION OSCILATIONS, DYNATRON
OSCILLATION.

ELECTRONIC PIANO : A piano without a sound board in which the
vibrations of each string affect the capacitance of a condenser
microphone and produce modulations in a loud speaker circuit.

ELECTRONIC RECTIFIER : See METAL RECIFIER.

ELECTRONIC RELAY : A Thermionic Valve used as a relay.

ELECTRONIC SWITCH : An electronic device, such as a transistor, that is used as a switch. These devices are usually operated as high speed switches where a very fast response is require, as for example in a computer.

ELECTRONICS : The science of the movements of Electrons.

ELECTRON IMAGE TUBE : An electron tube with a cathode of considerable surface covered with photoelectrically sensitive material upon which an optical image is thrown. This causes corresponding emission of electron streams which can be focused by systems of electron lenses upon a fluorescent screen. See also ELECTRON TELESCOPE and ELECTRON MICROSCOPE.

ELECTRON JET: A concentrated beam of Cathode Rays

ELECTRONJET INSTRUMENTS : Instruments, such as the Cathode Ray Oscillograph, employing an Electron Jet deflected by variable electric or magnetic fields a "pointer."

ELECTRON LENS: An appliance for spreading, concentrating or otherwise focusing a beam of projected electrons by the action of an electric or magnetic field. See also ELECTRON MICROSCOPE.

ELECTRON MICROSCOPE : An apparatus by which the structure of a body can be examined by spreading an electron beam emitted there form by electrostatic or electromagnetic deflection, capable of from 10 to 15 times the magnification of An optical microscope. Also called Super Microscope.

ELECTRON MULTIPLIER : An electronic tube in which current amplification is achieved by means of secondary emission of electrons. Primary electrons are released from the cathode by some means, such as the photoelectric effect. A highly exhausted tube with an anode in the centre and a flat

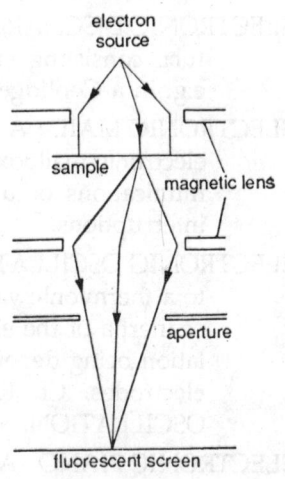

electron
source

sample

magnetic lens

aperture

fluorescent screen

Transmission
electron microscope

cold cathode at each end. An electron from a photoelectric or other source impinging upon one cathode causes emission of secondary electrodes ateed (due

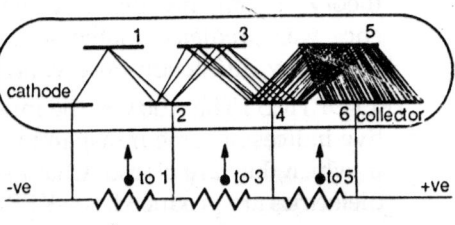

Electron multiplier

to the high Voltage maintained between the electrodes). These are caused by a focusing coil to miss the anode and to impinge upon the second cathode producing emission of further secondary electrons. Thus the actionis cumulative and a considerable amplifying effect is produced.

ELECTRONOMY : A name sometimes given to the study of atomic structure.

ELECTRON OPTICS : The study of the behaviour of electron beams under the influence of magnetic and electrostatic fields in a vacuum or very low pressure gas; he analogue of light beams passing through refractive media. The applied fields from electron lenses that are used for focusing or defocusing the electron beam.

ELECTRON POWER TUBE : See THERMIONIC OSCILLATOR.

ELECTRON TELESCOPE : An apparatus for seeing through haze, etc., by infrared rays in which an infrared ray image is formed by optical lenses on the cathode of an *Electron Image Tube* by which it is rendered visible.

ELECTRON TUBE : A Discharge Tube such as a Thermionic Valve or "X" Ray Tube in which there is a sufficiently high vacuum for the effect to be due to streams of electrons and not of gaseous ions, e.g. a Kenotron or a Coolidge Tube, often used indiscriminately for any Thermionic Valve, Cathode Ray Tube, etc. Cf. GAS DISCHARGE TUBE.

ELECTRON VALVE : An Electron Tube, the application of which depends upon its unilateral conductivity. See ELECTRON TUBE.

ELECTRON VOLTAIC EFFECT : A phenomen similar to the photovoltaic effect in which electrons striking the photocathode of a photocell with voltage to maximum and then decreases.

ELECTRON WAVES : Groups of waves which, according to recent

theory, accompany the movement of electrons. Their presence was predicted mathematically by de-Broglie and confirmed by diffraction observations.

ELECTRO OPTICS : The study of the interactions between the refractive indices of some transparent dielectrics and electric fields in which they are placed. Changes in the optical properties of dielectrics are produced. See KERR EFFECTS.

ELECTRO OSMOSIS : Syn. electroendosmosis. The movement of an electrolyte through a fine tube or membrane under the influence of an electric field.

ELECTRO PHONE : A name originally given to an early form (Ader's) of telephone transmitter, but later used to signify an auxiliary telephone service where by subscribers could listen to performances at theatres, etc., by means of special instruments over the ordinary exchange lines, and for certain forms of telephone for the use of the deaf.

ELECTROPHONIC EFFECT : The perception of sound when the human body is subjected to alternating currents of particular frequencies and magnitudes.

ELECTROSTATIC ADHESION : Adhesion between two substances or surfaces due to the presence of opposite charges, which attract each other.

ELECTROSTATIC MICROPHONE : A type of microphone depending for its action on electrostatic forces.

ELECTROSTATIC RETROACTION : Retroaction in a radio-receiving set in which the coupling between the plate and grid circuits is electrostatic. Cf. ELECTROMAGNETIC RETEROACTION.

ELECTROSTATIC SEPARATION : A method of separating fine powders of different permitivities. The powers undergo different deflections when placed in the intense electrostatic field between two highly charged electrodes.

ELECTROSTATIC UNITS (ESU) : See CGS system.

ELECTROSTATICS : The study of electric charges at rest and their associated phenomena.

ELECTROSTRICTION : A change in the dimensions of a body under the influence of an electric field in a medium of relative permittivty different from its own.

ELVERSION OSCILLOSCOPE : An apparatus for detection of vibrations and examination of movements of parts of machinery, etc, by viewing under recurrent flashes of short duration from a neon or other vacuum tube lamp, controlled by a special contact maker.

E-MAIL : Acronym for ELECTRONIC MAIL, which lets people send and receive messages stored solely on the computer.

EMISSION : The phenomenon of liberation of electrons or electro-magnetic radiation from the surface of a solid or liquid, usually electrons from a metal. The outer electrons of the atoms in a metal (conduction) move in a random manner among the lattice atoms with no net forces on them. The charge on the metal can be considered as an electric image located the same distance inside the metal as the electron is outside it. The force on the electron varies with distance x from the surface (fig. a). As the electron moves out from the surface work (W) is do force (fig. b). Where $W = \sqrt{F}dx$, the value W_1 represents a potential barrier that must be overcome by the electron. Electroheir energies are greater than W_1. W_1 is releated to the work function. ϕ by $\phi = W_1 - E_f$ where E_f is the Fermi level.

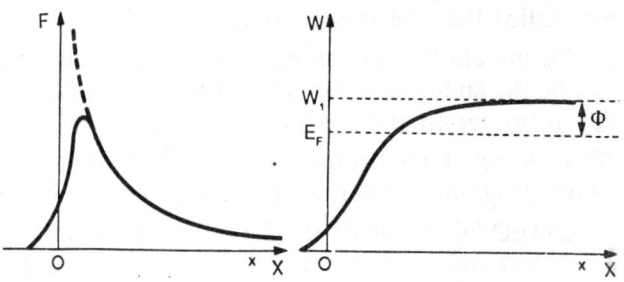

a Force on emmitted electron as function of distance from surface

b Work done to overcome force on emitted electron

EMITRON CAMERA : A camera used in television transmission in which an optical image is produced by a lens upon an insu-lating screen, with a vacuum tube, covered with a mosaic of separate deposits of photo-electrically active material, form-ing in connection with the back plate, a series of light sensi-tive cells. An electrostatically focused electron beam is caused by deflecting coils to scan the screen in parallel lines, and as it falls on each cell it releases the charge produced by the light falling on it, and sends through amplifiers and the transmitting apparatus, a series of signals of strength depend-ing upon the illumination of the part of the image being scanned.

EMITTER : Short for emitter *region*. The region of a bipolar junction transistor from which carriers flow, through the emitter junction into the base. The electrode attached to this region is the emitter electrode. See also TRANSISTOR; SEMICONDUCTOR.

EMITTER COUPLED LOGIC : A high-speed non-saturating logic family.

EMITTER FOLLOWER : An *amplifier* that consists of a bipolar junction transistor with common-collector connection, the output being taken from the emitter (see diagram). The transistor is suitably biased so that it is nonsaturated and conducting. The emitter voltage thus has a constant valve relative to the base at all times, and the emitter folllows the signal applied to the base. The voltage gain of the amplifier is therefore nearly unit but the current gain is high. The amplifier is often used as a *buffer* and is characterized by a high input impedance and low output impedance.

EMS : Acronym for Expanded Memory Specification, which is a special way for IBM computers to use more memory beyond 640 k. Developed jointly by Lotus, Intel, and Microsoft and sometimes called the LIM specification.

EMULATE : To imitate the way another item works. Often used to describe the ability of one pritner to act like another, such as "Epson Printer emulations."

EMULATION : A technique that uses both hardware and software to execute programs written for one computer on another.

ENCAPUSULATED POSTSCRIPT (EPS) : A special file format, containing text and graphics created using the Postscript programming language.

ENCHANCED SMALL DEVICE INTERFACE (ESDI) : An interface that speeds up disk drive access.

ENCODER : A device which takes information in one code and encodes it into another (e.g., BCD to binary encoder).

ENCRYPTION : The coding of a data stream to prevent unauthorized access to the data.

ENDODYNE (RADIO) RECEPTION : The arrangement of thermionic valve circuits for Beat Reception in which the source of the local oscillations of slightly different frequency is an actual part of the receiving aparatus; also called Auto-Heterodyne or Autodyne receiption. Cf. HETERODYNE.

ENDODYNE (RADIO) RECEPTION : RECEPTION END OF DATA; A signal identifying the termination of some quantity of data, as at the end of a reel of magnetic tape.

END OF FILE : The end of data for a file, abbreviated as EOF.

END OF FILE MARK (EOF) : A symbol or machine equivalent that indicates that the last record of a file has been read.

END USER PROGRAMMING : The growing trend for uses of information to employ very high level languages and other tools to access information without having a computer professional develop a program for them.

ENERGY BANDS : In a single atom, according to quantum theory, the orbiting electrons can only have certain discrete energies. The atom has a number of associated energy levels and the electrons occupy the lower levels and obey the Pauli exclusion principle, i.e. not more than two electrons can occupy each level.

ENERGY LEVELS : The possible values of energy of an atom or molecule. Quantum theory dictates that only certain fixed values are possible and these may only change by an integral multiple of some fixed amount. An electron orbiting the nucleus of the hydrogen atom may only occupy certain orbits of different energy. These allowed states of energy of the atom are called electronic energy levels. The ground state is the lowest possible energy level, higher states being called excited states. The total energy of an atom is composed of kinetic energy, as it translates through space, and the electronic energy.

ENHANCED GRAPHICS ADAPTER (EGA) : A colour graphics standard that defines the resolution of a program. EGA video boards and monitors display text with 640 x 350 resolution. Officially abandoned as a graphics standard in favour of VGA graphics. See also VIRTUAL GRAPHICS ARRAY.

ENHANCEMENT : The process of making changes and improvements in operational programs.

ENHANCEMNT MODE : A means of operating field effect transistors in which increasing the manitude of the gate increases the current. Compare DEPLETION MODE.

ENIAC : Electronic Numerical Integrator and Computer. Worlds first electronic digital mainframe type of computer (Ist generation) which utilized thermionic valves in their switchig circuits. This computer was based on decimal system rather than binary, on which all other digital computers are made.

ENIGMA : A German message coder. This was first german computer based on electric switching circuits utilizing relays. See COLOSSUS - 1.

ENTRANCE : A mean of beginning iterative action; the place in the routine which serves as the beginning point.

EPAM : (Elementary Perceiver and Memorizer) is a system developed by Feigenbaum and Simon for learning simple concepts and categories based on a discrimination network that grows as new positive training examples are presented.

EPITAXIAL D/MOS TRANSISTOR : Transistors that are formed in an n-epitaxial layer grown on a p-type substrate. Individual transistors on a chip may then be isolated by performing extra p-type isolating diffusions. See also V/MOS EPITAXIAL GROWTH : See EIPTAXY.

EPITAXIAL LAYER : See EPITAXY.

EPITAXIAL TRANSISTOR : See PLANAR PROCESS.

EPITAXY syn : Epitaxial Growth. A method of growing a thin layer of material upon a single-crystal substrate, such as *silicon*, so that the lattice structure of the layer is identical to that of the SUBSTRATE. The material, which may be the same as the substrate or a diferent one, is usually deposited from a gaseous mixture. The technique is extensively used in semiconductor technology when a layer (the epitaxial layer) of different conductivity to the substrate is required.

EPROM : Erasable Programmable Read Only Memory, a special PROM that can be erased under high-intensity ultraviolet light and reprogrammed.

EQUAL HETERODYNE : A hetryodyne receiving circuit in which the auxiliary oscillations and those due to the received waves are approximately equal in amplitude.

EQUISIGNAL RADIO-BEACON : A system of aerial navigation on a fixed route in which two modulated beams are projected from stations on each side of the route. When the aircraft is on its correct route, the received signals are equal. In one system, the signals are so interlocked that a continuous sound is heard on the route, but different discontinuous sounds are heard on either side. In a further development a visual signal is used showing a single illuminated line when on the course and *thick* and *thin* parallel lines when off the course, the thick line corresponding to that side of the route to which the machine has diverted.

EQUIVALENT SINE-WAVE: A true sin-wave form having the same R.M.S. valve and frequency as the actual wave form of an alternating current, voltage, etc.

ERASABLE DISK : An erasable optical disk capable of storing up to 2.6 gigabytes of information.

ERASE : To remove or replace magnetized spots from a storage medium.

ERROR : The general term expressing the deviation of the measured valve from the theoretically correct or true valve, or the part of the error due to a particular cause (e.g., trigger error, timebase error, etc).

ERROR CHECKING : The ability to check errors during data transfer, most commonly, while using a modern and communications software.

ERROR CORRECTING CODE : A code that detects and corrects its own errors.

ERROR DETECTING CODE : A code that will detect an error in the data.

ERROR MESSAGE : An audible or visual indication of hardware or software malfunction or of an illegal data entry attempt.

ERYTHEMETER : An instrument for measurement of that portion of the ultra-violet radiation given by a source of artificial sunlight which is effective in causing erythema (sunburn) of the skin. One method is by means of special photoelectric cells in conjunction with appropriate light filters. The range of wavelengths in question is from 2800 to 3200 A°.

ETHER : The hypothetical all-pervading medium by which light and all other electric waves are regarded as being propagated, and through which the force between electrons are exerted.

ETHEROPHONE : An electrical musical instrument, invented by Theremin, resembling a multi-valve radio-receiver in construction. The sounds are produced in a loud speaker by the beats due to the interaction of two oscillating circuits of fixed and variable frequencies respectively. The pitch is regulated by movement of the right hand of the performer in relation to an upright rod, thus making variations in the capacitance of the variable frequency circuit; while the volume is controlled by movement of the left hand relatively to a loop of wire in a part of the constant frequency circuit, altering the amplitude of the oscilations by regulating the degree of resonance of that part to the other part containing a *piezo-electric resonator*. A rectifying valve and one or more stages of low frequency amplification are used. Also called theremivox.

ETHONIUM : An electrical musical instrument on a similar principle to that of the Etherophone.

EUPATHEOSTAT : A form of thermostat for controlling room temperatures influenced by radiation and convection more nearly as is done by the human body that an ordinary thermostat controlled by air temperature alone. A blackened copper sphere communicates with a vessel in which a volatile liquid is heated. The amount of heat received by the sphere affects the vapour pressure within and this actuates a diaphragm controlled contact.

EUREKA WIRE : Wire made of an alloy of copper and nickel having a nearly constant temperature coefficient and not deteriorating to a great extend when kept at a high temperature.

EUTECTIC BOND : A thermo metallurgical bond used to provide contact between semiconductor chips. It is formed by heating a eutectic mixture, such as *gold* and *silicon*, and then allowing it to cool. A *eutectic mixture* is a mixture that solidifies as a whole when cooled without change in composition.

EVALUATION FUNCTION : A procedure, modelled as a mathematical function mapping states onto numbers, that computers are the desirability of a states in a state-space by estimating its distance to a goal state, or the cost of solving the problem from that starting point.

EXCESS-3 CODE : A code derived from the binary coded decimal system by adding the binary quantity 3 to each binary coded decimal representation. A binary-coded decimal code in which each word is equal to its binary value plus 0011.

EXCLUSIVE "OR" : A function which is valid, or its value is 1, if one and only one of the input variables is present. The exclusive "OR" applied to the case whence two variables are present, or 1, if the two binary input varibles are different.

EXCLUSIVE "OR" GATE : A two-input gate requiring either input to be true (but not both) to obtain a true output.

EXECUTE : To carry out an instruction or perform a routine. (1) The process of performing the task specified by an instruction; (2) The portion of an instruction cycle in which the task specified by the instruction is performed.

EXECUTE CYCLE : The inerpretation of an instruction and the execution of the operation it signifies, performed by the CPU of a digital computer.

EXECUTIVE PROGRAM : The control program that schedules and manages the computer's resources.

EXECUTIVE ROUTINE : A routine designed to control the execution of other routines.

EXIT : A means of stopping iterative action, as through the use of a test in a repeated loop of operations in a program; the place in the routine which serves as the stopping point.

EXPANDER GATE : A gate designed to increase the fan-in of a second gate.

EXPERT SYSTEM : An application that demonstrates significant human expertise; it involves a knowledge base and inference procedures is a knowledge-based sysem capable of performance in some restricted problem domain comparable to or better than that of human experts; it foten has the ability of explain its conclusions and interactively acquire new knowledge.

EXPERT SYSTEMS SHELL : An "empty" expert sytem with all the components by the domain-specific knowledge base, which can be created by a human expert without AI experience; it usually provides interactive guidance to simplify the initial knowledge acquisition process.

EXPLANATION FACILITY : Capability of an expert system's inference engine to explain to its users how it reaches its conclusions by describing *which* rules (in the case of a production system knowledge base) were fired, *when* and *why* or *why* not.

EXPONENT : A symbol written above a factor and on the right, telling how many times the factor is repeated. In the example of *A2*, *A* is the factor and 2 is the exponent. *A2* means A times A i.e. (A x A).

EXTENDED BINARY CODED DECIMAL INERCHANGE CODE : Abbreviated ABCDIC. A method of expressing alphanumeric characters in eight-bit words based upon binary decimal format.

EXTENSION : (1) Is the definition of a concept according to the examples of it in the world; extensionally, the "morning star" is the same as the "evening star" because the two terms denote the same physical object. (2) A one-to-three character set that follows a filename. The extension further defines or clarifies the filename. It is separated from the filename by a period (.)

EXTENSION LINES (TELEPHONE) : Lines radiating over a telephone installation from a private switchboad to the various extension instruments.

EXTENSION SET (Telephone) : An additional Subscriber's Set either connected permanently in bridge with the main instrument of a telephone subscriber's installation or connected when required.

EXTERNAL STORAGE : Storage devices physically separate from but connected to the automic computer, and holding information available to the automatic computer.

EXTRACT : To obtain parts of a word or fields as specified by a control (filter word or field) to unpack.

EXTRA HIGH VOLTAGE, PRESSURE OR TENSION : Any voltage above 3000 volts between any two conductors between one conductor and earth. Cf. HIGH PRESSURE.

EXTRA LARGE SCALE INTEGRATION (ELSI) : See INTEGRATED CIRCUIT.

EXTREMELY HIGH FREQUENCY (EHF) : See FREQUENCY BAND.

EXTRINSIC SEMICONDUCTOR : A semiconductor in which *impurities* or *imperfections* determine the charge-carrier concentration. See also SEMICONDUCTOR. Compare INSTRINSC SEMICONDUCTOR.

F

FACET : See SLOT.

FACSIMILE BANDWIDTH : A range of frequencies that is required for adequate transmission of the copy. A form of carrier wave is often used for facsimile transmission, the derived electrical signals-the facsimile baseband being used to modulate (see modulation) the carrier before transmission.

FACSIMILE TRANSMISSION: *Syn: Picture Telegraphy* : A method of transmitting any kind of graphic material to produce a pictorial likeness of the object. The system employs facsimile scanning the subject copy is scanned, so providing a successive analysis from which electrical signals are produced. These signals are transmitted to a receiver.

FACTOR OF MERIT : An expression sometimes used for the Sensitivity which a given galavnometer would have if its resistance were one ohm and its time of swing, ten seconds.

FADER : A device that maintains an electrical signal at a constant level while one signal is being faded out, i.e. smoothly reduced in amplitude, and another faded in, i.e. smoothly increased in amplitude.

FADING : Variations in signal strength at a receiver due to variations in the transmission medium. Destructive intereference between tow waves travelling by two different paths to the receiver is the most common cuase of fading; this is termed intereference fading.

FALL-TIME : The time required for a logic circuit to change "0" its output from a high level (logical 1) to a low level (logical 0).

FAN-IN : The number of inputs to a particular gate.

FAN-OUT : The number of inputs a gate output is capable of driving. The number of units loads an output can drive. Within a given family of logic circuits, the maximum number of inputs to other circuits that the output of a given circuit can drive.

FARADAY : A unit of quantity of electricity sometimes used in electrochemistry, being the quantity required to deposit a "gramme equivalent" of the substance, and equal to 96,500 coulombs.

FARADAY'S DARK SPACE : The dark space in the discharge in a moderately exhausted *Vacuum Tube* between the *Negative Glow* and the *Positive Column* (or the *Striae* into which it break up); varying in its position along the tube according to the degree of exhaustion, *Cf. Crooke's Dark Space.*

FAST-RECOVERY DIODE : A diode in which very little carrier storage occurs and that may therefore be used to give an ultrahigh speed of operation. Fast-recovery diodes may be formed as *p-n* junction diodes fabricated from a direct-gap semiconductor, such as *gallium arsenide,* in which the minority-carrier lifetimes are very much smaller than in silicon. The switching time from forward to reverse bias is of the other of 0.1 nanoseconds or less compared to about one to five nanosecods in silicon.

FAULT : An accidental discontinuity in a circuit, or a point of defective insulation, where there is a dead or partial earth, or short circuit, or leakage to earth or to the other pole in a circuit.

FDM : Frequency Division Multiplexing.

FEATURE : An attribute or distinguishing aspect of a concept or object.

FEATURE VECTOR : A representation for concepts in learning sytems that simply lists on object's feature names and their values; it is similar to a frame but strictly specifies only intensional features and not extensional properties like class membership.

FEDERAL COMMUNICATION COMMISSION [FCC] : A US government agency that sets standards and regulates communications in the United States.

FEEDBACK : To make the present output of a person or machine depend in part upon his or the machine's previous output.

FERMI DIRAC STATISTICS : A system of quantum statistics that is used to describe the behaviour of solids in terms of a free electron model. In this model the most weakly bound electrons of the constituent atoms are considered to behave as a gas subject to certain conditions.

FERMI LEVEL : The maximum electronic energy level that is occupied by an electron in a solid at a temperature of absolute zero. At higher temperatures some electrons are excited into higehr energy rates. The Fermi level then corresponds to the value of energy at whici the Fermi-Dirac distribution function has a value.

FERROELECTRIC CRYSTALS : Crystals that exhibit electrical properties analogous to certain magnetic properties, such as ferromagnetism.

FERROMAGNETIC : A substance is said to be "ferrormagnetic" when its permeability is very considerably above that of air, and varies at different values of the Flux Density; i.e. when it shows the very marked magnetic effects exhibited by iron, etc. The principal ferro-magnetic substances are iron (and certain steels), nickel and cobalt.

FET : Field Effect Transistor; a voltage-controlled semiconductor analogous to a triode. See FIELD EFFECT TRANSISTOR.

FETCH : The portion of an instruction cycle during which the instruction is read from memory.

FETCH CYCLE : The retrieval of data or instructions from memory and moving of them to the CPU.

FETRON : A junction field-effect transistor mounted in a suitable package so that it may directly replace a valve in a circuit without any modification to the circuit.

FICK'S LAW : If a concentration gradient of mobile impurity atoms (or ions) exists in a semiconductor, there will be a flow of such atoms (or ions) from the region of high concentration to the region of low concentration. This effect is utilized in producing a desired impurity profile in a particular specimen of semiconductor.

FIDELITY (As Applied To Sound Reproduction) : The degree of approach to equality of amplification of input at all audio-frequencies within the range required, on which the quality of the acoustic output depends.

FIELD : In a database record, a field is a specified area for specific types of data. Examples might *to* a *name field,* a *telephone field,* or a ZIP code field.

FIELD-EFFECT TRANSISTOR (FET) : A unipolar multielectrode semiconductor device in which current flows through a narrow conducting channel between two electrodes and is modulated by an electric field applied at a third electrode. The regions in the transistor are known as the *source, drain,* and *gate;* the narrow channel connects the source and drain regions and the modulating signal is applied to the gate. Application of a suitable bias across the transistor causes charge carriers to flow from soruce to drain. Devices with *n*-type source and drain regions and hence an *n*-type channel are referred to as *n*-channel device, those with *p*-type regions as *p*-channel devices.

FIELD-EMISSION : Syns *auto-emission; cold emission.* A type of emission in which the presence of a large external accelerating electric field reduces the potential barrier at the surface of the emitter (see chottky effect) and allows electrons to escape from the surface. The potential barrier is shown in the diagram as the work done, W for an electron to escape.

FIELD-EMISSION MICROSCOPE : An instrument for studying the surface of a solid, usually a metal, by causing it to undergo field emission.

FIELD-ION MICROSCOPE : An instrument for studying the surface of a solid, usually a metal, by subject it to field ionization. The form is identical to the field-emission microscope but the voltage is applied in the opposite direction.

FIELD-PLATES : The separate set of conductors used in some forms of Influence Machine to induce charges in the carriers.

FIFO : First-In First-Out.

FILAMENT EFFICIENCY (of a Thermionic Valve). The ratio of the Total Emission in milliamperes to the power used in heating the filament in watts.

FILE : A set of records on a common subject and usually organized or ordered on the basis of some combination of items of data uniformaly found in all the records of the file, such as a data.

FILE ALLOCATION TABLE FAT : A record, on a floppy or hard disk, that keeps track of each file's location on the disk.

FILE EXTENSION : A period followed by up to three characters, usually used to identify the file type for IBM computers.

FILE LOCKING : A method to prevent two or more user froms modifying the same file.

FILE MAKER PRO : File Maker Pro is simple for beginners to use. Currently, File Maker Pro is the leading flat-file data base for the Macintosh.

FILE NAME : A name to identify files on a floppy or hard disk. With IBM computers, file names can be up to 8 characters long. File names can be up to 31 characters long with Macintosh computers. The unique name, usually assigned by a user, that identifies one file for all subsequent operations that use that file.

FILE PROTECTION : The devices or procedures that prevent unintentional erasure of data on a storage device, such as a diskette.

FILE SERVER : A computer and program that controls a local area network (LAN).

FILE STRUCTURE : A conceptual representation of how data values, records and files are related to each other. The structure usually implies how the data is stored and how the data must be processed.

FILL : Arbitrary symbols or artificial data used to pad or complete a field record, block or file.

FILLER : Zeros, blanks, or other symbols used to fill or pad a word or field. Numeical quantities are in a specified number of places with the radix point implicity located at a constant position. A value that occupies a slot in a frame, often represented as a pointer to another frame in a frame system. See INTEGRAL, FRACTIONAL.

FILM RESISTOR : A type of resistor that uses a thin layer of resistive material deposited on an insulting core. For low-power applications film resistors are more stable than composition resistors and except for very high precision requirements are smaller and less expensive than accurate wire-wound resistors.

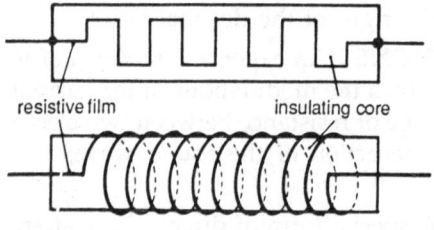

Types of film resistors

FILTER : An electrical network that will transmit signals with frequencies within certain designated ranges (pass bands) and supress signals of other frequencies (attenuation bands). The frequencies that separate the *pass* and *attenuation* bands are the cut-off frequencies.

FILTER CIRCUIT : See WAVE FILTER.

FILTERING : A process used by perception systems to modify its input, in content but not in format, in order to remove noise and highlight its important features and regularities; in vision, this consists of applying various mathematical functions to each point in Filtering a bitmap.

FINITE AUTOMATION (FA) : A simplest interesting computational device; it can model the decision procedure for any regular language but cannot handle all context-free languages.

FIRE : A rule that has already been triggered to be selected for application and have its action part executed.

FIRE ALARM : An apparatus for ringing an alarm bell, actuating indicators, etc., electrically, on the outbreak of a fire, either automatically or by hand. See CLOSED CIRCUIT SYSTEM and OPEN CIRCUIT SYSTEM.

FIRMWARE : A combination of software and microprogrammed hardware used to control the operations of particular computer.

FIRST DETECTOR : An expression formerly used for the Frequency changer in Super-heterodyne Reception.

FIXED DISK : A hard disk enclosed in a permanently-sealed housing that protects it form environmental interference. Used for storage of data.

FIXED LENGTH RECORD : A record in which the length and positon of each field is fixed for all processing.

FIXED POINT : The representation of numbers as integers with no digits to the right of the decimal point.

FLAME MICROPHONE : An experimental form of telephone transmitter in which the modulations of the current are controlled by the change of resistance between two electrodes in a flame owing to the action of the sound waves impinging on the flame.

FLAT AERIAL : A special form of directive aerial array used in the Beam Transmission, composed of a grid arrangement of wires, with a similar unexcited grid behind it acting as a reflector and giving a concentrated "bean" . one direction only.

FLAT PACK : The integrated circuit package with flat leads.

FLATTENER, WAVE : See WAVE FLATTENER.

FLAT TOPPED WAVE FORM : A wave form curve of e.m.f. current etc., the maxima of which are less pointed than are those of a true sine-curve.

FLAT TUNING : The existance of a considerable range of adjustment of the tuning range of adjustment of the tuning apparatus of a wireless receiver over which response to a particular wave can be obtained, often due to excessive self-capacitance or resistance.

FLEMING'S CYMOMETER : See CYMOMETER (Fleming's).

FLEMING'S RULE : If the thumb and first and second fingers be stretched out at right angels to one another, and the first finger represents the direction of the *magnetic lines of force*, the thumb the direction of *motion* and the second finger the direction of the *current* the right hand will give the correct relations between these directions for a conductor in the armature of a generator and the left hand for a conductor in the armature of a motor. (Sometimes called the Right and Left-Hand Rules.)

FLICKER EFFECT (In a Thermionic Valve) : A slight irregularity in the anode current probably mainly due to random emission of positive ions by the cathode.

FLIP-CHIP : A semiconductor chip with thickened and extended bonding pads enabling it to be flipped over and mounted upside down on a suitable *substrate;* such as a thin-film or thick-film circuit.

FLIP-FLOP : A storage device which can be used to retain I bit of information. A flip-flop can be in state or state. In the 1 state its Q output presents a high level and its Q output a low level. In the Q state, it Q output presents a low level and its Q output, a high level. An electronic circuit having two stable staes, two input lines, and two corresponding output lines, such that a current flows on either one or the other of the output lines if any only if the last pulse received by the flip-flop is on the corresponding input line; a modification on an Eccles-Jordan multivibrator circuit. Syn. *half-shift register.* A bistable multivibrator circuit that usually has two inputs corresponding to the two stable states. It is so called because application of a suitable input pulse causes the device to *flip* into the corresponding state and remain in that state *unit* a pulse on the other input causes it to *flop* into the other state.

FLOATING BATTERY : A storage battery connected permanently in parallel with the circuit, and normally subject to only slight charge and discharge currents, but ready at all times to acts as a stand-by on failure of the main supply. Cf. BALANCING BATTERY.

FLOATING CARRIER MODULATION : Syn. *controlled-carrier modulation.* A type of amplitude modultion in which the amplitude of the carrier wave does not remain constant but is automatically varied in a manner dependent upon the amplitude of the modulating wave whcih is averaged over a short time period. The modulation factor therefore remains substantially constant.

FLOATING CARRIER SYSTEM (Of Radio Transmission) : A method of transmission in whcih an approximately constant percentage of modulationis employed while the amplitude of the carrier wave is varied from the maximum down to a low value.

FLOATING POINT : A system of representing numerical quantities with variable number of places, a radix point being represented by a coded exponent of a power of the radix.

FLOATING POINT ARITHMETIC : A method of calculation in which the computer or program automatically records, and accounts for, the location of the *radix* point. The programmer need not consider the radix location.

FLOATING POINT ROUTINE : A set of program instructions that permit a floating-point mathematics operation ina computer which lacks the feature of automatically accounting for the radix point.

FLOPPY : See Diskette.

FLOPPY DISK : A small disk of flexible manetic recording material. It is usually enclosed in a cardboard jacket with an opening in it so that the head on the disk drive can read the information to and from the disk spins within the carboard jacket in the disk drives. A disk make of flexible plastic, coated with a magnetic surface to store data. Floppy disks come in 5¼nch and 3½inch sizes.

FLOW CHART : A system analaysis technique providing a graphic presentation of a procedure (not to be confused with a flow diagram).

FLOW DIAGRAM : In programming, a chart setting forth the particular sequence of operations to be done in a computer to handle a particular application. John von Neumann and other at the Institute for Advanced Study did much of the developed work on flow-diagraming.

FLUORESCENCE : The emission, by a body of energy absorbed from radiation at one wave length, at a different wave length, e.g. of visible light rays from ultraviolet or from "X" Rays. Also called Photo-luminescence.

FLUORESCENT SCREEN : A surface coated with a substance such as barium platino-cyanide, which fluroesces with a bright green light when such radiations as "X" Rays fall upon it; used to render visible the shadows cast by these rays, such as those of bullets or other metal objects embedded in the body.

FLUORESCENT X RAYS : See CHARFACTERISTIC RADIATION.

FLUOROSCOPY : The use of fluorescent screen acted on by "X" Rays.

FLUTTER : An unwanted type of frequency modulation in high-fidelity sound-reproduction systems resulting an audible variations of pitch above about 10 hertz. Compare WOW.

FLYING-SPOT SCANNER : A device that produces a video signal from an object, such as a film, by scanning the object with a spot of light, which is then focused on a photocell to produce corresponding electrical signals. The moving (or flying) spot of light is normally produced on the screen of a high-intensity cathode-ray tube used as a light source.

FLYWHEEL EFFECT : The continuation of oscillations in an oscillator during the intervals between exciting pulses. It results from electrical inertia, which is analogous to mechanical inertia of a flywheel.

FOCUSSING ELECTRODES : Electrodes provided in a Cathode Ray Tube to produce concentration of the electron beam.

FOIL CAPACITOR : A capacitor in which the electrodes are metal foil. The term is most commonly applied to paper capacitors but some polystyrene or polyester film capacitors use foil electrodes and one form of tantalum electrolytic capacitor uses a tantalum foil as one of the electrodes.

FOLDED DIPOLE : See DIPOLE AERIAL.

FONT : A set of letters, numbers, and symbols that appear in a particular tyepface size, and style.

FOOT PRINT : The amount of space a desktop computer uses.

FORBIDDEN CODE : An out-of-code representation for example, 1111 in the binary coded decimal system.

FORCED OSCILLATIONS : Oscillations, due to an external cause; of a frequency generally different from the natural frequency of the circuit in which they are set up.

FOREGROUND PROGRAM : The highest-priority program in a multi-programming environment. Contrast : BACKGROUND PROGRAM.

FORKED REPEATER : A telegraph repeating equipment which enables messages to be repeated to two stations in different directions from the repeating station without interfering with their ability to read each other's signals.

FORMAT : (1) Preparing a disk so that it can have information placed on it by a computer. (2) Placing words on a screen or on paper in a certain arrangement.

FORM FACTOR INDICATOR OR METER : An instrument for indicating the Form Factor of an alternating voltage by comparing the R.M.S. and Mean Values by a rectifier bridge method.

FORM FEED : To advance a page to a specified position in a printer.

FORTRAN : An automatic coding language used in programming computational applications (Formula Translator).

FORWARD CHAINING : An inference algorithm for production systems that operates in cycles, each time collecting all the rules that trigger, choosing one (conflict resolution), firing it, and repeating the process until no productions trigger.

FORWARD TRANSFER FUNCTION : See Feedback CONTROL LOOP

FOSDIC : An input method of photoelectrically reading microfilms of special documents to produce a magnetic tape. The special documents are business forms on which information has been coded as dark marks in certain positions. The method was developed by the Census.

FOUR CHANNEL DUPLEX TELEGRAPH SYSTEM : A variety of Voice Frequency system used on minor two-wire circuits in which some different frequencies are used in each direction.

FOURIER ANALYSIS : A mathematical method of analysing complex waveshapes or signals into a series of simple harmonic functions, the frequencies of which are intergral multiples of the fundamental frequency.

FOUR TERMINAL RESISTOR : A standard resistor that has four terminals. Two are used to connect the resistor to the current source and the other two to connect it to a measuring instrument. This arrangement ensures that the potential drops across the resistor is not affected by contact resistances at the terminals.

FOURTH-GENERATION LANGUAGE : A very high level language that produces a number of high level language statements for every statement in the fourth-generation language.

FOUR WIRE CIRCUIT : A circuit that consists of two pairs of conductors and that forms a simulataneous two-way communication channel between two points of a telecommunication system. One pair of conductors forms the go channel and the other the return channel.

FOX PRO AND FOX BASE +/ MAC (Macintosh) : FoxPro and FoxBase +/ Mac are completely file and command comptible with dBase III Plus and offer more commands and up to fifteen times faster speed. FoxBase +/ Mac lets you copy dBase III Plus files from an IBM to a Macintosh and use them without any modifications whatsoever.

FRACTIONAL : A type of fixed-point system in which quantities are represented as fractions of the numerical quantity 1; places to the right of the radix point.

FRAME : A data structure, similar to a record in a relational database, which represents a concept with a name and various properties arranged in slot/filler pairs.

FRAME AERIAL : A compact and portable form Aerial for receiving purposes composed of a flat coil of wire forming a rectangular (or polygona) frame, both ends of whcih are connected to the receiving apparatus but not to earth, having marked directional and selective quality, and responding to the magnetic rather than the electro-static compound of the wave.

FRAME FREQUENCY (In Television) : The number of times per second that the whole frame is coverred during Scanning. Cf. PICTURE FREQUENCY.

FRAME PROBLEM : A notorious representation and reasoning problem that asks how a system can know which elements of its knowledge become invalid as time passes, as when a robot moves into a new room and must somehow "unlearn" that it was at its previous location.

FRAME SYSTEM : An inheritance hierarchy of frames, including default values and procedure attachment, to make a complete system for representing declarative knowledge.

FRANKING AERIAL : A form of aerial used in short wave radio-communication in which several sections, each the length of half a wave, are used one above the other with Phasing Coils between them to reverse the phase in alternate sections.

FREE CHARGE : In the earlier theories, an electric charge was said to be "free" when not associaated with a neighbouring charge of opposite sign by which it is "bound".

FREE ELECTRON : An electron that is not bound to a specific atom or molecule and is therefore free to move when influenced by an applied electric field.

FREE LANCE PLUS : Freelnace Plus has sold on the strength of its close relationship and compatibility with the Lotus 1-2-3 spreadsheet. If you primarily use 1-2-3, then you may be more comfortable using Freelance Plus.

FREE OSCILLATIONS : Oscillations of the natural frequency of the circuit in which they take place. Cf. FORCED OSCILLATIONS.

FREE SPACE : The region used as an absolute standard and characterized by absence of gravitational and electromagnetic fields. Free space was formerly referred to as a vaccum. The electric constant and the magnetic constant are the formally defined values of the permittivity of free space and the permeability of free space, respectively. the velocity of light in free space is constant and is the maximum possible value.

FRENCH WIRE GAUGE : A wire gauge in use in France in which the higher numbers are given to the larger sizes of wire and the difference between diameters of successive sizes is an intergral number of tenths of millimetres increasing as the wire gets larger.

FRENOPHONE : A form of loud speaking telephone in which the diaphragm is actuated by variation in the frictional drag of a brake on a revolving cylinder of cork, or other material produced by control of the mechanical pressure by the telephone current. Cf. EDISON LOUD SPEAKING TELEPHONE, ELECTROMOTOGRAPH and ELECTROSTATIC ADHESION TELEPHONE.

FREQUENCY : The number of complete cycles (or double waves) which an alternating current, e.m.f. etc., executes per second. Symbol: f. The name Hertz has been proposed for a unit of frequency of one cycle per second. See also KILOCYCLE and MEGACYCLE.

FREQUENCY BAND : A particular range of frequencies that forms part of a larger continuous series of frequencies. The internationally agreed radiofrequency bands are shown in the Table. (a) Microwave frequencies, ranging from VHF to EHF bands (i.e. from 0.225 to 100 gigahertz are usually subdivided into bands designated by letters. These are not internationally agreed but the commonly used subdivisions are shown in Table (b) Limits for the bands may differ slightly from those shown. The entire electromagnetic spectrum is shwon in Table 10, backmatter.

Band	Frequency (GHz)	Wavelength(cm)
P	0.225-0.390	133.3-76.9
L	0.390-1.550	76.9-19.3
S	1.55-5.20	19.3-5.77
X	5.20-10.90	5.77-2.75
K	10.90-36.00	2.75-0.8834
Q	36.0-46.0	0.834-0.652
V	46.0-56.0	0.652-0.536
W	56.0-100.0	0.536-0.300

(b) Microwave bands

Wavelength	Band	Frequency
1 mm-1cm	extremely high frequency; EHF	300-30 GHz
1 cm-10 cm	superhigh frequency; SHF	30-3 GHz
1 cm- 1 m	ultrahigh frequency; UHF	3-0.3 GHz
1 m-10m	very high frequency;VHF	300-30 MHz
10 m-100 m	high frequency; HF	30-3 MHz
100 m-1000 m	medium frequency, MF	3-0.3 MHz
1 km-10 km	low frequency; LF	300-30 kHz
10 km-100 km	very low frequency; VLF	30-3 kHz

(a) Radiofrequency bands

FREQUENCY CHANGER : (1) An apparatus which converts a current at one frequency into a current at another frequency, such as an alternator driven by a synchronous motor, or an apparatus similar to an induction generator. (2) The additional apparatus including a local oscillator in a superheterodyne receiver for producing the intermediate frequency. See also STATIC FREQUENCY CHANGER.

FREQUENCY CONVERTER : A part of a frequency counter (often made as plug-in) which converts the frequency to be measured down to a value measurable by the counter.

FREQUENCY CYCLE : See FREQUENCY.

FREQUENCY DISCRIMINATOR : A discriminator that selects input signals of constant amplitude and produces an output voltage proportional to the amount that the input frequency differs from a fixed frequency.

FREQUENCY DISTORTION : Alternation of the quality of reproduced sound in radio telephony due to reproduction of different frequencies in different ratios. (Also called Tone Distortion.) Cf. AMPLITUDE DISTORTION.

FREQUENCY DIVERSITY : A diversity system that employs independent transmission channels on neighbouring frequencies.

FREQUENCY DIVIDER : A device that produces an output signal whose frequency is an exact integral submultiple of the input frequency.

FREQUENCY DIVISION MULTIPLEXING : A method of transmitting several channels over the same facility by assigning different frequency bands to each channel. A form of multiplex operation in whcih each user of the system is assigned a different frequency band. The transmitted signal contains several carrier waves each of a different frequency and seperately modulated with a different input signal. At the receiver a number of tuned circuits are used to separate the different carrier frequencies.

FREQUENCY DOUBLER : A frequency multiplier that produces an output signal with a frequency twice that of the input signal.

FREQUENCY FILTER : A network of inductance, capacitance and resistance designed to offer as little opposition as possible to currents at of a certain range of frequencies and as much as possible to currents of all other frequencies.

FREQUENCY GENERATOR : Apparatus for producing oscillations, usually of a standard radio frequency.

FREQUENCY INDICATOR (Or Frequency Meter) : An instrument for the measurement of Frequency; depending on the resonance of tuned reeds (See VIBRATING REED INSTRUMENTS). or on the fact that a current in a non-inductive circuit does not vary with the frequency whereas that in a circuit containing reacta but negligible resistance, is dependent upon the frequency. Other principles used are electrical resonance and the balancing effect of magnetic attraction and repulsion due to induced currents in a moving system containing iron.

FREQUENCY METER, INTEGRATING : See INTEGRATING FREQUENCY METER.

FREQUENCY MODE : The selected function of the counter/timer when the instrument functions as a digital frequency meter.

FREQUENCY MODULATED DIGITAL RECORDING : A method of recording digital data in which a 1 is transmitted as two transitions and a 0 as one transition per bit position.

FREQUENCY MODULATION : Modulation by variation of frequency of the carrier wave instead of variation of its amplitude.

FREQUENCY MODULATION (F.M. OR FM) : A type of modulation in which the frequency of the carrier wave is varied above and below its unmodulated value by an amount proportional to the amplitude of the signal wave and at the frequency of the modulating signal, the amplitude of the carrier wave remaining constant.

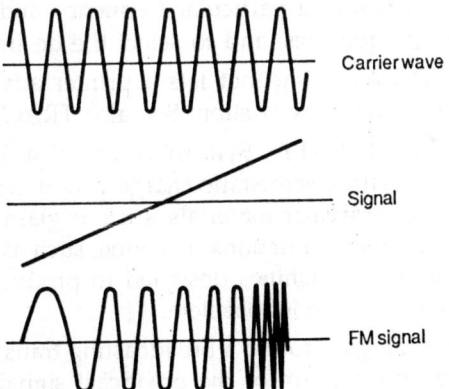

Carrier wave

Signal

FM signal

FREQUENCY MULTIPLIER : A circuit or device that produces an output signal whose frequency is an exact integral multiple of the frequency of the input. Particular cases of frequency multipliers are frequency doublers and triplers.

FREQUENCY MULTIPLIER : See STATIC FREQUENCY CHANGER.

FREQUENCY RANGE : The frequency range at which a circuit or device operates normally. The frequency range at which a particular device is useful depends on the particular operating conditions for the device.

FREQUENCY RATIO MODE : The selected function of the counter/timer where it measures the ratio between two input frequencies.

FREQUENCY RELAY : (1) A relay which comes into action when the frequency of a system departs from the normal by more than a certains margin. (2) A relay which comes into action when a current at a certain frequency is superposed upon a circuit. Sometimes used to actuate the change over switches of Two-Rate meters, or to connects or to disconnect a group of circuits when required.

FREQUENCY RESPONSE CHARACTERISTIC : The variation with frequency of the transmission loss or gain of any apparatus, circuit, or device. A series of tests at different frequencies designed to determine the frequency response characteristic is termed a frequency run.

FREQUENCY RUN : See FREQUENCY-RESPONSE CHARACTERISTIC.

FREQUENCY SELECTIVITY : The ability of any circuit or device to differentiate between signals at different frequencies or between a signal at a particular frequency and interference at different frequencies, and to select the desired signal.

FRICTION FEED : A setting that lets a printer advance paper between the platen by friction. See also TRACTOR FEED.

FRICTIONAL ELECTRICITY : Syn. triboelectricity. The phenomena associated with electrostatic charge produced by friction between two dissimilar materials, such as glass and silk or Ebonite and Catskin. Frictional machine, such as the Wimshurst machine, were machines designed to produce electricity by friction: they are now obsolete.

FRINGE AREA : A region round a broadcasting transmitter in which satisfactory reception of the broadcast signal is not always obtained.

FULL-ADDER : A device capable of adding two binary numbers to a carry in and providing a *sum* and a *carry out.*

FULLERPHONE : A telegraph apparatus used for military purposes in which Morse messages, consisting of very small direct currents in the line are read in a telephone receiver by breaking the current up by a local vibrator which owing to an arrangement of condensers and inductances, does not effect the continuity of the current in the line.

FULL-IMPACT : A software that offers nearly as many features as WINGZ, but has not been marketed as aggressively. Full Impact provides many presentation capabilites for creating reports.

FULL-WAVE DIODE : See DIPOLE AERIAL.

FULL WAVE RECTIFICATION : Rectification in which the half wave of an A.C. in both directions are utilized by reversing every alternate half wave. Also called TWO WAVE RETIFICATION. Cf. HALF WAVE RECTIFICATION.

FULTOGRAPH : A simplified form of Phototelegraphy for radio transmission and reception or "still pictures" in whcih the picture on the revolving drum in the receiver is produced by the action of a current between the stylus and the drum on chemically prepared paper. The synchronisation is effected by starting each revolution of the drum separately by a clutch on receipt of the snychronising signal.

FUNCTION : A computer action, as defined as a specific instruction. Some GW-BASIC functions are COS, EOF, INSTR, LEFTS and TAN.

FUNCTION KEYS : Specific keys on the keyboard that, when pressed, instruct the compuer to perform a particular operation. The function of keys is determined by the applications program being used.

FUNDAMENTAL COMPONENT : The component of the wave from of an alternating current, e.m.f., oscillation, or wave that is a pure sine wave of the principal frequency, uponwhcih harmonics at various higher frequencies may be superposed.

FUNDAMENTAL FREQUENCY : The frequency of the fundamental component of oscillations, etc.

FUNDAMENTAL OSCILLATIONS : Oscillations at the Fundamental Frequency.

FUNDAMENTAL UNITS : The independent units of length, mass

and time upon which a system of derived units is founded. See C.G.S. UNITS.

FUNDAMENTAL WAVE : See FUNDAMENTAL COMPONENT.

FUNDAMENTAL WAVE LENGTH : The wave length corresponding to the Fundamental frequency of the oscillations in a particular circuit, upon which Harmonics may or may not be superposed.

FUSE : (1) A wire or strip of metal, mounted in suitable fittings, designed to melt and to interrupt the circuit when a predetermined maximum current is exceeded. (2) An apparatus for electrically firing explosives, consisting of a case containing the necessary priming charge and a wire heated by the current which passes when the firing key is depressed.

FUZZY REASONING : See reasoning under uncertainly.

GAIN : The increase in the variation of power, voltage, or current obtained in an amplifier or repeater expressed in decibels or nepers; also called Amplification.

GAIN CONTROL Syn : Volume Control. A circuit or device that varies the ampltitude of the output signal or an amplifier.

GAIN CONTROL, AUTOMATIC : See AUTOMATIC GAIN CONTROL.

GALENA DETECTOR : A Crystal Detector employing galena (lead sulphide).

GALLIUM ARSENIDE : High resistivity that allows easier isolation between different areas of the crystal. The main uses for Ga-As have been as microwave devices, such as Gunn diodes or IMPATT diodes, but recently it has been used as MOSFET (a Ga-As junction field-effect transistor. A direct-gap 3-5 semiconductor that has a relatively large band gap and high carrier mobility. The relatively high carrier mobility allows the semiconductor to be used for high-speed applications.

GALVANIC BETTERY : Old-fashioned term for an ordinary primary battery.

GALVANIC CURRENT : A unidirectional uninterrupted current. See GALVANISM.

GALVANISATION : The curative application of uninterrupted unidirectional currents.

GALVANISM : (1) The science of electric currents. (2) In electrotherapeuties, curative treatment by uninterrupted unidirectional currents.

GALVANOLOGY : An old term for the science of electric currents.

GALVANO-MAGNETIC EFFECT : Any of various phenomena arising when a current is passed through a conductor or semiconductor in the presence of a magnetic field. The effects include the Ettinghausen effect, Hall effect, magnetoresistance, and the Nernst effect.

GAME TREE : A search tree used in adversarial game playing situations, in whcih the states are board positions and the operators legal moves.

GAMMA (y) RAYS : The most penetrating of the three principal kinds of rays given out by radio-active bodies; undeflected by a magnetic field, capable of exciting fluorewcence and consisting of other waves of very short wave length (from 0.06 to 1.4 Angstrom units), shortter than that of most "X" Rays, and therefore more penetrating. Cf. ALPHA and BETA RAYS.

GANG CONDENSERS (In Radio Receiving Apparatus) : Two or more variable condensers belonging to successive stage of amplification coupled together to be worked simultaneously by one dial. Usually with a small trimming condenser connected to each for fine adjustment.

GANG CONTROL : Control of a number of similar pieces of apparatus by one operation, e.g. the speed control of a number of motors driving printing telegraph instruments from one tuning fork apparatus.

GARBAGE : See Hash.

GAS DISCHARGE LAMP : See DISCHARGE LAMP.

GAS DISCHARGE TUBE : A Vacuum Tube in which the vacuum is not too high for a discharge to pass without special means of electron projection, such as a heated cathode, i.e. with a gas pressure not lower than about 2 millionths of an atmosphere. Cf. ELECTRON TUBE.

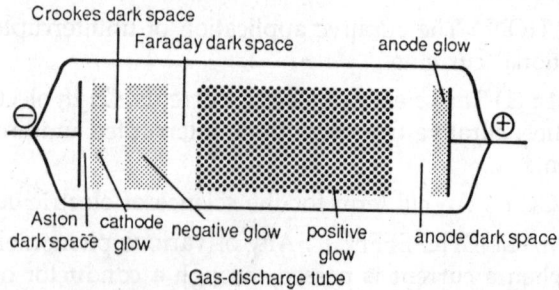

Gas-discharge tube

GAS ELECTRODE : An electrode that absorbs or adsorbs a gas so that when in contact with an electrolyte the gas effectively acts as the electrode. A gas cell is one that contains a gas electrode.

GASEOUS DISCHARGE : mass, consisting of gaseous ions. See also POSITIVE RAYS.

GAS FIELD RECTIFIER : A rectifying valve in which sufficient gaseous ions are present to facilitate passage of the current.

GAS FILLED RELAY : A valve including a certain amount of gas or vapour, provided with a control grid by variation of the voltage of which the discharge can be started or stopped as required.

GAS FILLED TUBE : An electron tube that contains a gas or vapour, such as mercury vapour, in sufficient quantity so that the electrical characteristics of the tube are determined entirely by the gas, once ionization has taken place.

GAS PLASMA SCREEN : A highly legible orange screen often used in laptop and portable computers.

GASSING : The copious evolution of gas in an accumulator cell when the charging current is continued after the charge is complete.

GAS TUBE : A term more particuarly applied to "X" Ray tubes without heated cathodes. See GAS DISCHARGE TUBE

GATE : A circuit with several inputs and one output which operates so that specified condition exists on the output line if and only if some specified combination on conditions is met on the input lines. The logic-operation AND is an example of this.

GEIGER COUNTER : A gas filled tube used to detect ionizing radiation especially alpha particles and to count particle, against radiation hazzard.

GEISSLER DISCHARGE : The broken-up luminous discharge seen in a Geissler Tube. See GEISSIER TUBE.

GEISSLER TUBE : A Vacuum Tube with a moderate degree of exhaustion sufficient for the brightly coloured luminous discharge to be broken up by dark spaces, i.e. with a gas pressure below about 1/100 of an atmosphere. Cf. GLOW DISCHARGE AND CROOKES TUBE, and see NEGATI-GEISSLER TUBE. VE GLOW, POSITIVE COLUMN, FARADAY DARK SPACE, CATHODE GLOW, and ANODE GLOW.

GENERALIZATION : An act of converting a specific concept into a new concept that covers, of subsumes, both the original concept as well as one or more additional concepts; it is an unsound form of logical reasoning, but an essential operation in learning from examples.

GENERALIZED CONES/CYLINDERS : The primitive elements in a
structural representation of three-dimensional objects in which
any element is composed of several connected axex, as in the
stick figure of a man, with closed figures of varying size surrounding them.

GENERATE AND TEST : A problem solving strategy that hypothe-
sizes entire solutions at once and then checks them for cor-
rectness, rather than building up a solutions that is known to
be correct the first time, as an optimal heuristic search.

GENERATING ROUTINE : A form of compling routine, capable of
handling less fully defined situations of more limited scope,
as for example, sorting situations.

GENERATOR : A sytem for use by analysts and programmers to
generate an application. The user works at a high level, and
the system produces code to execute the application.

GENESIS : A standard cell based VLSI Design System consisting of
two sets of modules-one for designing a cell library and another
for designing chips (systems) using library cells. It was designed to be used on simple, slow machines such as IMP-PC
with little memory and low resolution graphics, though
additional hardware can be used if available. Implementing
a complete design system on these machines imposes servere
constraints, particularly for interactive programs where response time is crucial. A software having many orginal algorithms incorporated in it, including a new circuits extraction
strategy, and original algorithms for placement, global routing and channel routing. It is being used as a teaching aid for
graduate and undergraduate students in design courses at
the Indian Institute of Technology Delhi. The modular nature
of GENESIS makes it useful as a vehicle for developing and
testing new algorithms in a design environment. It also makes
it facile to link new modules with it.

GENETIC ALGORITHMS : A set of adaptive search techniques based
on principles derived from natural population genetics, and
are currently being applied to various problems in science,
engineering, and AI.

GETTER : A name given in thermionic valve manufacture to a small
piece of magnesium, or similar metal, attached to the anode,
which is heated after the lamp is exhausted and sealed, so
that it voltalises combines with the residual gases, improves

the vacuum, and finally condenses on the bulb in a miror like deposit.

GHOST WORKING (In Telegraphy) : The superposition of telegraph working on Phantom telegraph workign on Phantom telephone Circuits to obtain an additional communication channel.

GIGABYTE (GB) : A unit of measurement equal to one billion (10^9) bytes.

GIGO : An informal term -that indicates sloppy data processing, an acronym for Garbage in Garbage Out. The term GIGO is normally used to make the point that if the input data is bad (garbage in) then the output data will also be bad (garbage out).

GILL : A measure of the arithmetic speed of an automatic computer. For a one-address computer with only one constant access storage device (such as magnetic cores), a gill is equal to ten times the average operations time. (Named after Stanley Gill of Cambridge University).

GIOGRI SYSTEM OF UNITS : See M.K.S. UNITS.

GLASS VALVE : *Syn. Glass Tube.* An electron tube in which the envelope is made entirely from glass without any different base material. The leads are taken in through special metal-glass seals.

GLOBAL SEARCH : Used in reference to a variable (character or command), a global search causes the computer to locate all occurrences of that variable.

GLOW DISCHARGE : A silent but continuous form of luminous discharge appearing at points on conductors at a high potential. Cf BRUSH DISCHARGE.

GLOW SWITCH : A switch that is formed from a glow-discharge tube containing heat-sensitivity contacts. When a glow discharge is produced in the tube the heat generated by it is sufficient to cause the contacts to operate.

GLOW TUBE : A vacuum tube, such as a neon tube emitting a visible glow when a suitable voltage is applied at its terminals.

GOAL DIRECTED : (Behavior/Reasoning) A term descriptive of a system that makes its decisions in problem solving on the basis of explicitly represented goal conditions it is trying to establish, as in means ends analysis or the proof metaphor of backward chaining inference.

GOALS/SUBGOALS : are objective states set up for or by a problem solver to guide its search.

GOLDSCHMIDT HIGH FREQUENCY ALTERNATOR : An alternator for producing currents at Radio-Frequencies, in which no attempt is made to provide a sufficiently high pole frequency but the fundamental frequency is multiplied by introducing harmonics by means of oscillating circuits in connection with both the field and the armature. A cumulative effect is produced by the interaction between the stator and the rotor, and very high frequencies can be obtained.

GONIOMETER : An instrument that measures angles, such as the angle between two radio-frequency waves (radiogoniometer); the direction of propagation of a wave may also be measured.

GRAMMAR : A formal framework that describes the syntactic structure of a language, usually by establishing various categories for words and sequences, as well as rules that restrict how those sequences may be combined in sentences in the language.

GRAMOPHONE Syn : *Record player.* An instrument that converts the undulations of the grooves of a gramophone record into sound. A stylus moves along the grooves and vibrates in sympathy with the modulations in the groove.

GRAPH : A structure consisting of nodes and arcs arranged with arbitrary connections, either *directed* or *undirected*, that can model the structure of AI concepts such as state-spaces or production systems used in backward chaining.

GRAPHIC INSTRUMENT : A measuring instrument arranged to make a continuous record of its indications on a travelling paper chart by means of a pen attached to its moving system or otherwise, also commonly called recording instrument.

GRAPHICAL USER INTERFACE (GUI) : A program that makes a computer easier to use by letting the user point to pictures representing files.

GRAPHICS : A hardware/software capability to display objects in pictures, rather than words, usually on a graphic (CRT) display terminal with line drawing capability and permitting interaction, such as the use of a light pen. Output involving figures, graphs, drawings, and/or animation.

GRAPHICS TABLET : An input device that translates two-dimensional (x,y) information into digital from which can be translated by the computer into a graphics display on the CRT.

GRATZ RECTIFIER : A method of rectifying alternating currents in which four electrolytic valves are employed per phase connected in the arms of a Bridge so that both waves are used, feeding the circuit taken off from the bridge points from different directions alternately.

GRAVITY CELL : A type of primary cell that contains two different electrolytes kept apart by their different densities.

GRAY CODE : A binary code in which each successive number differs in only one bit position.

GRENZ-RAYS : Soft X-rays produced when electrons are accelerated through 25 kilovolts or less. Grenz-rays are produced in many types of electronic equipment, such as colour television sets, but have an extremely low penetrating power. "X" Rays of the softer or less penetrating type employed in someforms of Radio-Therapy, produced by voltages from 3 to 12 KV.

GRID : (1) An electrode that has an open structure, such as a mesh or a plate with a hole in it, thus allowing an electron beam to pass through it. See *thermionic valve*. (2) The nationwide high-voltage transmission line system that interconnects many electricity power stations. It transmits voltages of up to 400 kilovolts. Voltages as high as 735 KV are used in some countries. (3) Anything made essentially of parallel wires or bars in one plane. Particularly a wire screen or auxiliary anode between the hot cathode and the plate anode in a Thermionic Valve. (4) The reticulated lead frame in a pasted type of accumulator plate in the interstices of which the active material is held.

GRID A. C. RESISTANCE : The ratio of a small change of grid voltage to a small change in grid current, other quantities being constant.

GRID BATTERY : A battery, sometimes consisting of a few dry cells, with or without a potentiometer, to apply a Grid Bias to a Thermionic Valve.

GRID BIAS : A low and usually adjustable steady voltage applied to the grid of a thermionic valve. A negative "Grid Bias" is often used to bring the action of an amplifier valve on the most suitable part of the characteristic.

GRID CIRCUITS : The circuit connected to the grid of a thermionic valve. Cf. PLATE CIRCUIT.

GRID CONDENSER (or Capacitor) : A capacitor in the grid circuit of a thermionic valve in connection with cumulative grid leak rectification, resistance capacitance coupling, or to maintain a negative bias on the grid of anoscilating valve.

GRID CONDUCTANCE (In A Thermionic Valve) : The reciprocal of Grid Resistance.

GRID CONTROL : (1) The control of the discharge in a discharge tube by varying the potential of a grid between the electrodes as in thermionic-valves, mercury arc rectifiers, etc. See GRID MODULATOR. GRID CONTROLLED PHASE-ADVANCER.

GRID CONTROLLED PHASE-ADVANCER : A Phase-Advancer consisting of apparatus similar to a grid-controlled mercury vapour rectifier in which the discharge through the main valve is arrested at a suitable point in the cycle by connection there to of a charged capacitor through auxiliary.

GRID CONTROLLED RECTIFIER : See GRID CONTROL and MERCURY VAPOUR RECTIFIER.

GRID CURRENT : The current flowing between the grid of a thermionic valve and another electrode; normally entering the valve by the grid and leaving by the filament. This will only flow if the grid has a slightly positive potential and can be prevented by a negative bias. See GRID POTENTIOMETER.

GRID CURRENT CHARACTERISTICS (of A Thermionic Valve) : Curves connecting simultaneous values of (i) grid current and grid voltages, and (ii) grid current and anode voltage, all other variables remaining constant. Cf. ANODE-CURRENT CHARACTERISTIC.

GRID CURRENT REVERSE : See REVERSE GRID CURRENT.

GRID CURRENT SURFACE (of A Thermionic Valve) : A surface connecting coordinates representing simultaneous values of Grid Current, Anode Voltage, and Grid Voltage. Cf. ANODE CURRENT SURFACE.

GRID D. C. RESISTANCE (of a thermionic valve) : The ratio of the grid voltage to the corresponding grid current under quantities being constant.

GRID EMISSION : A small amount of emission of electrons, liable to occur in some thermionic valves from the Grid, detrimental to the proper working of the valve.

GRID GLOW RELAY : A vacuum-tube relay with an anode, a cathode, and an intermediate grid in an inert gas, such as neon. The current passing is due to the glow discharge between the anode and the cathode, and its amount depends of the voltage conditions of the grid.

GRID LEAK : A high resistance connected between the grid and cathode of a thermionic valve that prevents an accumulation of charge on the grid and may also be used to develop the grid bias. A high resistance (of a value of megohm or more) in parallel with a Grid Condenser of a thermionic valve to relieve the grid from excessive static charges but allowing it to retain a small negative potential.

GRID LEAK RECTIFICATION : See CUMULATIVE GRID RECTIFICATION.

GRID MODULATOR : A Modulator in a radio-transmitting apparatus which acts by applying the variations of the microphone current to the grid of the oscillating valve, either direct or through an intermediate valve circuit.

GRID POLARISATION VOLTAGE : See GRID BIAS.

GRID POTENTIOMETER : A slide resistance carrying a current from a small battery for applying a variable Grid Bias, often used to given an adjustable positive grid potential to a high-frequency valve to produce just enough grid current to prevent self-oscillation.

GRID PRIMING VOLTAGE : See GRID BIAS.

GRID RECTIFICATION : Rectification by a thermionic valve in which the oscillations tube rectified are applied to the grid circuit of the valve. See CUMULATIVE GRID RECTIFICATION.

GRID RESISTANCE : See GRID D. C. RESISTANCE and GRID A. C. RESISTANCE.

GRID RESISTOR : (1) A resistor for tramway motor control or other purposes, made up of units, each composed of a cast metal grid, threaded on long bolts from which they are insulated. (2) A resistor in the grid circuit of a thermionic valve.

GRID SWEEP : The range of voltage change under working conditions of the grid of a thermionic valve.

GRID VOLTAGE : The voltage between the grid of a thermionic valve and the negative terminal of the filament or the centre of an indirectly heated cathdoe.

GRID VOLTAGE SWING : See GRID SWEEP.

GROUND : A term used (more especially in America) as a synonym to earth, e.g. "ground wire," grounded circuit," etc.

GROUND ABSORPTION : A loss of energy that occurs during transmission of radio waves due to absorption in the ground.

GROUND AERIAL : An aerial laid on or near the ground.

GROUND CLUTTER : See ground return.

GROUND CURRENT U.S. : syn for earth current.

GROUND EDBASE CONNECTION : See common-base connection.

GROUND PLATE (U.S.) : Syn. for earth electrode. see earth.

GROUND REFLECTION : Reflection of a transmitted radar wave by the ground before it reaches the target.

GROUND RETURN : Echoes received by a radar receiver due to reflections of the radar wave by the ground or by objects on the ground. Ground return from extraneous sources can lead to ground clutter on the radar screen. This type of noise can obscure the target.

GROUND STATE : The lowest possible energy level of a system. If a sytem has a configuration such that its energy is greater than its ground-state energy, it is said to be in an excited state. Many phenomena, including luminescence and semiconductor, depend upon system being in an excited state.

GROUND WAVE : A radiowave that travels between a transmitting and a receiving aerial situated above the earth. It has two main components: the space wave, which includes the direct wave and the ground-reflected wave, and the surface wave. Electromagnetic waves travelling along the ground in a way determined by the electrical constants thereof.

GROUNDED-COLLECTOR CONNECTION : See common-collector connection.

GROUNDED-REFLECTED WAVE : A radiowave that travels between a transmitting and a receiving aerial situated above the ground and that undergoes at least one reflection from the ground.

GROUP DRIVE : Electric driving in a factor, etc., where each motor drives a group of several machines.

GROUP FREQUENCY : The number of separate trains of waves per second in a spark or damped wave system of radio-transmission.

GROUP SELECTOR (In Automatic Telephony) : A selector which

makes connection with a desired group of links and then selects an idle link.

GROVE CELL : A two-fluid primary cell in which a zinc rod immersed in dilute silphuric acid forms the negative elements, which is separated from the positive element by a porous partition. The positive element consists of a platinum plate immersed in fuming nitric acid. E.m.f. : 1.93 volts.

GROWN JUNCTION : A p-function that is formed in a single crystal of semiconductor material while the crystal is being grown from a melt. The amount and type of inpurities added to the semiconductor are varied in a controlled manner as the crystal grows. A grown-diffused junction is produced by difussion of impurities into the semiconductor after a grown junction has been formed, in order to produce the precise doping profile required.

GUARD BAND : A frequency band that is left vacant during broadcasting to minimize mutual interence between two neighbourin frequency bands.

GUARDING : A method of shielding in DVMs for eliminating errors due to parasitic voltages from the surroundings.

GUARD RING : A device used to ensure uniform fields and to define the sensitive volume in absolute electrometers and standard capacitors. It consists of a metal plate surrounding and coplanar with a smaller plate; a narrow air gap separates the two.

GUARD RING CAPACITOR : A standard capacitor that uses a guard ring to reduce the edge effect (see diagram). The guard-well capacitor is a special type of guard-ring capacitor used for capacitance below 0-1 picofarads. In this type the guard ring forms a well on to which a pyrex disc is mounted in order to locate the electrode assembly accurately.

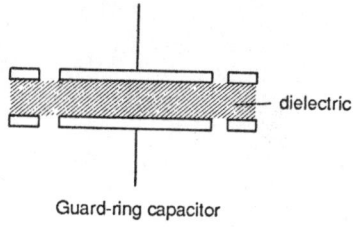

dielectric

Guard-ring capacitor

GUDDN-PHOL EFFECT : The transient luminescence observed in a phosphor previously exposed to ultraviolet radiation, when an electric field is applied to it.

GUILLEMIN LINE : A network that produces pulses with very sharp rise and fall times so that they are almost square.

GULSTAD RELAY : A sensitive form of telegraph relya in whcih the tongue is normally kept vibrating by an auxiliary winding at a speed equal to that of the reversals of current at the distant transmitter, so that the line currents only have to act in arresting the vibration of the tongue temporarily and holding it over against one stop.

GUMN DIODE : A negative-resistance microwave oscillator that operates by means of the Gunn effect. It is a diode formed from a sample of low-resistivity n-type gallium arsenide that produces coherrent microwave oscillations when a large electric field is applied across GUMN DIODE.

GUNDELACH TUBE : The type of "X" Ray Tube commonly employed having an auxiliary anode in another part of the tube but extermally connected to the Anti-Cathode or Target.

GUNN EFFECT : An effect that occurs when a large d.c. electric field is applied across a short sampole of n-type gallium arsenide. At values above the threshold value, typically several thousand volts per cm. coherent microwave oscillations are generated.

HACKER : Slang term for a someone who is particularly skilled and knowledgeable about computers. Often used to describe people who can perform unusual tricks with a computer, such as breaking into other computers.

HALF ADDER : A circuit having two-outputs and two inputs in which the output is related to the input. A logic device that adds two binary bits and provides a sum and carry out.

HALF-WAVE-LENGTH AERIAL : A transmitting aerial with an effective height equal to half the wave-length of the waves to be radiated.

HALF-WAVE RECTIFICATION: Rectification in which the half-waves of the alternating current in one direction only are made use of while those in the reverse direction are suppressed. (Also called One wave Retification.) Cf. FULL WAVE RECTIFICATION.

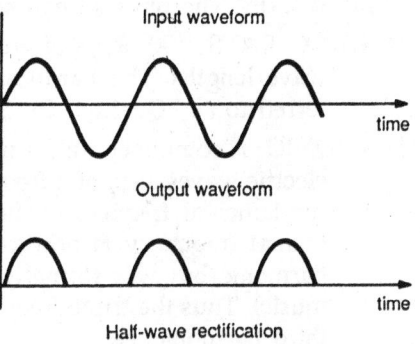

Input waveform

time

Output waveform

time

Half-wave rectification

HALF-WAVE TRANSMISSION : Long distance alternating current transmission in which line losses and pressure drop are lessened by arranging the natural period of oscillation of the line to be equal to four times the frequency so that resonance is produced. Cf. QUARTER WAVE TRANSMISSION.

HAMMING CODE : A particular type of error-correcting code invented by Hamming.

HANDLING : Communication and processing.

HAND MICRO-TELEPHONE : A combined microphone and receiver arranged to be conveniently held in the hand with the receiver to the ear and the microphonea convenient distance from the mouth.

HAND SHAKE : The sequence of events on the IEC bus during

transfer of a multiple remote message. The three hand shake lines are DAV, NRFD, and NDAC control this procedure.

HARD COPY : A printed copy of computer output in a readable form, such as reports, checks, or plotted graphs.

HARK DISK : A metal disk able to store more information that an floppy disk. Hard disks are usually installed internally or externally. An carry information worth 10, 20, 40, 80 or 120 m-bytes.

HARD TUBE : Vacuum tubes, such as "X" Ray tubes are said to be "hard" when the vacuum is very high. Hard "X" Ray tubes given more penetration than soft tubes. Cf. SOFT TUBE.

HARD VALVE : A thermionic valve with a hgih vacuum in which there are few gaseous ions, so that the discharge consists of a practically pure electron strem without gaseous ions. Cf. SOFT VALVE.

HARD WARE : The physical equipment that comprises a sytem.

HARD "X" RAYS : "X" Rays of great peneterative power, i.e. short wave lengths. The hardness of "X" Rays is sometimes referred to as "Quality" Cf. SOFT "X" RAYS.

HARMONIC : A component in the wave form of alternating currents, electric waves, etc., of a frequency which is a multiple of the fundamental frequency. The harmonic of twice the fundamental frequency is now commonly known as the seocnd harmonic but was formerly called the first harmonic (s in music). Thus the triple frequency harmonic is now called the third harmonic, etc.

HARMONIC ABSORBER : Apparatus used to absorb undesired harmonics in the wave form of currents in power circuits supplying transformers.

HARMONIC DETECTOR : A voltmeter arranged in a special circuit so that it measures only the component of the frequeny which is a harmonic of the fundamental sometimes provided with a multiway switch enabling the 3rd, 5th, etc., harmonics to be read separately.

HARMONIC DISTORTION : A distortion that is due to harmonics not present in the original wave form.

HARMONIC FILTER : A combination of inductance and capacitance designed to eliminate any particular undesirable harmonic in the wave form in a circuit. Also called Harmonic Stopper. See also REJECTOR CIRCUIT.

HARMONIC INTERFERENCE : Interference in radio-communication due to the presence of harmonics in the wave form of the transmitting oscillations.

HARMONIC POWER : The component of apparent power in the case of a non-sinusoidal wave due to harmonics only. Also called Distortive Power.

HARMONIC RINGING : A method of Selective or Party Line ringing for telephone circuits in which the different bells connected to one line are adjusted so that each only responds to ringing current of one frequency only.

HARMONIC SELECTIVE SIGNALLING : Signalling to one of a number of stations by a method similar to that described under Harmonic Ringing.

HARMONIC STOPPER : See HARMONIC FILTER.

HARP, TROLLEY : See TROLLEY HARP.

HARVARD PRESENTATION GRAPHICS : Harvard Presentation Graphics is the best-selling presentation graphics program for the IBM for its ease-of-use and ability to generate a wide variety of graphs.

HASH : Unwanted data or data left in a storage device from other applications or other runs; also known as garbage.

HASH CODING : The use of a mathematical calculation on a key to generate a storage address for a direct-access file.

HASH TOTAL : A check total computed by adding together all digits or all numbers comprised in a group of words of fields. The digits or number are added without respect to their identification, meaning, or significance.

HDLS : Abstraction of a digital system in computer manageable format. Typically an HDL supports the design, documentation and efficient simulation of hardware from the digitals system level to the gate level, and is designed to be independent of any underlying technology, design methodology, or environment tool.

HEAD : The device that reads or writes data to a floppy or hard disk, a part of corresponding disk drive.

HEAD (Magnetic) : A group of small electromagnets used for reading recording polarized spots on a magnetically sensitive surface.

HEAD PHONE HEAD RECEIVER, OR HAND SET : The telephone receiver worn round the bead by radio-telegraph and telephone operators.

HEADVISIDE UNITS : A sytem of electromagnetic units in which some of the units differ from those of the C.G.S. sytem by the factor 4 in order to simplify certain calculations.

HECTOMETRIC WAVES : Electromagnetic waves of a wave-length between 1000 and 100 metres.

HEISING MODULATOR : See CHOKE MODULATOR.

HELIUM VALVE : A rectifying valve consisting of a bulb containing helium at a low pressure which is easily ionised, with suitable electrodes, the cathode having a small and the anode a large surface.

HENRYMETER : An instrument for direct measurement of inductance in which an alternating current from a hand-driven magneto generator is passed through the inductance to be tested, in series with a fixed inductance, and the potential difference across its terminals, sensibly proportional to its inductance is read on a moving coil permanent magnet voltmeter working through a rectifying commutator driven off the same shaft as the alternator.

HEPTODE : A Thermionic Valve with seven electrodes, e.g. a transmitter amplifying valve for Push-Pull connection, in a single unit, provided with one cathode, two anodes, two control grids and two stabilishing grids. Used also as a frequency changer in super heterodyne receivers.

HERTZ : Symbol : Hz. The SI unit of frequency. It is the frequency of a periodic phenomenon that has a period of one second. It replac the cycle per second (c.p.s.).

HERTZ EFFECT : The facilitation of a spark discharge of ionisation caused by the incidence of ultra-violet rdiation.

HERTZ OSCILLATOR : A system of conductors in which electrical oscillations can take place, consisting of two insulated plates, spheres, etc., with as park gap between them, forming a condenser of very small capacity.

HERTZ WAVES (Or Hertzian Waves) : Electric waves of much greater wave length than light and including those as employed in radio-communication, so called because they were first experimentally observed by Hertz in 1988 although they has been theoretically predicted by Maxwell in 1864.

HETERODYNE : A term used as a verb in radio to signify producing Beats by the superposition upon received waves of oscillations from an independent source of slightly different frequency, either intentionally, as in Heterodyne Reception, or accidentally, owing to the interference of waves from antoher source of nearly equal wave length. It is also used as a noun both for the local oscillator producing the auxiliary oscillations and for the interference caused by other stations in this way.

HETERODYNE CAPACITANCE BRIDGE : An apparatus for comparing capacitances in which they are placed in oscillating circuits and a beat note produced by their combination; adjustement of the variable reference capacitance being made until a certain beat frequency as shown by a resonant indicator is attained.

HETERODYNE CONVERTER : A frequency converter working on the heterodyne principle.

HETERODYNE DETECTOR : A radio-receiver employing the Heterodyne system of Reception.

HETERODYNE FILTER : A filter consisting of condensers, and inductances, connected to the low frequency side of a receiving set with the object of by passing audio-frequency oscillations caused by the heterodyning of the carrier wave of the station next in order of frequency from that being received.

HETERODYNE INTERFERENCE : Sounds heard in a radio-receiver due to beats produced by the superposition of waves from extraneous sources of nearly equal wave-length upon those sought to be received, including those caused by neighbouring receivers in a state of self-oscillation.

HETERODYNE OSCILLATOR : A separate oscillator for providing the local oscillation required for producing the intermediate frequency in Superheterodyne Reception.

HETERODYNE RECEPTION : The arrangement of thermionic valve circuits for Beat Reception in which the local oscillations of slightly different frequency are provided by an independent source. Cf. ENDODYNE RECEPTION.

HETERODYNE SEPARATE : See SEPARATE HETERODYNE.

HETERODYNE WAVE METER : A wave meter employing the heterodyne principle to compare the waves to be measured with oscillations in a calibrated circuit.

HETODE : Syn. *pentagrid*. A thermionic valve that has seven electrodes, i.e. five grids between the anode and cathode. The most common use of the device is as a mixer.

HEURISTIC PROGRAMS : Programs that are not guaranteed to arrive to an optimal or even an acceptable solution; nonalgorithmic coding. Compare ALGORITHM.

HEURISTIC SEARCH : A search using heuristic knowledge to focus the exploration of the state space on more promising paths, avoiding dead ends, cycles, and blind alleys.

HEUSLER ALLOYS : Alloys showing marked ferro-magnetic properties containing manganese or manganese copper and one of the following ordinarily nonmagnetic elements, viz., aluminium, tin, arsenic, antimony, bismuth or boron. Of these the alminium-manganese bronzes show the strongest ferro-magnetic quality.

HEXADECIMAL : A number system whit a base or radix, of 16. The symbols used in this sytem are the decimal digits 0 through 9 and six additional digits which are generally represented as A, B, C, D, E and F of a processing procedure system whose radix is the quantity 16.

HEXADECIMAL NUMBER SYSTEM : A number system consistig of 16 unique symbols. See HEXADECIMAL.

HEXODE : A thermionic valve that has six electrodes, i.e. four grids between the *anode* and *cathdoe*. The most common use of the device is as a frequency changer.

HEXODE MULTIPLEX : See MORSE MULTIPLEX SYSTEM.

HIDEN FILES : Files that cannot be seen during normal directory searches.

HIERARCHICAL DIRECTORY : See TREE-STRUCTURED DIRECTORY.

H-FI : Acronym from high fidelity. See recording of sound; reproduction of sound.

HIGH DEFINITION TELEVISON : Televison in which the number of scanning lines to the picture is over two hundred.

HIGHER LEVEL LANGUAGES : User-oriented languages with single commands that would require many lines of code in a complier-level language.

HIGH FREQUENCY INDUCTION FURNACE : A form of induction furnace chiefly used for metal smelting in which frequencies from 400 to 20,000 cycles per second are employed, e.g. the Northrup Furnace.

HIGH FREQUENCY ALTERNATOR : An alternator giving a sufficiently high frequency to be employed for maintaining oscillators for the production of electric waves for radio-communication. See ALEXANDERSON and GOLGDSCHMIDT ALTERNATORS.

HIGH FREQUENCY AMPLIFICATION : See RADIO-FREQUENCY AMPLIFICATION.

HIGH FREQUENCY CURRENTS : Currents at frequencies of some thousands of cycles per second employed in radio-communication in various forms of research work and for medical purposes. Their properties are considerably different from those of alternating currents at the usual frequencies of supply, for example, ordinry conductors offter very much more resistance to their passage and they can be passed through the human body at pressures which would prove fatal at lower frequencies.

HIGH FREQUENCY RESISTANCE : The increased resistance of a conductor to currents at high frequencies owing to the current being concentrated near the surface, and other causes including dielectric absorption and radiation, See SKIN EFFECT.

HIGH FREQUENCY TREATMENT : Curative treatment involving the use of interrupted trains of damped high frequency oscillations. (Also called D' Arsonvalism.)

HIGH FREQUENCY WAVES : A term sometimes sued for electromagnetic waves of frequencies between 3000 and 30,000 kc. per sec.

HIGH LEVEL LANGUAGE : A language closer to English than assembler language that, when translated, produces many machine-language instruction for each input statement.

HIGH PASS FILTER : A waves filter which prevents the plassage of oscillations below a certain frequency.

HIGH POWER CHOKE MODULATOR : A Choke Modulator in which the modulating valve controls the oscillator directly and handles a power equal to or greater than that in the last high frequency stage. Cf. LOW POWER CHOKE MODULATION.

HIGH PRESSURE : Voltage exceeding 650 volts between two conductors.

HIGH SPEED (Telegraph) TRANSMISSION : Telegraph transmission at speed higher than are possible with keys manipulated by hand. See WHEATSTONE AUTOMATIC, POLLAK-VI-RAGE WRITING TELEGRAPH.

HILL CLIMBING : A search technique that is similar to best-first search in that it always tries to minimize the distance remaining to the goal (regardless of accumulated cost), but it only retains one node at a time in its OPEN queue; the effect is to have the search get stuck on local "hills," or maxima in the curve of negative remaining distance.

HITTORF TUBE : A form of Cathode Ray Tube with a concave cathode focussing the electron stream on to a platinum screen which becomes red not when the tube is in sue. Rontgen's discovery of "X" Rays was made with a tube of this kind.

HNIL : High-Noise Immunity Logic; closely resembling DTL, the basic differende being that HNIL uses a zener diode for standoff (raising the threshold voltage).

HOLLERITH CARDS : Cards in which holes are punched in specific places to represent data. Used for storage of information prior to invention of magnetic storage media.

HOLLERITH CODE : A binary code invented by Hollerith used for punched cards in expressing alphanumeric data.

HOLLERITH, HERMAN : Inventor of a punched card in 1889 and a device for reading it which was used in the 1890 cenus.

HOMODYNE RECEPTION : A system of reception for Suppressed Carrier systems of radio telephony in which oscillations of carrier frequency are provided from a separate source.

HOOK SWITCH : A switch included in a subscriber's or other telephone instrument, which automatically makes the necessary changes in the connections from ringing to speaking condition, when the receiver is taken off the hook; and, in the case of central battery working actuates the relay controlling the calling lamp at the exchange by closing the speaking circuit.

HORSE-POWER (H.P.) : The practical unit of mechanical power. The British Horse-power is equal to 33,000 ft. Ib. per minute or 746 watts. The *metric* Horse-Power is 0.986 of this, being 75 kg. metres per second.

HORSE-POWER, ELECTRICAL. (E.H.P.) : A unit of electric power equivalent to the amount of power in one mechanical "horse-power," equal to 746 watts; sometimes convenient in calculating efficiencies of motors, etc.

HOT CATHDOE : A cathode of an "X" Ray or other Cathode Ray Tube which is electrically heated to augment the disengagment of ions and electrons, i.e. to produce Thermionic Emis-

sion, with or without special oxides or other chemical substances to facilitate the action.

HOT CATHODE DISCAHRGE LAMP : A form of Discharge Lamp
in which a Hot Cathode is employed to increase its efficiency
and to improve the distribution of lighting the tube as well
as to facilitate starting.

HOT CATHODE MERCURY VAPOUR RECTIFYING VALVE : A
Thermionic Rectifier in a bulb containing mercury vapour
instead of a vacuum.

HOT CATHODE TUBE : See HOT CATHDOE.

HOT WIRE MICROPHONE : See THERMAL MICROPHONE.

HOT WIRE OSCILLOGRAPH : An oscillograph such as that of Irwin,
in which the mirror is deflected in accordance with the difference in the expansion of two fine wires carrying currents
proportional respectively to the sum and the difference of the
current to be measured (c) and a constant current (k). The
deflection is then proportional to $(c + k)^2 - (c - k)^2 = 4kc$.

HOT WIRE TELEPHONE : See THERMAL TELEPHONE.

HOUSE-KEEPING FUNCTIONS : Routine operations that must be
performed before the actual processing begins or after it is
complete.

HOWLER : Any apparatus for producing an oscillating current of
fixed or variable audio-frequency such as that sometimes
employed in telephone exchanges to pass such a current into
a subscriber's line if his receiver is left off the hook, in order
to make it give out a sound audible from some distance to
call attention to the fact.

HOWLING : If a telephone receiver is placed face to face with the
transmitter, connected up in the usual way, a continuous sound
or "howling" is produced by a cumulative effect setting up
oscillations of a electrical constants of the circuit as well as
the mechanical constants of the diaphragm. A similar effect
sometimes occurs in Loud Speakers when the natural frequency of the diaphragm coincides with that of the filament
of an amplifying valve by an action analogous to Microphonic
Noise. The term is applied rather indiscriminately to acc-
cidental loud continuous noises in radio receivers, due to
oscillations in radio-frequency or audio-frequency circuits. See
also SING.

HUGHES PRINTING TELEGRAPH SYSTEM : A type printing telegraph sytem which has been widely used on cables of short lengths, etc., depending upon synchronously revolving selectors at either end. These cause a current to be sent at that part of the cycle of operations when the receiving selector arm is in the position corresponding to the letter to be printer. The type wheel is thus stopped at the correct position. In the sending apparatus the depression of one of a number of keys raises a contact piece which touches the selector arm as it comes round.

HUNDREDS SELECTOR (In Automatic Telephony) : A selector responding to the impulses representing the second digit of a four figure number dialled in a call, the function of which is to make contact between a Thousands Selector and an appropriate Final Selector.

HYBRID CIRCUIT : A circuit made by a combination of different integration techniques.

HYBRID FUNCTION : A *sum of products* or *product-of-sums* expression that has been reduced by factoring; three or greater combinational levels of logic.

HYDROGEN ELECTRODE : A Normal Electrode formed by occluded hydrogen in a base of spongy platinum or some such substance.

HYDROPHONE : An arrangement of submerged microphones for detecting the proximity of submarines or other craft.

HYPERDYNE RECEPTION : A system of radio reception similar to the earlier forms of Superheterodyne Reception, but employing a higher intermediate frequency (about 2000 kc.), with Screen Grid Amplifiers.

HYPERFREQUENCY WAVES : See ULTRA-HIGH FREQUENCY WAVES.

HYPERTEXT : A way of linking text to other text or pictures. An application in which textual information is represented by frames; the system links related frames to facilities inquiry.

HYSTERESIS : The quality of a logic device to recognize a "1" when it is in the "O" state at a higher voltage than that necessary for recognizing a "O" when it is in the "1" state.

HYPERHETERODYNE RECEPTION : See HYPERDYNE RECEPTION.

I

IBM : International Business Machines (Corp.)

IC : Integrated Circuit.

ICL : International Computers Limited.

ICON : A graphical image that represents another object, such as file on disk.

ICONOSCOPE : A primitive television camera made on similar principle to that use in Emitran camera.

ICU : Intensive care unit.

IDEAL : Integrated Design Automation Language supports hiserarchical and modular description of digital systems at the architectural, register-transfer and logic levels. Descriptions such as behavorial, structural, procedural, functional and data-flow can be used at each level.

IDEAL TRANSDUCER : A hypothetical transducer that produces the maximum possible signal power output under specified input and load conditions; i.e. power loss in the transducer is negligible.

IDENTIFIERS : The menmonic symbols assigned to variables in a program.

IDFT : Inverse Discrete Fourier Transform.

IEC BUS : A standardized 16-line bus designed to connect IEC bus compatible devices. It can be divided into eight lines for the data bus, three lines for data-byte transfer control, and five lines for general interface management. The IEC bus is more correctly called the IEC interface bus.

IENOSECOND : One-billionth of a second; written as "ns".

IGNITION ELECTRICAL : (1) The firing of an explosive mixture of gases or vapours by an electric spark; either accidentally, as in explosions in collieries due to faulty electrical apparatus where open sparking takes place, or in internal combustion engines, where the charge is fired at the right part of the stroke electrically. The spark within the cylinder may be produced by an Induction Coil, or by mechanically breaking an inductive circuit, or, more usually, by a magneto generator itself, containing an induction coil, in combination with suitable current timing apparatus. (2) In a mercury vapour rectifier, the striking of an auxiliary arc to start the main arc. See MAGNETO.

IGNITION RATING : The rated ampere-hour capacity of small accumulstors used for ignition apparatus on a basis of intermittent working. Usually twice the continuous rating at a low discharge rate.

IGNITION VOLTAGE : The voltage required to start the arc in a mercury vapour rectifier, etc.

IGNITRON : A mercury vapour tube in which the time of formation of the arc is controlled by an auxiliary electrode of semi-conducting material from which a spark passes to the mercury when a certain potential is attained between them.

IKONOPHONE : An experimental apparatus combining Television by line currents with telephony to render the distant speaker's face "visible."

ILLIAC : Illinois Automatic Computer.

ILUMINATION : Lumen per square centimetre; and the Foot-Candle one lumen per square foot (or the illumination produced by one candle at a distance of one foot). One foot-candle is equivalent to 10.746 lux. Thus it is total amount of Luminous Flux (lumens) falling on a surface per unit area, i.e. the density of the luminous flux at the point in question. Several units have been employed for its measurement viz. (1) the lux, one lumen per square metre; the Photone.

ILLUMINATION PHOTOMETER : An instrument which measures the actual value of illumination at any point in a building, in the street, etc. Usually of a portable nature, and depending upon comparisons between a screen receiving the illumination to be measured and a second screen illuminated to a variable extent by a standard lamp within the instrument by tilting the screen or partly obstructing the light.

IMAGE : Processing of information that is in graphical rather than character mode; the image is represented by a bit pattern, for example, at 300 dopts per inch.

IMAGE UNDERSTANDING : Is the principal problem for computer vision systems, and consists of processing a perceived image of the external world to extract the important and relevant conceptual information and relationships that describe the contents of the world.

IMPEDANCE : The property of a circuit, depending upon *frequency, inductance, capacitance* and *resistance,* which determines the current produced by a given alternating voltage in that circuit, i.e. the ratio of the root-mean-square voltage to the R.M.S. current. Impedance, like resistance, is measured in ohms and its value for a circuit of resistance R ohms, inductance L henries, and capacitance C farads (not microfarads) at a frequency of f cycles per second, is $\sqrt{[R^2+(2\pi fL - 1/2\pi fc)^2]}$

IMPEDANCE BRIDGE : A Bridge method of measuring the combined resistance and inductance of a telephone receiver, etc., in which the current supply is alternating, of audiofrequency; a telephone is used in place of a galvanometer, and the instrument under test is compared with known variable resistances and inductances in the opposite bridge arm; a balance being obtained by adjusting both until there is no sound in the bridge telephone.

IMPEDANCE FACTOR : A term sometimes used for the ratio of the Impedance of a circuit to its Resistance.

IMPLICANT : A reduced Boolean term in tabular reduction.

IMPLICATION TABLE : A table used for reducing the number of asynchronous states to a minimum.

IMPORT : To load a file created by another program.

IMPULSE CIRCUIT : (1) A circuit in automatic telephone systems used only for the transmission of current Impulses, e.g. for calling signals and not telephone currents. (2) A circuit, built up of condensers, resistors, and spark-gaps to produce a sudden high voltage for insulator testing, etc., on the sparking over of the first or trigger spark-gap of the series.

IMPULSE FREQUENCY : The number of impulses occurring per second.

IMPULSE MACHINE (In Automatic Telephony) : An apparatus for sending impulses.

IMPULSE RADIATION : A term sometimes used for the "X" Rays of continuous spectrum given off by the anticathode of an "X" Ray Tube in addition to those of frequencies corresponding to the line spectrum of the material of the anticathode. Cf. SPONTANEOUS RADIATION.

IMPULSE RATIO : The ratio of the time of duration of an impulse to the Impulse Period.

IMPULSE REPEATER IN AUTOMATIC TELEPHONY : An appliance for repeating calling impulses in another circuit.

IMPULSE SPRING : A contact spring employed to control impulses in automatic telephony.

IN BRIDGE : An expression used in telephone engineering to signify in parallel.

INCREMENT KEY : A key employed in Quadruplex telegraphy which sends signals by varying the strength of a current but does not reverse it. Cf. REVERSING KEY.

INCREMENTER (Or Incrementer Relay) : An apparatus used in Quadruplex repeating stations to perform the functions of a remote controlled Increment Key.

INDEPENDENT DRIVE (In Radio Transmitters) : An arrangement for ensuring constancy of wavelength (independently of variation of capacitance of the aerial, etc.) consisting of an auxiliary oscillating circuit or Drive Oscillator coupled to the grid of the main oscillation generator, to control its frequency. (Also called Master Drive). See also TUNING FORK DRIVE CRYSTAL CONTROL, and VALVE DRIVE.

INDEX : To reset a computer or program to some starting values. When used to describe floppy or hard disks, the term means the same as format. See FORMAT.

INDEX REGISTERS : Computer registers used to hold data for address modification, subroutine linkage, etc.

INDIRECT ADDRESS : An address (formed by the contents of a particular storage location) that points to another storage location, which may contain either a direct or another indirect address.

INDIRECT LOGIC : Sequential logic in which the output is taken from gates, rather than the stee flip-flops.

INDIRECTLY HEATED CATHODE : A cathode in a thermionic valve or other discharge tube consisting of a plate coated with a special material, such as barium oxide, and heated by radiation from an independent filament which may be heated by alternating current.

INDIRECTLY HEATED VALVE : A thermionic valve with an Indirectly Heated Cathode.

INDIRECT RAYS (In Radio Communication) : That part of the radiation, at a considerable angle with the ground, which only reaches the receiving station by downward reflection or refraction from the lonsphere. Such rays are more readialy received in darkness. Cf. DIRECT RAYS.

INDUCED NEEDLE : A term sometimes used in telegraphy for a soft iron "needle" in a Single Needle Receiver, in which the required polarity is maintained by the field of fixed permanent magnets to prevent acidental reversal of polarity or gradual weakening due to the current in the coils.

INDUCTANCE COIL : A coil possessing considerable inductance compared with resistance or self-capacitance, such as the plug in coils used in the older radio-receiving sets.

INDUCTACE DROP : That part of the voltage drop along an alternating current transmission line, or in a transformer, etc., due to the Inductance of the line or apparatus.

INDUCTANCE FACTOR : A term occasionally used for the ratio of the wattless current to the total current in an alternating current circuit, i.e. the sine of the angle of lag. Cf. POWER FACTOR.

INDUCTION : Is the process of learning the formulating new concepts given only examples (positive and/or negative) of them.

INDUCTION COIL : (1) A coil with a primary winding of comparatively few turns of thick wire and a secondary of many turns of fine wire, used to obtain an intermittent high voltage in the secondary winding by supplying an intermittent current to the

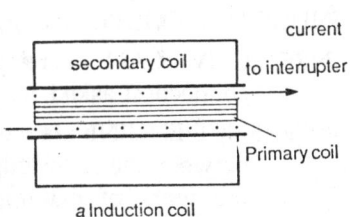

a Induction coil

primary coil. The primary current is broken by an Interrupter and the sharper the break the greater is the rate of change of the primary current and therefore the secondary voltage produced. Sufficiently high voltage are obtainable to give long spark discharged in air. Induction Coils were used in the earlier forms of radio telegraph transmitters, to excite Rontgen Ray and other Vacuum Tubes, and for electromedical purposes, and for Electric Ignition. (2) A form of transformer used in telephone speaking circuits is also called an *Induction Coil*. This is used to convert the fluctuating but unidirectional current of the microphone circuit into an alternating current which gives better effects in the receiver.

INDUCTIVE LEAK : A connection between some point in a telegraph circuit to earth having inductance as well as high resistance. Used to sharpen the signals in a similar way to a Signalling Condenser in Hughes fluplex and repeating stations, etc., and proposed as a method to alleviato generally the effect of capacitance in long lines.

INDUCTIVE LOAD : A load on a generating station, etc., containing a large proportion of apparatus such as induction motors causing a low power-factor, i.e. in which the current lags behind the voltage producing it.

INDUCTIVE LOADING : The provison of artifical or added inductance to a telephone line or cable to counteract the effect of capacitance, and to avoid distortion. See PUPIN COILS, CONTINUOUS LOADING, etc.

INDUCTORIUM : An old-fashioned name for an Induction Coil.

INDUCTOR LOUD SPEAKER : A type of Loud Speaker in which a pair of soft iron bars, on flexible parallel motion supports is directly connected to the coneor diaphragm and moves between laminated poles of magnets excited by the telephone current.

INDUCTIVE COUPLING : See COUPLING.

INEFFECTIVE CALL : A telephone call which has not resulted in the connection asked for.

INFERENCE ENGINE : Is the component of an expert sytem that lies between the knowledge base and the user interface and does the work of making deductions to solve problems; with production sytems; it typically uses either a forward or backward chaining algorithm.

INFLUENCE MACHINE : A machine for obtaining high electrostatic potentials from mechanical power by the relative movement of charged conductors having an inductive effect upon one another.

INFORMATION : Facts and knowledge derived from data. The computer operates on and generates data. The meaning derived from the data is information; that is, information results from data. The two words are not synonymous, although they are often used interchangeably. Knowledge about anything expressed in a symbolic from. In strictusage, information differs from data in its new value information is the part of data that is of most interest to the data user. In common usage, data and inforamtion are often interchangeable INFOMATION terms.

INFORMATION PROCESSING : Is an abstraction of the functionings of the brain, mind, and compuer to a high enough level to make them identical, not in detail but in input and output conditions; it views processes as "black boxes" whose inner working need not be precisely INFORMATION PROCESS-ING understood as long as their input-output mappings can be simulated.

INFRA-ACOUSTIC FREQUENCY : See SUB-AUDIO FREQUENCY.

INFRA-DYNE RECEPTION : A system of radio-reception similar to the Superheterodyne system, except that the "intermediate" frequency is higher than that of the received wave instead of lower so that Superretros action can be used.

INFRA-RED MICROSCOPE : An apparatus similar to an Electron Telescope in which the optical image on the cathode of the Image Tube is produced by a projecting microscope employing infrared rays.

INFRA-RED RAYS : Radiation of lower frequencies than those of the red end of the visible spectrum, including "radiant heat." Their wave-lengths range from 0.00008 to 0.04 cm.

INFRA-RED THERAPY : Treatment of disease by infrared radiation, of a wave-length over 7000 A.U.

INHERENT REGULATION : Voltage or speed regulation due to some feature of the apparatus itself such as the compounding of a generator or the differential winding of a motor, usually expressed as the percentage change between full and no-load under specified conditions. See REGULATION UP and REGU-LATION DOWN.

INHERITANCE : Enables of concept, such as "dog", in a semantic network of frame system to automatically acquire the properties of its superclass(es), such as "animal" and "mammal," without having them explicity attached to the concept itself; it works by scanning up the instance and subclass links to find ancestor concepts in the hierarchy.

INITIALIZATION : Setting variables and control indicators in a program to their starting values.

INITIALIZE : (1) To begin something. (2) To format a disk.

INK WRITER (or Inner) : A telegraph receiving instrument which marks the message in ink on a paper strip in the "longs" and "shorts" of the Morse Code.

IN-LINE : See ON-LINE (In-line is sometimes incorrectly used for real time).

INNER CONDUCTOR (or Main) : A term sometimes used for the *neutral* or *middle* conductor of a Three-Wire system, also sometimes used for the central conductor in a concentric cable. Cf. OUTER CONDUCTOR.

IN-PHASE : Two alternating currents, voltages, oscillation, etc., are said to be "in phase" when their wave forms are of the same shape and pass through their zero values at the same instants. Cf. INSTEP.

INPUT : (1) The process or device concerning the entry of data into a computer. (2) Actual data being entered into a computer. The information communicated to an automatic computer for the computer to process, or to direct the computer in doing data processing; to communicate information for one or both of these purpose. The power, measured mechanically in horsepower, etc., or electrically in kilowatts, etc., absorbed by a machine or apparatus and partly given out in a useful form and partly dissipated as Losses. See EFFICIENCY and OUTPUT.

INPUT/OUTPUT : A general term for devices that communicate with a computer. Input/output is usually abbreviated at I/O.

INPUT UNIT : The functional unit which takes into the automatic computer information from outside the computer.

INQUIRY AND POST SYSTEM : A system in which enquiries are made and data are entered and posted to a file for later updating.

INSTALLATION : The complete arrangement of wiring and apparatus for the production or utilisation of electrical energy in a building, etc.

INSTANTANEOUS POWER : In an alternating current circuit, the product of the e.m.f. and current at any instant which, in the case of currents out of phase with the voltage, may be negative for some parts of the cycle, meaning that power is being returned to the generating plant.

INSTANTANEOUS VALUES : The actual value at any instant of an alternating or otherwise varying current, e.m.f., etc., which may be anything from zero toits Peak Value. Cf. ROOT MEAN SQUARE VALUE, etc.

INSTEP : Two alternating currents, voltages, oscillation, etc., are said to be "in step" when they pass through their zero values at the same instants. Cf. IN PHASE.

INSTRUCTION : A program step that tells the computer what to do next. Instruction is often used incorrectly as a synonym for command. Also a word or field which directs the automatic computer to take a certain action. The instruction usually consists of a command together with one or more operand addresses, which, taken together, cause the computer to act upon the indicated operands.

INSTRUCTION LOCATION COUNTER : A register in the CPU that points to the next instruction to be fetched for execution.

INSTRUCTION MODIFICATION : The changing of one or more of the parts of any given instruction by means of computer operations specified in other instructions when they are executed by the computer. This is often called *address modification* becuase the operand addresses are the most commonly modified.

INSTRUCTION REGISTER : The register that holds the instruction currently being executed by the processor.

INSTRUCTION SET : The repertoire of instructions available on a computer.

INSTRUMENT : In electrotechnics generally the term instrument is used to mean Measuring Instrument, or appliances of similar nature actuating control apparatus, but in telegraphy and telephony includes a considerable class of transmitting and receiving apparatus.

INSTRUMENT CONTROL : The arrangements made for the provision of the force in measuring instruments tending to keep the moving system in its zero position and opposing the deflecting force due to the quantity to be measured. See GRAVITY, SPRING, and MAGNETIC CONTROL.

INSTRUMENT DAMPING : A widely used term in indicating instruments to bring the indicator to rest quickly. A device incorporated in an indicating instrument to provide damping is a damper, one that is *overdamped* so that the needle goes to its true position without wavering is a *dead-beat* instrument.

INSTRUMENT TRANSFORMERS : Small transformers used to couple ammeters, voltmeters, relays, etc., to high voltage circuits so that only low voltages and low currents are brought to the instruments themselves. See CURRENT TRANSFORMERS and VOLTAGE TRANSFORMERS.

INSULANT : A term proposed, but not extensively adopted, for Insulating Material as preferable to Insulator, Insulatin, or Dielectric.

INTEGER : A complete entitiy, having no fractional part. The whole or natural number. For example, 65 is an integer; 65.1 is not.

INTEGER VARIABLE : A quantity that can be equal to any *integer* and can taken on different values. There may be a range of values that it can take on.

INTEGRAL : A type of fixed-point system in which quantities are represented as integers; places of the left of the radix point.

INTEGRATED CIRCUIT : A complete electronic circuit contained in a small semiconductor component.

INTEGRATED DATA PROCESSING : Specifically, the use of paper-tape-producing equipment and paper-tape-actuated equipment for data handling; more generally, an approach to automatic data processing involving ideally only one manual transcription of input data into machine language.

INTEGRATED INJECTION LOGIC (IIL) : A high desnity bipolar logic consuming very little power. The latest polysilicon self-aligned IIL with sidewall base contact (1) compares favourably with CMOS in packing density, speed and power-delay product. The inherent wired logic capability makes gate arrays attractive for designing semi-custom ICs which are less sensitive to static chargs, hot electrons and radiation effects than CMOS circuits.

INTEGRATED PACKAGE : A program that combines the features of several programs such as word processor, spreadsheet, database, and communications program. Also called integrated program.

INTEGRATOR : An operational amplifier that outputs a ramp voltage when the input receives a *DC* value.

INTELLIGENT DEVICES : The addition of logic to a device or product, usually through the incorporation of a microcomputer.

INTENSION : Is the definition of a concept according to its innate properties; intensionally, the "morning star" is different from the "evening star" because the two appear at different characteristic times even though they are the same physical object.

INTENTIONALITY : A property, claimed by John Searle to be necessary for true intelligence but impossible to realize in a non-human system, that enables concepts to be directed at objects in the real world and underlies mental states like beliefs desires, goals and intentions.

INTERAL DRIVE : Another term for a hard drives. See also HARD DISK.

INTERCOMMUNICATION TELEPHONES : A telephone installation in a large office, etc., in which any station can call up and speak to any other without the medium of an exchange oprator.

INTER-CONNECTER, INTER-CONNECTING FEEDER : A main connecting two distributing centres for load equalisation or provisin of an alternative route in case of breakdown, or connecting two generating stations for mutual assistance.

INTERCONNECTION (In Automatic Telephony) : The connecting togethr of Level Multiples so that the links are available from different sections in a different order.

INTERELECTROLDE CAPACITANCE (of A Thermionic Valve) : The capacitance between any two electrodes, e.g. the grid and the anode; of greater importance the higher the frequency in avoiding self-oscillation. By suitable arrangements, its effect can be balanced out. See NEUTRODYNE RECEPTION.

INTERFACE : An information interchange path that allows parts of a computers and external equipment (such as printers, monitors, or moderns), or two or more computers to communicate or interact. (2) A link between two objects, such as a computer and a person. Such a link is often called a user inteface, with refers to the way a person gives commands to the computer.

INTERFERENCE : (1) The spoiling of received radio signals by the effect of extraneous waves including the effects of neighbouring electrical apparatus. See INTERFERENCE PREVENTER. (2) In general, the harmful reaction of power, traction, and other heavy current circuits on telegraph or telephone circuits. The expression is also used in a still more general sense to include effects of neighbouring and associated telephone circuits.

INTERFERENCE PREVENTER : Additional radio-receiving circuits, slightly out of tune with the main receiving circuit, arranged to have opposing actions upon the main receiver, when practically equal forced oscillations are produced in both main and auxiliary circuits, so that the effects of all waves not producing a markedly preponderating resonance in the main circuit cancel out.

INTERFERENCE SUPRESSOR : A system of capacitance and inductance applied to plant to prevent interference with neighbouring radio-receiving apparatus.

INTERLACE : To assign addresses, codes, etc., in a nonsequential pattern selected usually to increase the speed of computer operation.

INTERLACED SCANNING : A system of Scanning in television in which the whole field is covered in several stages known as frames, each consisting of lines not included in the previous frame. The spaces left out are covered in subsequent frames.

INTER-LEAVE : The relationship between the rate at which a hard disk spins and the way it organizes files on the disk.

INTERMEDIATE FREQUENCY AMPLIFIER : An Amplifier in a Supersonic Hetrodyne receiver working at the beat frequency.

INTERMEDIATE RELAY : See REPEATER.

INTERMEDIATE TELEGRAPH STATION : One of a number of telegraph stations in series on a main telegraph line where no connection is made to earth and messages can be sent or received in either direction.

INTERMEDIATE WAVES : An expression to represent waves between 50 to 200 metres (6000, 1500) cf. LONG MEDIUM, SHORT and VERY SHORT WAVE'S and see HECTOMETRIC and DECAMETRIC WAVES.

INTERMITTENT CURRENT : A unidirectional current continually interrupted at regular or irregular intervals.

INTERMITTENT INTEGRATION : The action of an integrating meter which does not make a true continuous integration of the current, etc., but only sums up values taken at regular intervals. Cf. CONTINUOUS INTEGRATION.

INTERMITENT RATING : The rating of a motor or other piece of apparatus according to the output it can give when working for alternate periods of load and rest having a definite ratio to one another, or when running for a stated period only, insufficient to cause it to reach its final temperature, followed by an indefinite period of cooling. See LOAD FACTOR RATING.

INTER-MODULATION DISTORTION : A distortion that results form spurisous combination-frequency components in the output of nonlonear transmission system when two or more sinusoidal voltages, applied simultaneously, form the input. Intermodulation distortion of a complex wave form arises from intermodulation (see modulation) within the waveform.

INTERNAL IMPEDANCE : The impedance of a thermionic valve between the cathode and the anode, as defined by the ratio of change of anode voltage to change of anode current on the straight part of the characteristic with constant grid potential. The internal impedance is approximate constant over a large range of frequency.

INTERNAL MEMORY : Storage devices that are in integral physical part of the automatic computer. These are usually high-speed storage devices.

INTERNAL SELF-INDUCTANCE : An expression used in connection with cables and other conductors for the inductance due to the field within the conductor. Considerably greater in iron or steel than in copper conductors. Cf. EXTERNAL SELF-IN-DUCTANCE.

INTERNAL VOLTAGE : The actual e.m.f. generated in the armature of a machine without deduction of the potential drop when a current is flowing through it.

INTERNATIONAL COULOMB, FARAD, GILBERT, HENRY, JOULE AND MAXWELL : Units based on the *International Ampere* and *International Ohm*, and having values respectively of 0.99991, 0.99948, 0.9999991, 1.00052, 1.00034 and 1.00043 of the true values of the coulomb, and farad, etc.

INTERNATIONAL OHM : The resistance offered to an unvarying current by a column of mercury at the temperature of melting ic, 14.4521 grammes in mass, of constant cross section, and of a length of 106.300 centimetres. Equal to 1.00052 true Ohm.

INTERNATIONAL UNITS : The sytem of units of electrical measurement recognised by law in most countries, based upon the values of the standards representing the *International Ampere* and *International Ohm* agreed upon at the International Conference in London in 1908.

INTERNATIONAL VOLT : The e.m.f. or electrical pressure which when applied to a conductor, the resistance of which is one International Ohm, will produce a current of one International Ampere. Equal to 1.00043 true Volt.

INTERNATIONAL WATT : The energy expended in one second by an unvarying current of one International Ampere under an electrical pressure of one International Volt. Equal to 1.00043 true Valt.

INTERNODE : See ANTINDOE.

INTERPOLATOR : A form of Regenerative Repeater for retransmitting signals over a long cable received through a relay in a clearer form.

INTERPRET : To translate interior restate in a human language.

INTERPRETER : A program that *reads, translates* and *executes* a user's program, such as one written in the BASIC language, one line at a time. A complier, on the other hand, reads and translates the entire user's program before executing it. A program for a computer that translates English input into assembly language. The translation is done one line at a time, and the computer runs the instructions for each line before the next line is translated.

INTERPRETIVE ROUTINE : An executive routine which, during the course of data-handling operations, translates a stored macroceded program into a machine code and at once performs the indicated operations by means of sub-routines.

INTERROCORD GAP : A 3/8 in length of bland tape between records that is used as a starting and stopping point for the transport.

INTERRUPT : An event that causes the computer to temporarily stop, and perform another task. A signal that causes the current program in the CPU to terminate execution. Depending on the nature of the interrupt, a different program may be loaded and exeucted.

INTERRUPTED CONTINUOUS WAVES : Modulated Keyed Continuous Waves in which the audio-frequency modulation consists of periodic interrution.

INTERRUPTED CURRENT : A current which is periodically interrupted, but does not change its direction and is in that sense a "direct current".

INTERVALVE COUPLING : Any method of coupling thermionic valves in cascade so that the current variations in the plate circuit of one valve reappear as voltage variatious on the grid of the next. See INTERVALVE TRANSFORMER, RESISTANCE-CAPACITANCE COUPLING and TUNED ANODE COUPLING.

INTRANSIC BRILLIANCY : The Brightness (candle power emitted per unit projected area in a particular direction), of a source itself emitting light.

INTRANSIC SEMICONDUCTOR : A pure semiconductor that has equal concentrations of electrons and holes under condifitons of thermal equilibrium. Absolute purity is unobtainable in practice and nearly pure materials are called intrinsic.

INVERSE CURRENT : A current induced by the cessation or diminution of a neighbouring current or field; i.e. in a contrary direction from that induced by the establishment or increase of such a current or field.

INVERTED VACUUM TUBE : A thermionic valve used for voltage reduction power amplification by using the anode circuit as a high voltage input circuit, and the grid circuit as the output circuit.

INVERTER : (1) A logic device that outputs a "1" with a "0" input and outputs a "0" with a "1" input. A device (NOT gate) which performs the operation of inversion. It will present at its output the inverse or complement of the information at its input. (2) A term sometimes used for an Inverted Rectifier particularly of the thermionic or mercury vapour type.

I/O : An acronym that stands for Input/Output.

IODIDE ACCUMULATOR : See SPEECH INVERTER.

ION : A minute dissociated portion of a substance carrying a positive or a negative charge. Used (1) of the dissociated portions into which molecules split up during electrolysis. See ANIONS and CATIONS; and (2) of constituent dissociated corpuscles into which ION gases split up when ionised; the positive and negative ions having respectively a deficiency or an excess of Electrons. A current through an ionised gas consists of processions or streams of electrons detached in this way. See POSITIVE and NEGATIVE IONS, ION. IONISATION, etc.

IONIC CURRENT : A current through a gas carried by ions or electrons.

IONIC VALVE : Any apparatus using the properties of ionised gases or liquids for the conversion of the nature of a current, e.g. as a Rectifier or Inverter.

IONISATION : Dissociation of molecules of a gas on an electrolyte into ions, causing it to become more conducting. In the case of a gas, this can be done by the action of ultraviolet light and certain rays from radioactive meterials, by dielectric stress and some forms of electric discharges, and by flames and in can descence.

IONISATION CHAMBER : An instrument for gauging the intensity of "X": Rays by measuring the degree of ionisation of a gas.

IONISATION POTENTIAL : The minimum energy in electron-volts required to remove an electron from a given atom or molecule to infinity. Earlier the ionization potential was defined as the minimum potential through which an electron would fall in order to ionize the atom and was measured in Volts.

IONOSPHERE : The whole ionised zone of the upper atmosphere including the kennelly-Heaviside (E1 and E2) and the Appleton (F1 and F2) Layers. This ionisation is probably due to ultra-violet radiation from the sun. It is reflection from the ionosphere which renders reception by long and medium waves over much longer distances possible by night than by day. The structure of the ionosphere is considerably affected by magnetic storm conditions. See also FADING and SKIP DISTANCE.

IR : Instructions Register. Information Retrieval (also, Instruction Registor).

IRG : Inter Record Gap.

IRREGULARITY : This term is sometimes used in connection with réctified or other pulsating current for the ratio of the alternating to the mean direct component.

IRREGULARITY FACTOR : The ratio of the speed variation to the mean speed, during one revolution, of a generator driven by an engine of unequal crank effort.

IRWIN HOT WIRE OSCILLOGRAPH : See HOT WIRE OSCILLO-GRAPH.

ISAM : Index Sequential Access Method.

ISDN (Intergrated Services Digital Network) : A network from common carrier that lets the subscriber send data and voice traffic over a ditital network; the network provides a number of services to the user.

ISO : International Standards Organization.

ISOCHRONISM : The running, exactly in step at equal speed, of two shafts, etc., such as those of the selecting devices in some forms of printing telegraph systems, where the sending and receiving apparatus is kept in step by periodical correcting current impulses, or in general the performance of related functions at the same speed.

ISODYNE RECEPTION : A system of radio-reception employing bigrid h.f. amplifying valves in such a way that the currents in the two grid circuits compensate any tendency to self-oscillation.

ISOPERM : One of a group of special alloys of *nickel, iron* and *copper* or aluminium used for cores of inductance coils on account of the small variation of permeability and good magnetic stability.

ISOTOPES : Varieties of an element of the same chemical properties and atomic number, but with a nucleus containing a different number of mass units but the same atomic charge. Many elements, as usually occurring, have been found to consist of a mixture of isotypes. with different integral mass numbers.

ITERATION : A single cycle of a repetitively executed series of steps.

ITERATIVE-LOOP : A repeated group of instructions in a routine.

J

JACK : The receptacle on a telephone or telegraph switchboard, etc., provided with a metallic bush and several contact springs behind the board for making contact with a plug attached to a flexible cord, the various parts of the plug making contact with the different springs and also separating contacts normally existing between the springs or closing contacts normally open. (Formerly sometimes called Switch-Springs).

JACKSON TUBE : An early form of "X" Ray Tube with inclined anticathode in which a concave cathode was introduced to focus the cathode rays, but an auxiliary anode was not employed. Cf. GUNDELACH TUBE.

JACOBI'S UNIT OF RESISTANCE JACOBIETALON : A resistance unit proposed in 1848, being that of a copper wire weighing 22.4932 grammes; 761.975 mm long and 0.667 mm. in diameter.

JACQUARD, JOSEPH-MARIE : French Inventor who, in 1801, invented a set of cards that were used on looms to weave intricate patterns, Jacquad also desigend the attachment for the loom that would use these cards to produce the designs. Not only were geometric designs in wool, linen or cotton woven on these "computerized" machines but also very fine portraits in silk thread were made.

JAMMING : (1) The sending of distrubing radio signals to interfere with the reception of another message, as in the operations of fleets in war time. (2) Interference due to any signals other than those it is desired to receive

JCL : Job Control Language.

JET WAVE RECTIFIER : The equivalent of a commutator rectifier formed by pairs of stationary commutator segments upon which a jet of mercury impinges. Part of the jet carries an alternating current of the same frequency of the supply, and is in a strong D.C. field so that it.

JET WAVE RECTIFIER : Is deflected from side to side and thus makes contact alternately with the two segments.

JIGGER : A high frequency transformer as used in the aerial circuit of a radio-transmitting equipment.

JK FLIP-FLOP : A flip-flop capable of operating as a type T or type D, depending on the J and K leads. Flip-flop will go to the 1 state on receipt of the clock pulse, when J is low K is high, the flip-flop will switch to the 0 position at the clock pulse. When both J and K are high, the flip-flop will complement its initial state on receipt of the clock pulse. JK FLIP-FLOP is a flip-flop whose inputs are designated J and K. These devices are almost invariably clocked and their outputs are the same as the R-S type except when logical is appear together at the inputs. In these circumstances the evice changes state.

JOB : A collection of tasks viewed by the computer as a unit.

JOCKEY RELAY : A delicate relay of the moving coil type, used in cable telegraphy, automatically corrected for stray erth currents by an arrangement analogous to a shifting zero in which the "fixed" contacts are caused to drift or change their position according to the earth currents.

JOHNNIAC : John von Neumann's Integrator and Automatic Computer.

JOHNSON COUNTER : A complementing ring counter.

JOINT TESTING : Tests of joints in conductors, cables, etc., required to be made to ensure that the joint is not inferior to the rest of the cable in (1) mechanical strength, (2) electrical conductivity, and (3) insulation.

JOYSTICK : A device that uses a lever to control to screen display. Often used for playing games.

JUMP : See transfer.

JUMPER WIRE (in Automatic Telephony) : A temporary connecting wire in a Cross-Connection Field during repair or rearrangement of permanent connections.

JUNCTION CALLS : Calls made or received at a telephone exchange over a Junction Line. (2) Such a circuit together with a Junction Liner.

JUNCTION KEY : A key on a telephone switchboard for signalling through a Junction Line.

JUNCTION LINE : A telephone line from one exchange to another, in the same district. Cf. TOLL LINE and TRUNK LINE.

JUNCTION SELECTOR (in Automatic Telephony) : A selector re-

sponding to the current impulses of the exchange code in a
call dialled, the function of which is to make connection through
a vacant junction line between a First Selector and an appropriate Thousands Selector in the wanted exchange.

JUNGNER ACCUMULATOR : A variety of nickel-cadmium Accumulator with alkaline electrolyte.

JUSTIFICATION : The placement of text of graphics between the left
and right margins.

JUSTIFY : To line up with the *left* or *right* edge of something, as the
significant contents of an accumulator with one end of the
accumulator.

K : The symbol signifying the quantity 2^{10}, which is equal to 1024. K is sometimes confused with the symbol k (kilo), which is equal to 1000. K represents one thousand in scientific terms. In computers, it actually replaces 1024 or 2 to the tenth power.

KAPPA : κ symbol for Permittivity (dielectric constant).

KARNAUGH MAP : A system of reducing Boolean expressions using a matrix of 1's and 0's invented by Karnaugh.

KAROLUS SYSTEM OF PHOTO-TELEGRAPHY : A system of phototelegraphy in use on the Continent in which a photoelectric cell is used in the transmitter and a Kerr Cell in the receiver, considerable use being made of thermionic amplifications.

KATHIONS, KATIONS, KATHODE ETC : See CATHIONS, CATIONS, CATHODE, etc.

KAY-STONE DISTORTION : A distortion that is due to the length of the horizontal scan line varying with the vertical displacement of the line. It is most pronounced when the electron beam is at an acute angle to the screen and results in a trapezoidal image instead of a rectangular.

KC : Abbreviation for Kilocycles per Second.

KELVIN : A name proposed by the Board of Trade of U.K. (in 1892), for the commercial unit of energy, one Kilowatt-Hour or one Board of Trade Unit. (Named after William Thomson, Lord Kelvin, 1824-1907).

KELVIN BRDIGE : A method of measuring low resistances in which the potential drops produced by the same current in the resistance under test and in a standard low resistance slide wire are balanced against each other, also called Double Bridge.

KENNELLY HEAVISIDE LAYER : The lower section of the Ionosphere or ionised layer in the upper atmosphere, between which and the earth's surface the waves used in radio-communication follow the curvature of the earth, owing to being reflected or refracted thereby. Now regraded as being divided into the E_1 layer about 100 km. up and the E_2 layer existing during daytime about 130 km. up. There is, however, some variation in the heights. See also IONOSPHERE.

KENOTRON : The name given to an early form of two-electrode thermionic valve tube for use as a rectifier, having a high enough vacuum to depend on a full electron discharge.

KERNEL : A series cycle of programmed instructions that are timed from published timings and used to evaluate the performance of a specific computer or system.

KERR CELL : A light relay used in phototelegraphy, etc., depending for its action on the Electrostatic Kerr Effect.

KERR EFFECT : The rotation of the plane of polarisation of plane polarised light when reflected from the pole of magnet. The angle of rotation is proportional to the intensity of magnetisation.

KERR'S, CONSTANT : The number, varying for different wavelengths and for different materials, by which the intensity of magnetisation has to be multiplied to give the angle of rotation forming the Kerr Effect.

KEY : A computer word or field which determines the position of a record in a sorted sequence; also, a lever on a manually operated machine, as a typewriter.

KEY BOARD : A device that generates an electronic signal through the pressing of a key. An ANSI keyboard has keys placed in a pattern similar to that of a standard typewriter.

KEY COMBINATIONS : The pressings of the two keys at the same time. For example, Alt-X means to press the Alt key and hold it down, then press the X key, then release both keys at the same time.

KEYING : Arrangements in radio-telegraphy for breaking up trains of waves into morse signals either by interrupting them altogether when the signalling key is released, making an alteration in frequency or otherwise.

KEY LESS RINGING : Ringing of the telephone bell at a distant station by making connection to the line without the use of a separate key.

KEY PUNCH : A machine that punches holes in cards to represent data.

KEY SENDER : An appliance for the control of automatic telephone apparatus by depressing keys.

KEY SET CALL SENDER : A calling Device actuated by keys.

KEY STONE DISTORTION GULAR ONE : It can be removed using suitable transmitter circuits.

KEY VERIFY : To verify the punching on punched cards by the operation of punched card verifier.

KEY WORD ANALYSIS : A primitive technique of natural language processing, whereby input sentences are scanned for certain special words and patterns that signal actions for the programs to take.

KICKOUT : A rejection under the control of the program of erroneous or invalid input data.

KILO : Prefix meaning a thousand times, e.g. kilovolt, a thousand volts. Abbreviation : k.

KILOBYTE [K] : A unit of measurement that refers to 1,024 bytes.

KILOCYCLE (kc) : One thousand cycles. In radio-communication it is often more convenient to specify frequency in kilocycles per second instead of wave-length in metres, especially when considering questions of interference of stations of only slightly different wave-length. Their frequencies should differ by at least 10 to 15 kilocycles per second. The frequency in kilocycles per second equals 300,000 divided by the wave length in metres.

KILOHERTZ : A proposed unit of frequency equal to one kilocycle per second.

KILOHM : Abbreviation of Kilo-ohm, 1000 ohms.

KILOMETRIC WAVES : Electromagnetic waves of a wave-length between 10,000 and 1000 metres.

KILOVOLT (kv) : One thousand volts. A unit often used for expressing lines, test pressures for insulators, etc.

KILOVOLTAMPERE (kVA) : A unit of Apparent Power used in rating alternators, etc., being 1,000 Volt Amperes.

KILOWATT (kw) : The unit of electric power generally used for rating electrical machinery and for other practical purposes; equal to 1,000 watts. Equivalent to 1.34 h.p.

KILOWATT HOUR METER : A meter integrating the power supplied to a circuit, i.e. measuring the energy supplied in a given time in kilowatt hours.

KINESCOPE : A form of cathode ray television reproducing apparatus in which the amplified signal impulses vary the strength of an electrostatically focused electron beam which is caused by suitable deflecting coils to scan a fluorescent screen in synchronism with the scanning beam in the transmitter, rendering it luminous in proportion to the light falling on to the transmitting screen.

KIRCHHOFF'S LAWS : (1) At any point where circuits branch from one another the algebraic sum of the currents meeting at that point is zero. (2) The total e.m.f. in a circuit is equal to the sum of the resistances of its various parts multiplied by the current in the circuit.

KIRKIFIER : Popular name for a thermionic valve detector system, using the valve as an anode rectifier with the space charge neutralised by a positive Grid Bias.

KIRR FACTOR : The ratio between the R.M.S. values of the fundamental wave in an oscillation to that of the harmonics present. See NON-LINEAR DISTORTION FACTOR.

KLYDONOGRAPH : An instrument for the recording of surges in transmission systems, etc., depending on the photographic reproduction of Lichtenberg figures on a moving film.

KLYSTRON : A form of thermionic apparatus suitable for use as an oscillator, amplifier or frequency multiplier in which the Rhumbatrib principle is employed to produce velocity modulation of the electronstream.

KNOWLEDGE : Is any information that is useful for the task being performed.

KNOWLDGE ACQUISITION : A learning problem for expert systems that involves augmenting the knowledge of a running system, either by receiving it from human experts or by creating it automatically, as in chunking.

KNOWLEDGE BASE : Is the store of domain-specific knowledge of an AI program; in a rule-based expert system it is the production system being used by the inference engine.

KNOWLEDGE BASE : Data and rules about the data from the knowledge base of expert systems.

KNOWLEDGE BASED APPROACH : (to intelligence) says that intelligence arises from vast bodies of knowledge rather than any particularly clever algorithms, and that the right way to design intelligent programs is to build-in large quantities of domain-specific knowledge.

KNOWLEDGE BASED SYSTEM (KBS) : Is a system that achieves intelligent behaviour by using fairly simple algorithms with a large body of specific and commonsense knowledge, rather than by applying powerful but complex algorithms with little domain-specific information.

KNOWLEDGE ENGINEERING : Is the practice of building expert systems by extracting from expert their domain and problem-specific knowledge and representing it in a suitable from for computer implementation.

KNOWLEDGE REPRESENTATION : Is the science of encoding real-world knowledge in an efficient format that makes if easy for programs to use it.

KOHLRAUSCH BRIDGE : A form of Bridge with alternating current supply and a telephone receiver instead of a galvanometer. Suitable for measuring resistance in whcih a back e.m.f. is present, etc.

KORH SYSTEM OF PHOTOTELEGRAPHY : A system of Phototele-graphy, employing, in its original, every part of which is traversed by a beam of light modifying the resistance of a selenium or other light sensitive cell and thus causing modu-lation of the line current according to the density aratus. In another form of the system a metallic original in relief is employed with which a metal stylus makes contact when it passes over a raised portion of the protion of the original passing. These modulations control a shutter by an arrange-ment similar to an Einthoven String Galvanometer which causes corresponding variations in a beam falling upon the synchro-nously moving film in the receiving apparatus.

KR : A symbol formerly frequently employed in telegraphy for the product of the capacitance and resistance of a cable or line, upon the reciprocal of which the working speed (words per minute) depends. The expression CR is now more commonly used.

KYLSTRON : An electron tube that is used as a microwave amplifier or oscillator. It is a linear-beam microwave tube in which velocity modulation is applied to an electron beam in order to produce amplification of a microwave-frequency field.

KYMOGRAPHY : A system of recording by "X": rays the movement of internal organs in which the film is exposed through a series of slits and moved periodically by a distance equal to that between them. By this means the equivalent of a direct record of the movements is automatically made.

L

LABEL : An identifier, in either human or machine language, eg. a magnetic tape lable identifying the tape or an address identifying a storage location.

LADDER NETWORK : See QUADRIPOLE.

LADDER NETWORK : (1) A series/parallel resistor network used for D/A conversion. (2) See QUADRIPOLE.

LADDER SECTION : See QUADRIPOLE.

LAGGING CURRENT : An alternating current is said to *lag* behind the voltage which produces it when its wave is out of phase with that of the voltage in a direction such that it reaches its maximum after that of the voltage wave. Cf. LEADING CURRENT.

LAGGING PHASE : (in the two-watt-meter method of power measurement). The circuit in which the phase of the current lags behind that of the voltage taken to the wattmeter connected therewith.

LAMBERT : A unit of brightness.

LAMP SYNCHROSCOPE : A Synchroscope in which the moment when synchronism is reached is indicated by the cessation of flicker of incandescent lamps.

LAND LINE : A telegraph line across the land as opposed to a submarine cable circuit.

LANGUAGE : Is, formally, a set of strings; if the set is finite, it can be simply enumerated, but if it is infinite it must be represented as a finite decision procedure.

LAPTOP : A computer that can run off batteries and weighs less than 12 pounds.

LARC : Livermore Automatic Research Computer

LARGE-SIZE : Of an automatic computer with a combined purchase price of about $ 700,000 or more. (Year 1990)

LARYNGAPHONE : A telephone transmitter in which the voice vibrations are taken up by a pad pressed on to the exterior of the throat instead of from the sounds issuing situations in that it does not pick up extraneous sounds.

LASER PRINTER : A printer that uses light to print text and images onto paper, using the same techniques as photocopying machine.

LAST PARTY RELEASE : A system whereby a telephone circuit is cleared when both parties have hung up their receivers.

LATCH : A DG or RS flip-flop.

LATENCY : The time required for a mechanical storage device to begin transmitting data after a request. For a movable-head disk drive, the seek time to position the read-write heads plus the rotational delay time.

LATERAL INHIBITION : In connectionist networks in the existence of bidirectional, negatively weighted links between mutually inconsistent, competing features as the same level in a recognition hierarchy.

LATTICE NETWORKS : Networks having the input and output terminals at a junction between two or more conductors a bridge network is a particular type of lattice network.

LAY : The axial pitch of one complete turn of the strand or of the core of a cable.

LCD : Acronym for Liquid Crystal Display, most commonly used to describes screens that come with laptop computers.

LCS : Large Core Storage

LEADING CURRENT : An alternating current is said to "lead," when owing to the effect if capacitance, etc., it reaches the maximum of its wave sooner than the voltage which produces it.

LEADING EDGE (of a pulse) : See PULSE.

LEADING LOAD : Load in which the current leads in front of the voltage, due to the presence of capacitance or the effect of synchronous motors,etc. (Also called Capacitance Load.) Cf. LAGGING LOAD.

LEAF Node : A node in a tree with no children.

LEAK : An intentionally or accidentally provided path, through which part of the current can return through the earth or to the oppo-site conductor. See also LEAK COIL and LEAK CIRCUIT.

LEAKAGE CONDUCTANCE : The reciprocal of Leakage Resistance or Insulation Resistance.

LEAKAGE DETECTOR : See LEAKAGE INDICATOR.

LEAKAGE INDICATOR : An instrument for indicating or measuring th presence of a leakage current to earth on a system; usually by indications of the potential difference between the respective poles of the system (if normally insulated) and earth, or in systems earthed at one point, consisting of an ammeter in the earth wire.

LEAK CIRCUIT : A very high resistance shunt circuit such as that used in Repeater Stations with a galvanometer or receiver to observe the character of the transmitted signals without diverting enough current to affect them.

LEAK COIL : A high resistance coil used in telegraphy for shunting a condenser.

LEARNING : Is the improvement of performance with experience over time, and includes becoming more efficient, rule learning, category, induction, knowledge acquisition, and scientific discovery.

LEARNING ELEMENT : is the portion of a learning AI system that decides how to modify the performance element and implements those modifications; it normally includes any learning algorithms employed by the system.

LEAST SIGNIFICANT BIT (L.S.B.) : The bit of a logic word that represents that least weighted value.

LED : Light-Emitting Diode, when the minority carriers recombine in a forward-biased junction diode, they can give up their energy in the form of quanta of light (photons). A number of these diodes can be placed in an array to form α (alpha) numerical indicator Light-Emitting Diode

LED-RELAY : See SOLID STATE RELAY

LEGAL OHM : A unit of resistance agreed upon in 1884 by an international congress at Paris, but not actually recognised by law; defined as the resistance of a column of mercury 1 sq. mm. cross section and 106 cm. long at 0 C; equal to 0.99718 International Ohm.

LEMSTROM MACHINE : A form of electrostatic Influence Machine in which the tinfoil carriers are mounted upon concentric cylinders revolving in oppsite directions

LENARD RAYS : The name given to cathode rays which have bombarded and succeeded in passing through an aluminmum "window" out of the tube in which they were produced.

LENGTHENING COIL or LENGTHENING INDUCTANCE : Additional inductance in an aerial circuit to make it respond to a longer wave length (Sometimes called Loading Coil.).

LEPEL DISCHARGER : A Quenched Spark discharger used in early radio telegraph systems; employing shock excitation; used in conjunction with an audio-frequency oscillating circuit to obtain a muscial spark.

LESS LETTER QUALITY : Refers to print quality that looks as if a typewriter printed it.

LEVEL : (1) The number of bits positions in a binary word. (2) A term used for the value of a variable quantity, particularly when measured according to a logarithmic (decibel) scale. (3) (in Automatic Telephony). One row in a Rank of fixed Contacts in a Selector.

LEVEL OF ADDRESS : Any of three main ways for specifying operands in an instruction, as follows : FIRST LEVEL : The operand specification symbols appearing in the instruction are the address of the operand ("direct addressing); SECOND LEVEL: The operand specification symbols appearing in the instruction are the address of the address of the operand ("indirect" addressing).

L.F. : Low Frequency

LIBRARY ROUTINES : A collection of standard and fully debugged programs, routines, and subroutines by means of which many types of problems and parts of problems can be processed or handled.

LIEBEN REISS VALVE : An early form of three-electrode thermionic valve with oxide-coated filament practically contemporary with the deForest Audion.

LIFO : Last-In First Out

LIGHT CURRENT ENGINEERING : A branch of engineering dealing with electrical and electronic circuits below 110 volts and currents of very low values mostly using semiconducter devices. See ELECTRONICS and ELECTRICAL ENGINEERING.

LIGHT EMITTING DIODE (LED): A semiconductor two terminal "light bulb "made of semiconductor material (such as gallium phosphide) that make light when electric current is passed through it in a particular direction.

LIGHT PEN : A device that looks like a fountain pen which is sensitive to light and transmits signals to the computer based on light input to the pen. The computer will translate this signal into information and store it.

LIGHT RELAY : An apparatus used in Phototelegraphy. etc., in which current variations are converted into variations in the intensity of a beam of light, e.g. by the displacement of a beam falling upon a special shaped aperture. See also KAROLUS CELL.

LILIENFELD TUBE : A form of "X" Ray Tube with a heated filament forming the cathode of an auxiliary circuit and projecting electrons through a hollow working cathode connected to the main high voltage circuit.

LIM : Lotus-Intel-Microsoft used to describe a memory specification called Expanded Memory Specification. See also EMS.

LIME SPOT CATHODE : A Hot Cathode in an "X" Ray or other Cathode Ray Tube in which the action is augmented by the provision of a small quantity of lime which emits electrons freely when incandescent.

LIMIT DEVICE IN RADIO COMMUNICATION ETC. : An arrangement for limiting the response of a receiver however great the strength of the incoming signals.

LIMITER (In Radio Communication) : A device for keeping the response of a receiver or transmitter below a certain maximum value, whatever the value of the input.

LINE : (1) Any electric circuit for telegraph, telephone, power or lighting purposes, etc. (2) An unofficial name formerly used indis-criminately for the C.G.S. electromagnetic units of magnetic force, field, or flux.

LINEAR AMPLIFICATION : Amplification by a Thermionic valve in which the change in the anode current is directly proportional to the change in the applied potential: i.e. when the valve is being worked on the straight line portion of its characteristic.

LINEAR DISTORTION : Amplitude, Distortion or Phase Distortion where the alternation of the quantities in question is not dependent upon the voltage or current amplitudes.

LINEARITY : The digital output of an A/D converter compared with what the output should be.

LINEAR MODULATION : Modulation in which the amplitude is stricity proportional to that of the impressed sound wave at all audio frequencies.

LINEAR NETWORKS : Networks having a linear relationsip between the voltages and currents; otherwise they are nonlinear.

LINEAR PROGRAMMING : An operations Research mathematical technique not related to comput·r·programming techniques.

LINE DRIVER : A logic element used to transfer data to a bus.

LINE FREQUENCY (In Television) : The number of scanning lines traversed per second.

LINE LAMP : A lamp on a telephone switchboard which glows when a call is being made through the particular subscriber's or junction line to which it belongs.

LINE NOISE : Disturbing noise in a telephone due to causes inherent in the line and to variations in resistance due to had contacts, or inducttive interference of power lines, etc. Cf. TERMINAL NOISE.

LINE PRINTER : A very high speed printer output device that can type text at several hundred lines per minute. A very expensive computer peripheral, but also a very powerful one.

LINE RECEIVER : A logic element with input hystersis designed to reject noise.

LINE RELAY : A relay which controls a Line Lamp on a telephone exchange switch board.

LINE TELEGRAPHY, TELEPHONY and TELEVISION : Telegraphy, telephony and television over metallic circuits as distinguished from radio-communication.

LINK : (1) A solid removable connection, usually a flat strip fixed by bolts only to be disconnected when the circuit is dead. (2) In automatic telephony, a wire connecting selectors, etc., in an ex-change. (3) A communication channel or circuit that is used to connect other channels or circuiits. (4) A path between two switches that from part of a central control system in automatic switching.

LIQUID CRYSTAL : An organic liquid consisting of long-chain modecules that line up under the influence of an applied electric field to give a quasi-crystalline structure of the liquid. A change in the applied field causes a change in the reflectivity advices.

LIQUID CRYSTAL DISPLAY (LDC) : A type of passive display that uses liquid crystals, as in the seven-segment numerical display of digital watches and pocket calculators.

LISP : List Processing, an intelligent computer programming language developed by John McCarthy in the late 1950s, and is the dominant programming language for AI research and development because of its built-in facilities for manipulating symbols and lists in sophisticated structures.

LISSAJOU'S FIGURES : Any closed figures traversed by a point moving with the resultant of two periodic oscillatory motions at right angles. Originally applied to certain experiments in connection with pendulums and sound, but now used for a class of records of this nature by such instruments as the Cathode Ray Oscillograph. See also CYCLOGRAPH.

LIST : A group of logically related items that are stored with points to the next item on the list. Also, a series of pointers running through a storage file.

LISTDE WIRE : A wire of uniform resistance provided with a sliding contact that can make a connection at any desired point along the length. Slide wires are used to provide a variable resistance, as a potentiometer, or to provide a desired resistance ratio. (see Wheats)

LISTENER : A device that is able to receive device-dependent data from the IEC bus after it is addressed with its *listen* address by the controller in charge.

LISTENING KEY : A key on a telephone switchboard by means of which an operator connects her head set to a line, to communicate with the calling subscriber or to ascertain if a line is in use.

LITERAL : (1) A letter in a position of a number. (2) An operand; a zero level address. See LEVEL OF ADDRESS.

LIZENDRAHT WIRE : A multistranded wire formed from many fine conducting filaments. It is employed for high-frequency applications, as in coils usd in radio, in order to reduce the high-frequency reistance.

LLO : Local Lockout; universal bus command.

L-NETWORK : See QUADRIPOLE.

LOAD : (1) To retrieve a file into a program such as word processor or spread sheet, or BASICA. To put a program or data into the computer's main memory from auxiliary storage such as disk or tape. (2) The output (kilowatts) that a generating plant is being called upon to give at any time, or the horse-power which is being exterted by an electric motor. (3) Inductance added to a telephone or radio-circuit. See LOADING.

LOAD AND GO : The use of an interpretive routine.

LOAD CARD : A punched card with machine language instruction which are executed immediately by the automatic computer. The load Card often serves as a boot-strap.

LOAD DESPATCHER : An official in a special office connected by telephone to all parts of a large power distribution systsm, traction system, etc., who is responsible for the distribution of the load in various parts of the system according to requirements.

LOADED CABLE : A cable, such as a submarine telephone cable, provided with extra inductance to counteract the effect of its capacitance and thus to diminish attenuation. See LOADING COIL, CONTINUOUS LOADING, etc.

LOADED CIRCUIT or LINE : A telephone circuit containing extra inductance to improve its transmission efficiency. See LOADING COILS.

LOADER : A program that places a translated computer program in primay memory before its execution.

LOAD FACTOR : The ratio of the maximum to the average load during a given period.

LOAD IMPEDANCE : The impedance presented by a load to the driver circuit that supplies power to it. The effect of variations in the load impedance on the performance of an oscillator is seen in a load impedance diagram in which the oscillator output is plotted against load impedance.

LOADING : The addition of inductance to telephone lines or cables to counteract the effect of capacitance in order to diminish Distortion.

LOADING COIL : (1) An inductance coil with an iron core for insertion in a telephone line or cable to lessen the effect of its capacitance and to reduce distortion. (Also called Pupin Coil.) (2) Inductance added to a radio aerial.

LOAD ROUTINE : A routine which causes a computer to read a program into storage, some times punched a load card.

LOCAL AREA NETWORK (LAN) : A communications network connecting devices in a local area, such as on one floor of a building. Used for communications and to share common devices such as a printer. A group of computers connected through cables, able to transfer files to other computers connected through the cables. CF: WAN.

LOCAL BATTERY : A battery in a local circuit such as that opened and closed by a Relay.

LOCAL BATTERY TELEPHONE SYSTEM : A system employing a

battery at the subsriber's station to provide the speaking current.
Cf. CENTRAL BATTERY SYSTEM.

LOCAL CIRCUIT : An auxiliary circuit connected by a relay or otherwise
to the main circuit.

LOCAL OSCILLATIONS : Oscillations in a radio-receiving circuit
due to the action of the apparatus itself,e.g. from a separate
oscillator valve for heterodyne reception, or due to self
oscillation.

LOCAL OSCILLATOR : An oscillator producing auxiliary oscillations
to obtain a heterodyne effect.

LOCATION : Any physical place in a storage device which is used
for storing a word or field and which usually has an address.

LOCKOUT : The time interval in the cycle of an input or output
device during which the rest of the automatic computer is
prevented from access to the contents of part or all of storage.

LODGE-MUIRHEAD SYSTEM OF RADIO TELEGRAPHY : An early
system distinguished principally by the use of accurately tuned
sending and receiving aerials with symmetrical upper and
lower capacitance areas with no earth connection, and a multiple spark gap in the aerial circuit.

LODGE VALVE : A rectifying discharge tube, with a gas pressure of
about one millionth of an atmosphere, with one coiled and
one plate electrode in a special bulb arranged so that it is only
when the plate electrode is the cathode that the projected
electrons reach the anode and cause a current to pass.

LOGARITHM : A logarithm of a given number is the value of the
exponent indicating the power required to raise a specified
constant, known as the *base*, to produce that given number.
That is, if B is the base, N is the given number and L is the
logarithm, then $B^L = N$. Since $10^3 = 1000$, the logarithm to the
base 10 of 1000 is 3.

LOGIC : (1) Any precisely defined formal reasoning system that includes
wellformed-formulae, axioms, and rules of inference that allow
new theorems to be deduced from the axioms and other existing
theorems; it is used explicitly as a representation and reasoning
tool by many, and implicity in the operation of most, AI
systems. (2) A system of representing the validity of an output
by analyzing inputs. (3) In computer language, logic is a form
of mathematics based upon two-state truth tables. Electronic
logic uses two-state gates and flip-flops to perform decision-
making functions.

LOGICAL DESIGN : Automatic computer design that deals with the logical and mathematical inter-relationships that must be implemented by the hardware.

LOGICAL RECORD : A collection or an association of fields on the basis of their relationships to each other.

LOGIC CIRCUIT : A circuit designed to perform a particular logical function based on the concepts of 'and', 'either-or' 'neither-nor', etc. Normally these circuits operate between two discrete voltage levels, i.e. *high* and *low logic levels,* and are described as *binary logic circuits.* Logic using three or more logic levels is possible but not common. Binary circuits are extensively used in computers to carry out instructions and arithmetical processes, and may be formed from discrete components or, more commonly, from integrated circuits. Families of integrated logic circuits exist based on bipolar transistors (*see* diode-transistor logic (DTL), emitter-coupled logic (ECL), resistor-transistor logic (RTL), I²L, and transistor-transistor logic (TTL)). MOS logic circuits are based on field-effect transistors. The basic digital gates are:

AND gate. A circuit with two or more inputs and one output in which the output signal is high if and only if (sometimes written *iff*) all the inputs are high simultaneously.

Inverter (NOT gate). A circuit with one input whose output is *high* if the input is *low* and vice versa.

NAND gate. A circuit with two or more inputs and one output, whose output is high if any one or more of the inputs

Popular (formerly BSI) symbol	Binary logic circuit	IEC approved symbol	Popular (formerly BSI) symbol	Binary logic circuit	IEC approved symbol
AND gate	&		NOR gate, negated output	≥1	
NAND gate, negated output	&		Nor gate negated inputs	≥1	
NAND gate, negated inputs	&		Exclusive-OR gate	= 1	
OR gate	≥1		Inverter (NOT gate)		

is low, and low if all the inputs are high.

NOR gate. A circuit with two or more inputs and one output, whose output is *high* if and only if all the inputs are *low*.

OR gate. A circuit with two or more inputs and one output whose output is *high* if any one or more of the inputs are *high*.

Exclusive OR gate. A circuit with two or more inputs and one output whose output is high iff one input is high.

LOGIC DIAGRAM : A schematic diagram using logic symbols.

LOGIC GATE : See AND, OR, NAND, NOR, NOT, and Exclusive-OR.

LOGIC OPERATIONS : The operations of comparing selecting extracting etc.

LOGIC PROGRAMMING (LP) : Is the use of logic as a programming language, in which programs consist of axioms and control is exercised by a theorem proving algorithm.

LOGIC SYMBOL : A symbol used to represent a logic element graphically.

LONGITUDINAL VOLTAGE : A term used in connection with Interference in communication circuits for a voltage induced between different parts of the sameline, or between a conductor and earth. Cf. TRANSVERSE VOLTAGE.

LONG WAVES : A term formerly in common use for wave-lengths over 1,000 metres, but lately limited to waves over 3,000 metres (below 100 kcs.)See also KILOMETRIC and MYRIAMETRIC Waves.

LOOP : A series of computer instruction that are executed repeatedly until a desired result is obtained or a predetermined condition is met. The ability to loop and reuse instructions eliminates countless repetitious instructions and is one of the most important attributes of stored programs. An interative rountine, usually subject to instruction modification in a progrressive manner unit an exit condition is reached. A set of instructions that causes the same set of action to be repeated a specific number of times. (1) In a radio direction finder: one of the two aerial systems used at right angles to each other. (2) See ANTINODE. (3) Of an armature coil: the sharp bend at the end of the end connection portion particularly in a Former Wound Coil.

LOOP AERIAL : Originally a receiving aerial composed of a large rectangle of wire, similar to a Frame Aerial of one turn, used particularly in Direction Finders. Now used for a frame aerial generally.

LOOP DIALLING : A system of Dialling in automatic telephony employing Break Impulses in a loop or double metallic circuit.

LOOP TUNING ERROR (of Radio Direction Finder) : Inaccurate reading caused by differences in the natural frequencies of the two loops.

"L" OPERATIONAL AMPLIFIER : An amplifier with a very high open-loop gain.

LORIMER AUTOMATIC TELEPHONE SYSTEM : An early automatic telephone exchange system distinguished by the use of power-driven line selectors instead of selectors actuated by the calling current impulses.

LOTUS 1-2-3 : Lotus 1-2-3 version 3.0 is designed for AT or compatible computers and includes the ability to create three-dimensional spreadsheets, Lotus 1-2-3 version 2.2 is designed for the older IBM PC/XT computers but cannot create three-dimensional spreadsheets.

LOUD SPEAKER : A receiver for radio or line telephony or sound reproduction generally of sufficient power to be heard from some distance, indoors or in the open air.

LOW DEFINITION TELEVISION : Television in which the number of scanning lines to the picture is under 200.

LOWER SIDE BAND : A Side Band composed of frequencies below the carpier wave frequency.

LOW FREQUENCY AMPLIFICATION : See ADUIO-FRQUENCY AMPLIFICATION.

LOW FREQUENCY RESISTANCE : A term sometimes used for ordinary ohmic resistance at frequencies where skin effect etc., is negligible and the resistance is practically the same as for direct current.

LOW FREQUENCY TRANSFORMER : A transformer for audio-frequencies. See also INTERVALVE TRANSFORMER.

LOW FREQUENCY WAVES : A term sometimes used for electromagnetic waves of frequencies between 30 and 300 kc. per sec.

LOW PASS FILTER : A Wave Filter which prevents the passage of oscillations above acertain frequency.

LOW POWER MODULATOR : A Coks Modulator in which the modulating valve controls the oscillations before their amplification to full power. Cf. HIGH POWER MODULATOR.

LOW PRESSURE or Voltage : An official description of a voltage below 250 volts, between any two conductors or between one conductor and earth. Cf. MEDIUM, HIGH and EXTRA HIGH PRESSURE.

LOW RESISTANCE BRIDGE : A Wheatstone's Bridge or similar apparatus arranged to be suitable for the measurement of low resistances, e.g. Kelvin Bridge, Carey Foster Bridge, etc.

LP : Linear Programming

LRC : Longitudinal Redundancy Check

LSB : Least Significant Bit or digit. The bit or digit at the end of a number which has the smaller numerical value.

L-SECTION NETWORK : See QUARDRIPOLE

LSI : Large Scale Integration. A way of making denser and more complex integrated circuits. Eventually will be replaced by VLSI (Very Large Scale Integration) and VLSI by ELSI (Extremely Large Scale Integration).

LT : Logic Theorist

LU : Logic Unit

LUGGABLE : A term describing a computer that weighs less than 30 pounds and can be carried in one unit. See also LAPTOP.

LUMINESCENCE : Emission of visible radiation due to casuses other than high temperature, e.g. the glow in Discharge Tubes, etc., *Fluorescence* and *Phosphorescence*.

LUMPED CHARACTERISTIC (In a Thermionic Valve) : The characteristic showing the relation between Lumped Voltage and Anode Current.

LUMPED VOLTAGE (In a Thermionic Valve) : The anode voltage which would have to be applied when a certain grid voltage is present to bring the anode current back to its valve for zero grid voltage.

LUXEMBURG EFFECT : The transfer of the effect of modulations of the waves sent out from a powerful radio-transmitter to the carrier wave of anotherstation of different frequency passing through the same region of the atmosphere, causing signals of the former station to be distinctly audible during reception of the latter.

LX (Liquid-crystal displays) (LCD) : These are displays which consist of a very thin layer of liquid crystal fluid sealed between two glass plates. On the inside of these plates electrodes are etched in the form of characters. When an electric feild is applied the molecular alignment of the fluid is distrubed. Due to the difference in reflection, the activated characters become visible.

M : The symbol signifying the quantity 1,000,000 (10^6). When used to denote storage, it mere precisely refers to 1,048,576 (2^{20}).

MACHINE : A term somewhat loosely employed in electrical engineering usually to signify an apparatus with moving parts for converting mechanical into electrical energy or vice versa such as a generator, or a motor, acting either on electromagnetic or electrostatic principles, but not including stationary apparatus, such as transformers.

MACHINE LANGUAGE : A language consisting of binary data that only a computer can understand. The actual string of digits the computer hardware interprets and executes. Information represented in a form which an automatic computer can handle; for example, magnetic tape is in machine language whereas spoken words and number are not. The lowest level of computer programming; this is the only language the computer can understand without the aid of another program. Looks like:........ FF FF FF FF B8 C9 D9 CO.........

MACHINE LEARNING : Is the study of making computer systems exhibit learning behavior.

MACHINE RINGING : Telephone ringing by interrupted current from a power-driven generator started by a key or relay and stopped by a relay on the called party lifting his receiver.

MACHINE SWITCHING TELEPHONE SYSTEM : See AUTOMATIC TELEPHONE SYSTEM.

MACHINE TELEGRAPH SYSTEMS : An expression usually comprising such printing and other telegraph systems as involve the synchronous running of mechanisms at the sending and receiving stations.

MACRO : A set of instruction that can by typed automatically by pressing a specified key combination. Short for macroinstruction. An instruction that generates a larger sequence of instructions.

MACROINSTRUCTION : A symbolic command or instruction that is translated into two or more machine language instructions (often an entire subroutine) by the translator program.

MAC-TOOLS DELUXE PC : Tools Deluxe offers as many features as The Norton Utilites, but with the addition of a simple word processor, database, and telecommunication program combined with easy-to use menus.

MAC-TOOLS DELUXE MACINTOSH : Mac-Tools Deluxe is the Macintosh version of PC Tools Deluxe but without the word processor, database, and telecommunication portion.

MAGNETIC AMPLIFIER : Any form of amplifier depending upon the properties of ferro-magnetic materials.

MAGNETIC CORE : A small piece of magnetic material that can be used to represent a 0 or 1. A memory device in which information is represented by the magnetic polarity of a wire-sensed permeable ring.

MAGNETIC CROSSED-FIELD MODULATOR : A type of frequency doubler consisting of an annular coil surrounded by a hollow toroid of magnetic material. The toroid forms the core of an external toroidal winding (see diagram). The toroidal winding carries a direct voltage and the alternating current is input to the annular winding. The output voltage developed in the toroidal winding has a frequency twice that of the input signal due to the interaction of the orthogonal magnetic fields in the coils.

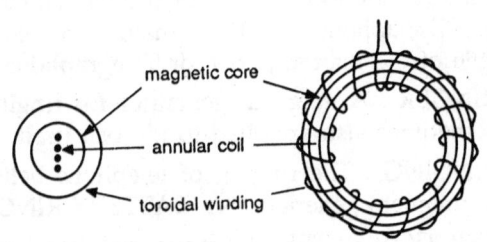

magnetic core

annular coil

toroidal winding

Magnetic crossed-field modulator

MAGNETIC DIAPHRAGM RELAY : A relay having a coil wound around a central coore with a thin metal diaphragm plate mounted close to its and (Fig. c). When the coil is energized the central portion of the diaphragm moves towards the core and makes contact with it.

MAGNETIC DRUM : A rotating cylinder; the surface of which is coated with a magnetic material on which information may be stored as a small potarised spots.

MAGNETIC HYSTERESIS : See HYSTERESIS

MAGNETIC MODULATOR : A Modulator for radio-transmitting stations in which the variations of the microphone current cause changes in the inductance of iron-cored coils in a circuit in parallel with the aerial inductance. No amplifying valves are employed.

MAGNETIC REED RELAY : A coil wound around a glass envelope containing fixed contacts and a centrally placed reed contact in the form of a thin flat metal strip. When the coil is energized the reed is deflected and can either make or break contact with the fixed leads. See REED RELAY.

MAGNETIC TAPE : Tape made of plastic (rarely metal) coated with magnetic material on which information may be stored as a small polarised spots.

MAGNETIC TAPE UNIT : A sequential storage medium that operates as does a home tape recorder.

MAGNETO BELL : A telephone bell constructed to work with the current, usually alternating, from a magneto generator; with a vibrating armature which causes the hammer to strike two gongs alternately.

MAGNETOPHONE : (1) One form of Moving Coil Microphone. (2) An instrument similar in construction to a telephone receiver for producing a loud sound when supplied with an undulatory current. (3) A sound recording instrument similar in purpose to the "Dictaphone," but Magnatophone employing the principle of the Blattnerphone or Telegraphone.

MAGNETO RINGER : A magneto generator for ringing telephone bell; sometimes also apoplied to the bell itself.

MAGNETO RINGING : The ringing of telephone bells by current from a magneto generator. Cf. BATTERY RINGING.

MAGNETOSTRICTION FILTER : A Frequency Filter depening on the principles of the Magnetostriction Oscillator.

MAGNETO TELEPHONE SYSTEM : A telephone system in which the call signals are actuated by magneto generators at the subscribers' stations.

MAGNETRON : Syn. *magnetron oscillator*. A crossed-field microwave tube that produces radio frequency (v.f.) oscillations in the microwave region. An early magnetron was used as a rectifier— the rectifier magnetron-but all modern magnetrons are designed as oscillators. (1) A Thermionic Valve in which the stream of the electrons is controlled by a magnetic field. In large magnetrons producing continuous oscillations, the high frequency alternating field for the purpose is provided by the filament current and causes oscillations in the anode circuit of twice the frequency, without the use of a Grid. Such tubes can deal with powers up to about 1,000 kw.

MAGNETRON EFFECT : The weakening of electron emission in a thermionic valve, due to the magnetic field of the filament current which imposes a limit on the output of large valves.

MAGNETRON OSCILLATIONS : Oscillations produced in Thermionic Valves circuits by the interaction of magnetic fields and electron streams.

MAGNETRON RECTIFIER : A gas tube rectifier in which the electrodes are not heated and the electron stream is controlled by a magnetic field.

MAGNIFIER : A term used in cable telegraphy in preference to relay or amplifier, for an apparatus for amplifying signals received over a submarine cable by controlling a circuit otherwise than by opening and closing contacts (see below); also sometimes used for an amplifier in radio telephony.

MAIN : A line that brings power from a generator, converter, or service cut off switch to the main distribution center, or service panel for power lines inside a building from electric supply company.

MAIN CONTACT SPRING In Telephony : A contact spring making contact between two other contact springs.

MAIN DISTRIBUTION FRAME : The frame in a telephone exchange to which the circuits from the cables entering the building are brought, usually carrying fuses, for every incoming circuit. Connections are taken thence to the intermediate distribution frames, and temporary connections. can be made for testing.

MAIN GATE : The input gate to the DCU, controlled by the input signal and by the time-base generator.

MAIN FRAME : A larger computer system used for multiple purposes such as batch processing, on-line applications, and time sharing. The main part of the computer, not including the peripherals.

MAIN MEMORY : Another term for random-access memory (RAM). With IBM computers, this term is used to differentiate between the first 640 K in your computer (main memory) expanded memory, and extended memory.

MAINS RECEIVER RADIO : See ALL MAINS RECEIVER.

MAINTENANCE : (1) The process of modifying operational programs to fix errors or add enhancements. (2) The activity of keeping an instrument, circuit or a system in perfect working order. This is of three types Preventive maintenance (P.M.), Maintenance Prevention (M.P.), and Breakdown maintenance.

MAJORANA LIQUID MICROPHONE : A type of microphone used in early experiments on radio-telephony in which the diaphragm is attached to a nozzle and causes vibrations in a jet of liquid, and consequent alteration of the surface upon which the jet is falling. The resulting variation of resistance between electrodes in this surface produces the required modulation of the current.

MAJOR SORT : The controlling or general order of items in a sequence. For example, a month's invoices could be sorted by customer and within each customer group, by date. The sort by customer is the *major* sort; the sort by data is the *minor* sort.

MAKE-BEFORE-BREAK CONTACT SPRING IN TELEPHONY : A contact spring arranged not to break its normal (back) contact before contact is established with the operating (front) contact.

MALFUNCTION : Hardware failure, i.e. incorrect operation of some physical equipment.

MALTER EFFECT : An effect observed when a semiconductor of high secondary-emission ratio, such as caesium oxide, is subjected to electron bombardment. The semiconductor can become strongly positively charged and if a layer of such a material is separated by a thin insulating layer from a metal plate it can be used to develop a potential difference. With an insulting layer of about 0.1 micrometre thickness a potential of up to 100 volts can be produced.

MANAGERIAL CONTROL DECISIONS : Decisions primarily involving personnel and financial control, concerned with ensuring that resources are applied to achieving the goals of the organization.

MANIAC : Mechanical and Numerical Integrator and Computer

MANTISSA : The fractional or decimal part of a logarithm of a number. For example, the logarithm of 163 is 2.212 in which the mantissa is 0.212, and the characteristic is 2.0. In floating-point numbers, the mantissa is the number part. For example, the number 24 can be written as 24,2 where 24 is the mantissa and 2 is the exponent. The floating-point number is read as $.24 \times 10^2$, or 24.

MANUAL : A book that contains instruction for using equipment or software. Usually loaded with typos, hard to read, and poorly organized.

MANUAL TELEPHONE SYSTEM : A telephone system in which the calls are given verbally, and all switching operations are carried out by hand.

MAPPING : A system of reducing Boolean expressions using a matrix of minterms.

MARCONI DIRECTIVE AERIAL : A type of partially directive aerial consisting of an inverted L aerial with a long horizontal portion.

MARCONI--REISS MICROPHONE : A form of carbon microphone used for broadcasting, etc.,in which the sound waves act directly upon the carbon granules without a diaphragm.

MARCONI SYSTEM OF RADIO COMMUNICATION. The systems of radio-communication associated with the name of G. Marconi have undergone so much alteration and development that no one method can be defined under this heading. Originally employing an earthed aerial containing the spark gap and coherent in its own circuit, the system was later improved by the addition of coupled tuned circuits production and detection, leading up to the valve and other methods of transmission and reception, and the short wave (Beam) system with highly-directive aerials now employed for the longest systems.

MARKING CURRENT : A telegraph signal current in the direction which causes the Morse instrument to mark the paper. Cf. SPACING CURRENT.

MARKING WAVE : A term used for the wavelength used during the actual Morse signal in wireless telegraph systems where signalling is affected by a change in frequency of the continuous waves employed and not by their interruption. Cf. SPACING WAVE.

MASER : Microwave Amplification by Stimulated Emission of Radiation. A source of intense coherent monochromatic radiation in the microwave region of the electromagnetic spectrum The maser can be used as a microwave amplifier or oscillator.

MASKING : A method used for rejecting noise in the input signal. The input signal is masked during a certain time slot that no miscounting due to noise can occur.

MASS CORE : A core for electromagnetic apparatus, particularly in telegraphy,composed of pulverised iron or other magnetic material with a binder of non-magnetic material.

MASS SPECTRUM : A spectrum obtained by deflecting a beam of positive rays emitted from a tube, containing a residual gas to be investigated, byelectric or magnetic fields. The extent of the deflection depends upon m/e (the ratio of the mass of the projected positively charged particles of which rays are composed to the atomic charge). Thus, every element has its characteristic spectrum lines like those of the light spectrum. Isotopes were discovered by investigations into mass spectra, which are now applicable to ionised metallic atoms as well as to gases.

MASTER DISK : The floppy disks that store programs you buy from a company.

MASTER FILE : A data file composed of records having similar characteristics that rarely change. A good example of a master file would be an employee name and address file that also contains social security numbers and hiring dates.

MASTER TAPE : A paper tape or magnet tape carrying semi-permanent information (such as a file of information).

MASTER SLAVE : A flip-flop which contains two flip-flops, a master flip-flop and a slave flip-flop. The master flip-flop receives its informatin during the leading edge of a clock pulse, and the slave or ouptut flip-flop receives its information from the master during the trailing edge of the pulse.

MASTER SLAVE FLIP-FLOP : A flip-flop composed of two internal FF'S one to receive the inputs (the master) and one to drive the outputs (the slave).

MATRIX : An arrary of arrangement having properties that depends on the ordering of the elements arid upon the nature.of the elements.

MAXTERM : An OR of single-lettered terms that includes all the variables

MAXWELL BRIDGE : A four-arm bridge for measuring inductance in terms of a capacitance and resistances *(see diagram)*. At balance, as indicated by a null response on the instrument I.

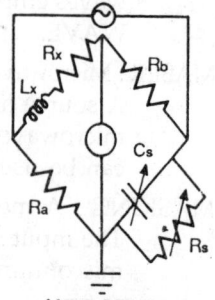

Maxwell bridge

$R_sR_x = R_bR_a$

$L_x = R_bR_aC_s$

MBQ : Modified Biquinary Code.

MC : Magnetic Core.

MCP : Master Control Program

MCST : Magnetic Card Selectric Typewriter.

MD : The memory data register, also called the memory buffer. MD Magnetic drum or magnetic disc.

MDAC : Multiplying Digital-to-Analog Converter

MEANS ENDS ANALYSIS : A problem-solving strategy that works by analyzing the difference between the current state and the desired goal state and choosing the operator that is though to be best at reducing that difference; it was pioneered in the GPS system.

MEDIA : The plural of medium.

MEDICAL ELECTROLOGY : The application of electrical methods to medical practice.

MEDIUM : The physical material on which data is recorded and stored. Magnetic tape, punched cards, and diskettes are examples of media.

MEDIUM FREQUENCY : Abbreviated MF, any frequency in the region from 300 kHz to 3 MHz.

MEDIUM FREQUENCY WAVES : A term sometimes used for electromagnetic waves of frequency between 300 and 3,000 kc. per sec.

MEDIUM PRESSURE or VOLTAGE : The official description of a voltage above 250 volts but not exceeding 650 volts between any two conductors or between any conductor and earth. Cf. LOW, HIGH and EXTRA HIGH PRESSURE.

MEDIUM SIZE : Often automatic computer with a combined purchase price between about $100,000 and about $ 700,000.

MEDIUM WAVES. : Expression formerly in common use for wavelengths between 100 and 1,000 metres, but lately defined as applying to waves between 200 and 3,000 metres (1,500-100 kc). Cf. LONG, INTERMEDIATE, SHORT,and VERY SHORT WAVES.See also HECTOMETRIC and KILOMETRIC WAVES.

MEDLARS : Medical Literature Analysis and Retrieval System.

MEDLINE : Medlars On-Line System.

MEGA or MEG : Prefix meaning one million times. (Abbreviation : M.)

MEGACYCLE : One million cycles. Radio frequencies are sometimes expressed in megacycles per second. Cf. KILOCYCLES.

MEGAHERTZ : A measures of transmission frequencies; megacycle or millions of cycles per second.

MEGGER : An instrument of the Ohmmeter type for measurement of insulation resistance by direct reading; distinguished by its compact form and employing a hand-driven generator in the same case, and using the same field magnets, as the moving coil indicating instrument.

MEMORY : The high-speed work area in the computer where data can be held, copied, and retrieved. Storage inside the computer for data. A device for storing and retrieving binary words at locations called address. An (electronic) device in which data can be stored and from which data can be obtained when necessary (also called storage).

MEMORY ADDRESS REGISTER : The register that contains the address of the memory location currently being *read* or *written* into.

MEMORY CAPACITY : The amount of information which a memory device can store.

MEMORY DATA REGISTER : The register that contains the data to be written into the memory or the data that were just read from the memory. Abbreviated as MDR

MEMORY DUMP : See STORAGE DUMP.

MEMORY MATE : Memory Mate works as a memory-resident program, available to use from within any other program. It lets you store random information including letters, database records, or spreadsheets stored as text or ASCII characters.

MEMORY REFERENCE INSTRUCTION : An instruction that requires an operand from memory or one that places data into memory.

MEMORY-RESIDENT : Term describing certain programs that copy themselves into memory and remain ready to run.

MEMORY UNIT : Storage unit, the computer unit where information can be introduced and later extracted.

MENU : A list of choices from which an operator can select a task or operation to be performed by the computer. A list of commands or instruction displayed on the screen. Menus are supposed to organize commands and make a program easier to use.

MENU-DRIVEN : Describes a program that provides menus for choosing commands.

MERCADIER MAGUNNA MULTIPLEX TELEGRAPH SYSTEM : A

Multiplex system in which each message is sent on a different audio-frequency, and is selected at the receiving station by a Monotelephone which responds only to that frequency.

MERCURY AIR VALVE REGULATOR : An appliance to regulate the vacuum in an "X" Ray Tube connected to the main tube by apporous plug normally sealed by mercury which can be displaced by application of external air pressure, to allow a little air to enter the main tube. Cf. PILON and OSMO MERCURY AIR VALVE REGULATORS.

MERCURY-ARC RECTIFIER : A cold-cathode arc-discharge tube that has a mercury-pool cathode. When a sufficiently high voltage is applied to the anode an arc is struck across the tube. The tube is used as shown in the diagram. As the alternating-current input is applied the arc oscillates between anode 1 and anode 2 as each becomes positive in turn. Direct current is output to the load.

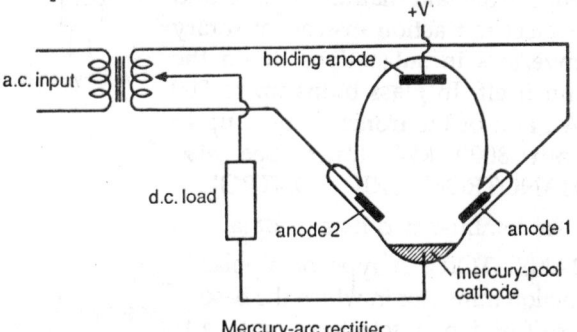

Mercury-arc rectifier

Many types of mercury-arc rectifier contain an extra anode known as the *holding anode*. This anode is supplied with a direct voltage and is used to maintain the arc when the *main anode(s)*, i.e. the anode(s) supplying the load, draw no current. Use of such an electrode to maintain the arc prevents *misfire*, i.e. a failure to produce the arc during an intended conduction interval.

The main anode is often surrounded by a metal shield—the *anode shield*—to protect it from massive ionization or radiation. One type of mercury-arc rectifier has an anode in the form of *steel tank*. This type of anode is very robust. An *arc baffle* is often used to prevent mercury splashing from the cathode on the anode. One version of the mercury-arc rectifier uses a small auxiliary electrode dipped into the cathode to start the arc.

After the arc has been struck a typical voltage droop across the tube is 20 to 25 volts; such tubes are thus more efficient than high-vacuum rectifier tubes.

MERCURY VAPOUR FREQUENCY CHANGER : An apparatus is like a Mercury Vapour rectifier of the reversible design with the control grid excited at a different frequency from that of the supply.

MERCURY VAPOUR RECTIFIER : A rectifier depending for its action upon the unidirectional conductivity of the mercury arc in an exhausted vessel. Used in small sizes for accumulator charging from alternating current circuits,and in larger sizes for running cinema and other direct current arc lamps from alternating circuits and for electric traction instead of rotary convertors in substations or on the train itself. In glass bulbs up to 200 kW., and of the ironclad type up to about 8000 kW. units. See also PHANOTRON, GRID CONTROL.

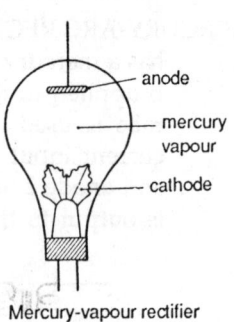

Mercury-vapour rectifier

MERGE : To combine two lists of data.

MESA TRANSISTOR : A type of bipolar junction transistor in which the base is diffused non selectively into a suitable substrate and the portions around the base then etched away to leave a plateau above the substate. A double diffusion technique may be used to form the emitter inside the base region or the emitter may be alloyed into the base. The collector is the region of the substrate below the base.

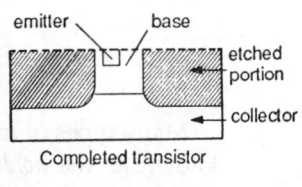

Completed transistor

MESA Transistor

MESFET : Acronym from Metal Semiconductor Field-Effect Transistor. Syn. Schottky gate field-effect transistor. A type of junction field-effect transistor that has a Schottky barrier as the gate electrode rather than a semiconductor junction. The voltage current characteristic is similar to that of a junction FET. The Schottky barrier gate electrode can be formed on semiconductors, such as gallium arsenide, in which doping cannot be easily effected, and is most often used to form gallium arsenide FETs. It has the added advantage that it can be made at much lower temperatures than are required to form a *p-n* junction.

MESH CONNECTION : Also known as Delta Connection in case of three phase circuits, the connection of phases in the polyphase circuit so that they form a closed figure. cf: STAR CONNECTION.

MESTRON : A form of "heavy electron" with a mass of about 170 times that of an ordinary electron, produced by the impingement of cosmic radiation upon gaseous molecules.

METABOLONS : The products of sucessive disintegration of radioactive bodies.

META-LEVEL KNOWLEDGE : Is knowledge about knowledge, or higher-order information in a knowledge base about its organization, structure, and usage.

METALLIC CRYSTAL : A crystal that consists of a regular array of positive ions together with a cloud of free electrons.

METALLISED VALVE : A Thermionic Valve, the bulb of which is sprayed with a metal film to produce a screening effect to counteract the effect of stray capacitance couplings.

METALLIZING : Syn. *silvering.* Depositing thin films of a metal (originally silver) on a glass, semiconductor, or other substrate in order to render it electrically conducting. The metal is then etched into the required metallization pattern using a sepcially designed mask. The technique is widely used in solid-state electronics for the formation of interconnections on integrated circuits or thin-film circuits and to form bonding pads on integrated circuits or discrete components.

METAL OXIDE RECTIFIER : See METAL RECTIFIER.

METAL OXIDE SEMICONDUCTOR : Abbreviated MOS, a metal-insulator-semiconductor structure in which the insulating layer is an oxide of the substrate material, such as silicon dioxide. Compare MESFET, MOSFET (see below).

METAL OXIDE SEMICONDUCTOR FIELD EFFECT TRANSISTOR (MOSFET) : Abbreviated MOSFET a field effect transistor in which the gate is insulated from the channel by a metal-oxide dielectric film. Consequently, this allows for a higher input impedance by making the gate forward biased to enhance the conductivity of the channel. Also called an insulated gate field effect transistor, or IGFET.

METAL RECTIFIER or Metal Oxide Rectifier : A form of rectifier depending on the unidirectional conductivity of a junction of such materials as copper and copper oxide not apparently involving any chemical or electrolytic action. (Also called Electronic, couprox, or Barrier Film Rectifier.)

METAL VALVE : A thermionic valve in which a metal container replaces the glass bulb, and in some cases constitutes the anode.

METAL "X" RAY TUBE : An "X" Ray Tube partly of glass and partly of metal, the latter for the prevention of egress of rays where not required.

META-RULE : Is a rule in knowledge base whose conditions and actions mention both elements in working memory and other rules in the system; it is most commonly used to control the search conducted by the inference engine in ways the standard algorithms cannot.

METER-AMPERE : See METRE-AMPERE

METER KEY : A key on a telephone switchboard which closes a circuit to actuate a Telephone Meter.

METER RACK : A frame in a telephone exchange on which are mounted rows of call meters.

METHOD : A way of doing something.

METRE-AMPERE : An expression used in connection with radio aerials for the product of the mean height and the rated current.

METRIC WAVES : Electromagnetic waves of a wave length between ten metre and one metre.

MFT : Multiprogramming with a Fixed Number of Tasks.

MHO : The unit of Conductance or Admittance, being that of a conductor having a resistance or impedance of *one* ohm.

MICARTA : An insulating material capable of replacing ebonite, etc., made chiefly of mica.

MICR : Magnetic-Ink Character Recognition. The machine reading of characters printed in magnetic ink; primarily used in check processing.

MICRO : Prefix meaning one millionth, e.g. microampere, one millionth of an ampere; a unit sometimes used for very small currents.

MICRO-ALLOY TRANSISTOR : A type of bipolar junction transistor in which the semiconductor substrate forms the base; the *emitter* and *collector* are formed by recrystalization from a suitable impuritymetal/semiconductor alloy in small pits etched in the surface of the base. Micro-alloy diffused transistors are formed in a similar manner but a gaseous diffusion is carried out first to produce a nonuniform base before the emitter and collector are formed. These transistors are now obsolete.

MICRO CHANNEL ARCHITECTURE [MCA] IBM's : Proprietary design for transferring information through a computer.

MICROCOMPUTER : A small computer, often for a single user or a small number of simultaneous users. Developed from advances in chip fabrication such that only a few chips are needed to produce an entire computer.

MICROFARAD : (Abbreviation: μf.) The most widely used practical unit of capacitance; being one millionth of a *Farad*. The latter although used in many formulae on account of its correspondence with other practical units, is inconveniently large for use in defining the capacitances met with in practice.

MICRO HENRY : Abbreviation: H. One *millionth* of a henry.

MICROPHONE : An instrument founded on the discoveries of Hughes (1878) and Edison (1877), devised for magnifying small sounds, and now used in a slightly different form as a telephone transmitter. The ordinary carbon microphone consists of a diaphragm set in vibration by the sound waves and causing by its motion variation in the resistance of a mass of lossely packed carbon granules and corresponding undulations in the current through the instrument faithfully copying the form of the original sound waves. See also TELEPHONE RECEIVER. A number of other forms are used to a limited extent for radio-telephony and for experimental purposes, and the term has come to be applied to any variety of sensitive telephone transmitter, whether dependant upon variation of resistance or not on.

MICROPHONE AMPLIFIER : (1) An amplifier used in conjunction with a broadcasting microphone to amplify the variations of current caused thereby before their transmission to the Control or Modulating System. (2) A microphone acting on the same principles as a Microphone Relay.

MICROPHONE BATTERY : The battery which supplies the current in a microphone circuit.

MICROPHONE CIRCUIT : The local circuit containing a microphone, a battery, and the primary of an Induction Coil.

MICROPHONE RELAY : A relay in which the slight movement of the armature of an electromagnet varies the resistance of a loose carbon contact and thereby controls the variations of current in the local circuit with a similar action to that of a Microphone.

MICROPHONE TRANSMITTER : See MICROPHONE.

MICROPHONIC NOISE : Valve Noise in a radio-receiving apparatus, consisting of a more or less musical note, due to the effect of mechanical vibration of the filament.

MICROPHONIC VALVE : A valve with insufficiently rigid electrodes liable to take up vibrations which produce Microphonic Noise.

MICROPROCESSOR : A semiconductor central processing unit (CPU) in a computer. Sometimes called the Central Processing Unit (CPU) or processor, this chip does all the calculations for the computer. An integrated circuit capable of interpreting and, with the help of RAM and input/output devices, executing instruction. Thus it is a essential electronics of a computer miniaturized to a signal chip the size of a pin head. Contained usually in an IC package with 18 to 40 leads.

MICRO-PROGRAM : A special command repretoire for an automatic computer that consists only of basic elemental operation which the programmer can combine into commands to suit his all convenience and in terms of which he would then program.

MICRO-PROGRAMMING : The combining of series of elementary hardware functions (invisible to the programmer) to make a single instruction.

MICRO RAYS : See MICROWAVE.

MICRO-SECOND : One one-millionth of a second, written "μs".

MICRO-SOFT BASIC (IBM) : See QUICK C.

MICRO-SOFT C (IBM) : See QUICK C.

MICRO-SOFT EXCEL : (IBM and Macintosh) Microsoft Excel has long been considered superior to LOTUS 1-2-3 for its large number of features and excellent graphing capabilities. Available for both the IBM and Macintosh. Microsoft Excel lets you share data files between two computers.

MICRO-SOFT WORD (IBM) : See MICROSOFT WORD [MICINTOSH] MICROSOFT WORK (MACINTOSH) Microsoft Word and

Word for windows are rival programs to Word Perfect. Microsoft Word is also available for the IBM and Macintosh, letting you easily share data between the two computers. Microsoft Words remains popular for its low cost, simplicity and powerful features including a spell checker, graphics and communication program.

MICRO-TELEPHONE : A name sometimes given to a hand combination telephone set in which the transmitter (Microphone) and Receiver are joined together and held in the hand when in use.

MICRO-WAVE : Pertaining to wavelengths ranging from 0.3 to 30 cm, correspondig to frequencies from 1 to 100 GHz. An electromagnetic wave with a wavelength in the range three millimetres to 1.3 metres, i.e. between infrared radiation and radiowaves on the frequency spectrum. Microwaves are used in radar and telecommunications and are also used commercially for extremely rapid cooking.

MICRO-WAVE TUBE : An electron tube that is suitable for use as an amplifier or oscillator at microwave frequencies (see frequency band). These tubes usually employ velocity modulation of the electron beam rather than density modulation as in the values used in audio-frequency valve amplifiers or oscillators.

MIDI : Musical Instrument Digital Interface.

MID-RANGE FREQUENCY : Any audio frequency between 600 Hz and 6 kHz.

MILLER BRIDGE : An AC bridge circuit that measures the amplification factor or transconductance of vacuum tubes.

MILLER EFFECT : The gain degeneration of an amplifier due to collector-base capacitance. The effect whereby the input capacitance of a triode vacuum tube amplifier is effectively the sum of the grid-cathode capacitance plus the product of the grid—plate capacitance and one plus the voltage gain of the triode stage. The phenomenon by which an effective feedback path between the *input* and *output* of an electronic device is provided by the interctrode capacitance of the device. This can affect the total input admittance of the device, which results in the total dynamic input capacitance of the device being always equal to or greater than the sum of the static electrode capacitances.

MILLER INTEGRATOR : An integrator circuit consisting of an

operational amplifier with a series resistor and a capacitor in its feedback path. Also called a Miller time base.

MILLER TIME BASE : See MILLER INTEGRATOR.

MILLER'S THEOREM : A network transform that states that an amplifier circuit with a feedback impedance network connected from its output terminal to its input terminal may be replaced by an equivalent network having transformed impedances between its input and output terminals to ground, both of which depend on the amplifier's voltage gain and the feedback impedance. In addition the feedback element is removed.

MILLI : Prefix meaning one thousandth, e.g. millivolt, one thousandth of a volt. (Abbreviation: *m*.)

MILLI-AMPERE : One thousandth of an ampere. A unit used by telegraph and telephone engineers who are therefore sometimes spoken of as "milli-ampere men."

MILLIONS OF INSTRUCTIONS PER SECOND (MIPS) : Unit of measuremnt to determine how fast a computer can process instructions.

MILLISECOND : One one-thousand of a second; written *"ms"*.

MIMR : Magnetic-Ink Mark Recognition

MINICOMPUTER : A computer between the micro and mainframe; often used for a dedicated on-line application or for general time sharing. Computers that are larger than a microcomputer but smaller than the larger computers in terms of physical size, price, and amount of memory.

MINIMAX : A method for searching game trees that deals with an adversary by assuming that he will always choose moves to minimize the result or static evaluation (towards) at his turn; if the player on move always tries to maximize, the line of best play can be found.

MINIMIZATION : The process of reducing a Boolean expression so that a minimum number of gates will be required for implementation.

MINIMUM ACCESS PROGRAMMING : See MINIMUM LATENCY PROGRAMMING.

MINIMUM LATENCY PROGRAMMING : Arranging instructions and other data in a cyclio storage device so that the average realized access time is close to the minimum access time. This is usually achieved by interlacing instructions and operand data, both separately and together.

MINOR SORT : The order of items within the homogeneous groups formed by a major sort (See major sort).

MINORITY CARRIER : In a semiconductor, current carriers that are in the minority. In a p-type semiconductor, minority carriers are electrons; in an n-type semiconductor, they are holes.

MINTERM : A product of single-lettered terms that contains all the variables.

MINTERM DESIGNATOR : The decimal or binary number formed by assigning 0's to the complemented literals of a minterm and 1's to uncomplemented literals.

MIS : Management Information System or Metal-Insulator-Semiconductor

MIT : Massachusetts Institute of Technology.

MIXED RETROACTION : Retroaction in a radio-receiving apparatus in which both electrostatic and electromagnetic coupling are employed.

MIXER : A term sometimes used for a valve in a superheterodyne receiver in which the frequency of a local oscillator, which may be incorporated in the valve itself, is superposed upon that of the received signal to produce the required intermediate frequency by beats.

MIXES : The weighting of a representative series of instructions for the purpose of evaluating machine performance.

MIXING : The process of creating intermediate frequency in super heterodyne receiver. See MIXER.

MKS SYSTEM : A practical international system of electrical units (amperes, volts etc. dependent on forming all units based on *metre, kilogram* and *second* if the pearmeability of space is taken as 10^{-9} instead of 1. India adopted this system instead of FPS system in 1956, although the system was internationally adopted in 1935 by the International Electro-technical Commission. See MKS UNITS below.

MKS UNITS : The practical system of electrical units (amperes, volts, etc.) which may be taken as being founded on the metre, kilogram, and second if the permeability of space is taken as 10^{-9} instead of 1 in the same way as the C.G.S. system is based on the centimetre, gramme and second. Adopted in 1935 by the International Electro-technical Commission as superseding the C.G.S. absolute electro-magnetic and electrostatic systems.

MLC : Multiline Controller.

MNEMONIC : Pronounced "new-monic." A short word or group of letters that stands for another word, and is easyto remember. For instance, "Add X to the Accumulator" might be abbreviated "ADDX." Alphabetic symbols used in place of numeric codes to facilities the recognition and use of computer instructions.

MNEMONIC SYMBOLS : Symbols chosen for the aid they afford to the human memory.

MODE : The selected operational function of measuring instrument.

MODEL (Modulate and Demodulate) : A device that converts *digital* computer signals into analog from and modulates them for transmission. Demodulation is the reverse process that occurs at the receiving point.

MODEM : Acronym for *modulator demodulator*. A modem converts data from a computer to analog signals that can be transmitted through telephone lines, or converts the signals from telephone lines into a form the computer can use. A modem lets a computer send an receive data through an ordinary telephone line. Usually connects between the computer and telephone line and converts digital signals to audio tonesand vice-versa.

MODULA-2 : Programming language designed for writing complicated programs. Similar to Pascal.

MODULAR PROGRAMMING : The subdivision of a system and of programming requirements into small building blocks to reduce programming complexity and take advantage of common routines.

MODULATING VALVE. : A thermionic valve used in a radio-transmitter to amplify the audio frquency oscillations produced by the microphone and to superpose them upon the carrier wave by suitable action upon the anode circuit of that oscillating valve or an amplifier connected there to. (Also called control valve), See also CHOKE MODULATOR.

MODULATION : The coding of a digital signal onto an analog one, for example, by changing the amplitude of the carrier signal to represent a 0 or a 1. The process by which a characteristic of one signal is varied in proportion to the information contained in another signal. Types include amplitude, frequency, phase, pulse-amplitude, frequency, phase, pulse-amplitude, pulse-code, pulsewidth, pulse-frequency, pulse-position, and pulse-time modulation. The periodic alternation of amplitude, frequency or phase of the Carrier Wave according to the sound waves by the action of the microphone, or in some other way,on some part of the transmitting circuit. Cf. DEMODULATION.

MODULATION ENVELOPE : An imaginary curve connecting the peaks of a modulated waveform in order to more clearly show its shape. The envelope then represents the waveform of the intelligence carried by the signal.

MODULATION FACTOR : The ratio of half the difference between the maximum and minimum amplitude of a modulated wave to its mean amplitude.

MODULATION METER : An apparatus for measuring the depth of modulation of transmitted waves.

MODULATION PERCENTAGE : Modulation Factor expressed as a percentage.

MODULATION SYSTEM : See CONTROL SYSTEM.

MODULATOR : A transmitter circuit or device that varies the amplitude, frequency, phase, or other characteristic of a carrier signal in proportion to the waveform of the modulating signal that contains useful information. The apparatus which superposes the modulations on the carrier current or waves in a line or radio-telephone system.

MODULATING GRID : The remainder after division by the specified number, as "modulo 5".

MODULO-N ARITHMETIC : Arithmetic in which all carries beyond n-1 are discarded.

MODULUS : The maximum possible number in an arithmetic system or register.

MONITOR : (1) A television-like display that lets the computer show information to you. (2) The control program that schedules and manages the computer's resources.

MONOCHROME DISPLAY : A monitor that can only display one colour, such as green, amber, or black and white.

MONOCHROMATIC RADIATION : Electromagnetic radiation of a single frequency. In practice radiation of a single frequency is never achieved and the term is applied to a narrow range of frequencies. The term is also applied to particulate radiation when the particles are all of the same type and energy but in this case the description *homogeneous* or *monoenergetic* is usually preferred. *Compare* POLYCHROMATIC RADIATION.

MONO-STABLE MULTIVIBRATOR : A device that will output a pulse of preset length every theme it receives an input.

MONOTELEPHONE : A telephone receiver which has a selective action in responding to only one audiofrequency. See MERCADIER MAGUNNA MULTIPLEX SYSTEM.

MONOTONICITY : The property of a D/A converter to maintain consistent analog voltage increments between binary increments.

MORASE INKER : A telegraph receiving instrument which marks the message of a paper strip by means of an inky wheel attached to a lever worked by the armature of an electromagnet, in the Morse Code.

MORPHEME : A combination of phonemes that is a valid word in the language being used.

MORSE KEY : A single lever key for signalling in the Morse Code. See SINGLE CURRENT KEY, MARKING CURRENT, SPACING CURRENT.

MORSE MULTIPLEX TELEGRAPH SYSTEM : A Multiplex System for signalling in the Morse Codel called Multiplex Diode, Triode, Tetrode, Pentode, or Hexode, according as it provides for 1, 2, 3, 4, 5, or 6 ways.

MORSE SOUNDER : A telegraph receiving instrument which gives a distinctly audible sound on the attraction and release of the armature, owing to its striking against stops, at the beginning and end of each Dot and Dash, from which sounds the message can be read by a trained operator.

MORSE TELEGRAPH SYSTEM : Any telegraph system employing the Morse Code.

MOS : Metal Oxide Semiconductor.

MOS-FET : Metal-oxide semiconductor field-effect transistor; consists of source and drain regions on either side of a P-type or N-type channel, plus gate electrode insulated from the channel by silicon dioxide. See METAL OXIDE SEMICONDUCTOR FIELD EFFECT TRANSISTOR.

MOST SIGNIFICANT BIT (MSB) : The bit of a binary word having the most weight.

MOTHER-BOARD : The main circuit board inside the computer where every part plugs into.

MOTOGRAPH RELAY : An early Relay working on the same principle as the Edison Loud Speaking Telephone in accordance with the signals received.

MOUSE : A hand-held device that moves a cursor, called a *mouse* or *pointer*, on the screen.

MOUSE BUTTON : A button on top of a mouse that performs a specific action depending on the location of the mouse pointer on the screen.

MOVING COIL MICROPHONE : A Moving -Conductor Microphone in which the moving system is a coil.

MOVING CONDUCTOR MICROPHONE : Any pattern of microphone in which e.m.f.'s are produced by the movement of a conductor, mechanically connected to the diaphragm, in a magnetic field.

MOVING IRON MICROPHONE : A microphone depending on the e.m.f.'s produced by the movement of a mass of iron in a variable magnet field.

MPU : Micro Processor Unit. Another way to say micro processor. Motorola makes an MPU called the 6800.

MQ : Multiplier-Quotient.

MQR : Multiplier-Quotient Register.

MS : Milli Second.

MSB : Most Significant Bit.

MSDOS : Acronym for Microsoft Disk Operating System, installed in IBM or IBM compatible personal computers.

MSI : Medium-Scale Integration.

MSM : Message-Swithcing Multiplexing.

MT : Magnetic Tapes.

MTBF : Mean Time Between Failures.

MTS : Michigan Terminal System.

MTST : Magnetic Tape Selectric Typewriter.

MU GREEK LETTER "MU," EQUIVALENT TO "M" : A symbol used (1) for Permeability, and for Amplification Factor, and (2) for the millionth part of a unit, e.g. μf (microfarad)

MULLER TUBE : A variety of hot cathode tube with an auxiliary cathode, or grid, connected to the main cathode through a leak resistance.

MULTI ADDRESS : Often automic computer that has a built-in ability to use more than one address in the address part of each instruction. In some computer, one of the additional addresses may taken over of the control counter.

MULTI-VALVE AMPLIFIER : See MULTISTAGE AMPLIFIER.

MULTI-CHANNEL TELEVISION : A system of Television in which the picture is built up of a very large number of separate elements divided into several sections, each worked independently on a different frquency band.

MULTICHANNEL VOICE FREQUENCY TELEGRAPH SYSTEM : See VOICE FREQUENCY TELEGRAPH SYSTEM.

MULTICS : Multiplexed Information and Computing Service.

MULTIGRID VALVE : A Thermionic Valve with more than one grid,e.g. a Pentode or a Screen Grid Valve.

MULTIPLE : In telephone exchange switchboards, the jacks connected to every subscriber's line which are provided on every section of the junction operators' boards (or in small exchanges the subscriber's operators' boards), so that any operator can make connection to any subscriber's line are spoken of as "multiples." In the same way, in automatic telephony, the groups of connections by which lines are brought to the Banks of a number of selectors. As a verb, to make a telephone circuit available at a number of points.

MULTIPLE FOLDED DIPLOE : See DIPOLE AERIAL.

MULTIPLE MICROPHONE : A microphone with a number of carbon elements all acted upon by one diaphragm.

MULTIPLE-OUTPUT MINIZATION : The process of reducing multiple equations to take advantage of terms that could be shared between the equations.

MULTIPLE PERIOD MODE : The function where the counter/timer measures a selected number of successive periods.

MULTIPLE RECEPTION : Reception of several separate radio signals simultaneously on the same aerial or with the same carrier-frequency.

MULTIPLE TUNED AERIAL : An aerial to which several separately tuned circuits are connected.

MULTIPLE TUNER MARCONI : An early tuning apparatus for use with the Marconi Magnetic Detector.

MULTIPLE TWIN CABLE : A telephone cable containing a number of cores each consisting of two twisted pairs twisted together. Cf. QUADCABLE.

MULTIPLE UNIT STEERABLE ANTENNA (MUSA) : A receiving aerial system in which the vertical angle of greatest sensitivity can be adjusted to obtain optimum selectivity, and to reduce interference.

MULTIPLE VALVE : A Thermionic Valve with several sets of electrodes in one bulb acting as independent valves.

MULTIPLE WAY TELEGRAPH SYSTEM : Any system in which more than one message is sent out simultaneously.

MULTIPLEX : To combine two or more signals so they can be transmitted over a single channel.

MULTIPLEX SYSTEM : A system where by more than two messages can be sent over the same circuit simultaneously. CF. DUPLEX, QUADRUPLEX SYSTEM, and MULTIPLEX TELEGRAPH SYSTEM.

MULTIPLEX TELEGRAPH SYSTEM : A system whereby more than two messages can be sent over the same circuit simultaneously. One such system depends upon the use of rapidly revolving contact arms at the two stations driven at exactly the same speed and in step with each other whereby successive contact is made with a number of local circuits. Cf. DUPLEX, DIPLEX and QUADRUPLEX.

MULTIPLEXER : A buffer capable of coordinsting, within limits, several inputs or outputs of a data movements.

MULTIPLEXER, BINARY : A multiple-input device that receives data on the input lead determined by the control leads and transmit them to the output.

MULTIPLEXING : The combination of several low-speed signals onto a higher-speed line for communications. The process of transmitting multiple data channels over a single facility. The transmission of different data simultaneously or sequentially over a single line or a group of lines.

MULTIPLEXOR : A device that combines signals received from a series of low-speed lines and transmits them over a high-speed line. No storage is provided, and signals must be demultiplexed on the receiving end.

MULTIPLIER : (1) In a voltmeter, a resistor used in series to increase its full-scale range. (2) A circuit or device whose output is proportional to the product of both its inputs. (3) A device whose output frequency is a multiple of its input frequency.

MULTI-PROCESSING : A technique for executing two or more instruction sequences simultaneously in one computer system by the use of more than one processing unit.

MULTI-PROGRAMMING : The presence of more than one semiactive program in primary memory at the same time; by switching from program, the computer appears to be executing all concurrently.

MULTI-STAGE AMPLIFIER : A Thermionic Amplifier consisting of several valves in Cascade to produce a cumulative effect.

MULTI-TASKING : The ability to run more than one program simultaneously.

MULTI-TONE TRANSMITTER : A radio transmitting apparatus of the singing spark class in which by variations of the inductance of a circuit resonating with the spark or audio-frequency, variations can be made at will in the tone or note of the signals.

MULTI-USER : The ability to let two or more people use a computer simultaneously.

MULTI-VALUED LOGIC : Acronym for (MVL), as against binary logic, is suited for applications requiring large global interconnections, as each MVL line carries more information that a binary line.

MULTIVIBRATOR : An oscillator that contains two linear inverters coupled in such a way that the *output* of one provides the *input* for the other. There are several types of multivibrator, the action of which depends on the type of coupling used.

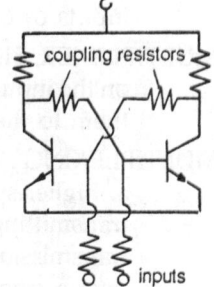

Bistable multivibrator

Capacitive coupling produces an *astable multivibrator* that has two quasi-stand states; once the oscillations are established the device is free-running, i.e. a continous waveform is generated without the application of a trigger.

MULTI WAY TELEGRAPH SYSTEM : See MULTIPLE WAY SYSTEM.

MULTI WIRE AERIAL : An Aerial in which several parallel wires, fixed a short distance apart, are used to obtain greater capacitance than to obtain greater capacitance than would be produced by a single wire,e.g.a CAGE AERIAL.

MUMETAL ("μ" METAL) : A magnetic alloy of specially high permeability and low hysteresis, used in loaded submarine cables, for moving iron instruments and other purposes, consisting of about 75 per cent nickel, 25 per cent iron, and small quantities of copper and manganese.

MURRAY PRINTING TELEGRAPH SYSTEM : A type printing telegraphy system in which the message is first punched in a special code on a paper strip which is put through a transmitting instrument similar to a Wheatstone transmitter. The received signal currents actuate a punching mechanism which multiplex system so as to actuate several instruments printing different messages simultaneously, by synchronous revolving selectors reproduces the original strip, and this new strip is passed through the printing instrument where a mechanical selecting devicecauses the correct letter to be printed. The system is adapted to a very high speed working, and can be arranged to work on multiplex system so as to actuate several instruments printing different messages simultaneously by synchronus revolving selectors CF. BAUDOT CREED, HUGHES, and STELJES PRINTING TELEGRAPH SYSTEM .

MUSH : A term used in wireless *telegraphy* and *telephony* for irregular disturbing radiation, due to various causes, and, in particular, produced by arc transmitters, causing a rushing sound in receiving telephones.

MUSICAL INSTRUCTION DIGITIAL INTERFACE (MIDI) : A standard specification for connection sound-producing equipment, such as synthesizers, to computer.

MUSICAL SPARK : A spark discharge of a regular frequency within the limits of audition; used in some systems of radio transmission to produce a singing note of contant pitch in the receiving telephone; also called Singing Spark.

MUTATOR : A term proposed to cover all applications of a grid-controlled mercury vapour discharge apparatus whether used as a Rectifier,Inverter, Frequency-Changer, or Phase Changer.

MUTUAL A.C. CONDUCTANCE (of a Thermionic Valve) : The ratio of a small change of anode current to the corresponding change in grid voltage, all other quantities remaining constant; also called slope.

MUTUAL IMPEDANCE : A property of two neighbouring circuit analogous to Mutual Inducatance which determines the effect of the rate of change of current in one on the rate of change of current in the other taking into account all other contributory components as well as Inducatance.

MVT : Multiprogramming with a Variable Number of Tasks

MYRIA-HERTZ : A unit of frequency equal to 10,000 hertz (i.e. 10 kilo. cycles per sec). Suggested by the Union Internationale de Radiophonie, as the minimum difference of frequency for broadcasting stations

N

NAND and NAND GATE : A Nand Gate to which has been added an inverted output.

NAND GATE : A gate that is enabled when both its *inputs* are present or *high*. When a *nand* gate is enabled, its output is *low*. The term *nand* is a contraction of the words, and not.

NANO-SECOND : A unit of measurement representing one-billionth of a second.

NAPIER : The transmission unit now usually known as the Neper.

NASA : National Aeronautics and Space Agency

NATURAL BCD : A binary-coded decimal system using weighted values of 8 : 4 : 2 : 1

NATURAL FREQUENCY : The frequency at which free oscillations occur in an electrical or mechanical system. It is the frequency at which resonance occurs in such a system in response to a periodic driving force. (1) The frequency of natural oscillations in a circuit, equal, when the effect of resistance is negligible, to $1/2\pi\sqrt{(LC)}$, when L= the inductance and C-the capacitance of the circuit. (2) The fundamental frequency of an aerial without the addition of extra inductance or capacitance.

NATURAL LANGUAGE INTERFACE (NLI) : A program used as a front-end to an application program, such as a database manager or expert system, that enables the user to communicate with the application in a subset of a natural language like English instead of the special language and commands of the software.

NATURAL LANGUAGE PROCESSING (NLP) : The problem of constructing internal representations of input in a human language like English or a subset of it, to convert information from a language convenient for the user into one convenient for the computer system.

NATURAL OSCILLATIONS : Oscillations in a circuit at a frequency entirely dependent on the electrical constants of the circuit. Cf. FORCED OSCILLATIONS.

NATURAL RECTIFIER : A mineral or chemical substance having a much greater resistance to currents in one direction than to those in the other, e.g. certain crystals used as detectors in radio-reception.

NBCD : Natural Binary-Coded Decimal

NBCH : Natural Binary-Coded Hexadecimal

NBS : National Bureau of Standards (USA)

NCC : Network Control Center or National Computer Conference (USA)

NCHS : National Center for Health Statistics (USA)

NCR : National Cash Register (USA)

NDAC : Not data accepted, handshake bus line.

NDRO : Non-Destructive Read Out

NEEDLE INSTRUMENT : A telegraph receiving instrument giving visual signals by the deflection of one or more pivoted needles. See SINGLE NEEDLE and DOUBLE NEEDLE INSTRUMENTS.

NEGATIVE ATTENUATION : A term used in connection with the strength of radio signals for the increase in the product of amplitude and distance with increasing distance which sometimes occurs within 10 miles or so of a transmitting station.

NEGATIVE AUTOMATIC VOLUME CONTROL : A similar arrangement to Automatic Volume Control but acting in the opposite direction, i.e. to increase the amplification of loud signals used in radiogramophones to increase contrast when playing from records.

NEGATIVE FEEDBACK : An output-to-input signal path where the output signal is 180 degrees out of phase with the input, leading to a reduction in circuit gain and distortion. Also called inverse feedback, stabilized feedback, or degeneration.

NEGATIVE LOGIC : A logic system in which low levels are considered 1's and high levels 0's.

NEGATIVE MODULATION (In Television) : Modulation of the carrier wave in a manner whereby the zero value corresponds to the full white of the picture and 70 per cent modulation being available for synchronising signals. Cf. POSITIVE MODULATION.

NEGATIVE PHOTORESIST : See PHOTORESIST.

NEGATIVE RESISTANCE : A property of certain devices whereby a portion of the current-voltage characteristic has a negative slope, i.e. the current decreases with increasing applied voltage. Devices that exhibit negaive resistance include the *thyristor*, the *tunnel diode,* and the *magnetron.* A term sometimes applied to the state of affairs in an arc, vacuum tube, or other apparatus when a rise of e.m.f. produces a fall in current.

NEGATIVE RETROACTION : The effect of coupling between valve circuits in a direction tending to reduce the amplitude of the output, and to supress self-oscillation.

NEGATIVE WIRE In Telephony : The wire forming that side of a telephone line, within the exchange, normally connected to the negative side of the battery.

NEGATRON : (1)A Negative Electron. (2) A form of Thermionic Valve with four electrodes. So called because it has the effect of a negative resistance.

NEON TUBE RECTIFIER : A rectifier consisting of a tube containing neon at a low pressure with special shaped electrodes not independently heated, in some cases provided with grid control.

NEPER : A unit used principally on the European Continent in the comparison of currents. The relation between current on the neper scale is expressed as the naperian logarithm of this ratio. On neper is equivalent to 0.8686 Bel.

NESTED PROGRAMS (or subroutines) : A program or subroutine that is incorporated into a larger routine to permit ready execution or access or eachy level of the routine. For example, nesting loop involves incorporating one loop of instructions into another loop.

NET LOSS : The difference between the attenuation and the gain in any circuit, device, network, or transmission line.

NETWORK : See GRAPH. A collection of computers connected through cables. Networks are often used to share equipment, such as laser printers or hard disks. A combination of communications lines tying various locations together. A number of impedances connected together to form a system that consists of a set of inter-related circuits and that performs specific functions. The behaviour of the network depends on the values of the components, such as the resistances, capacitances and inductances, from which it is formed and the manner in which they are interconnected.

NETWORK (Electrical) : (1) A Distributing Network. (2) A combination of inductance, capacitance, and resistance usually arranged in a formation of closed cells for the purpose of acting as a Frequency Filter or otherwise modifying the constants of an electrical circuit. Networks are described as resistive, resistance-capacitance (R-C), inductance-capacitance (L-C), inductance (L) networks, etc., depending on their components.

NEURAL NETWORK : In biology is a interconnected web of nerve cells; in AI it denotes a parallel distributed processing network modelled closely on the property of real neurons.

NEUTRAL POINT : Of a supply system. A point normally at earth potential when the branches of the system are symmetrically loaded and often connected to earth.

NEUTRALISING CONDENSER : A condenser used to compensate for the effect of valve and stray capacitance and to produce negative retroaction, in order to secure freedom from oscillation.

NEUTRODYNE RECEPTION : A form of radio-reception with high-frequency amplication in which oscillation is prevented by the compensation of interelectrode capacitance.

NEUTROSONIC RECEPTION : A system of radio-reception incorporating the principles of *Superheterodyne Reception* with neutralisation of the intermediate frequency stages as in Neutrodyne Reception.

NIBBLE : A portion of a byte, usually four bits.

NIC : Network Information Center (USA)

NICHROME : An alloy containing nickel and about 20 percent of chromium, used for heating elements, which can remain at a bright red heat for extended periods.

NICKEL CADMIUM ACCUMULATOR : An Alkaline Accumulator with nickel hydroxide and cadmium oxide for the positive and negative active materials respectively.

NICORE : A material similar to that described under Dust Core but with nickel as the magnetic material.

NIT : Numerical indicator tube; a gas-filled, cold-cathode display tube with a common anode and 10 individual cathodes, formed in the shape of the numerals 0 to 9.

NIT DRIVER : Generally a combined circuit consisting of a BCD-to-decimal decoder a low to high-level interface.

NLM : National Library of Medicine. A curve exhibiting the relation between excitation and voltage, or any other two quantities relating to a machine or apparatus, under No-Load Conditions. Also called open circuit characteristic.

NOCTOVISION : A process similar to Television, employing invisible infra-red rays in the transmitting apparatus, forming the equivalent of "seeing in the dark."

NODE : (1) Any point, line, or surface in a distributed field at which some specified variable of a standing wave, such as voltage or current, attains a minimum value, usually zero. A partial node has a non zero minimum. A point at which maximum magnitude is attained is an *antinode*.

NODES : The zero points of a series of Stationary Waves or oscillations in a conductor. Cf. ANTINODES.

NODON VALVE : An Electrolytic Rectifier with a cathode consisting of an aluminium rod, an anode formed by a lead containing vessel and an electrolyte of ammonium phosphate. A high current density is employed at the cathode.

NOISE ANALYSIS : Determination of the frequencies of the constituent harmonics of a telephone Noise.

NOISE IMMUNITY : The amount of noise a system can receive without being amplified beyond unity gain.

NOISE (In Telephone Engineering) : All disturbing sounds audible in a receiver other than the speech required to be transmitted.

NOISE MARGIN : The voltage difference between the minimum voltage input guaranteed to be a logical 1 and minimum voltage output guaranteed to be a. logical 1 ; also, the voltage difference between the maximum voltage input guaranteed to be a logical 0 and the maximum logical 0 and the maximum logical 0 voltage output. Also called immunity; a critical IC parameter. It is the difference between the normal operating logic levels and the threshold voltage.

NOISE MEASUREMENT : The quantitative estimation of telephone Noise in arbitrary units, by means of a "Noise Measuring Set," in which a comparison is made by means of a potentiometer arrangement with a standard Noise Generator.

NO LOAD CONDITIONS : The conditions when a machine, transformer, etc., is giving its voltage but no current is being taken from it other than that required to make up losses, or when a motor is running but not giving useful power.

NO OPERATION : An instruction that causes the automatic computer to take no action except to go to the next instruction.

NON--LINEAR DISTORTION : A distortion that is produced in a system when the instaneous transmission properties depend on the magnitude of the input. Amplitude, harmonic, and intermodulation distortion are all results of nonlinear distortion. Distortion of received signals in which the alterations of the quantities in question are not independent of the amplitudes, including the introduction of parasitic frequencies by apparatus in the circuit.

NON--LINEAR DISTORTION FACTOR : A measure of the degree of Nonlinear Distortion in a telephone circuit, etc., being the ratio of the total power in the circuit due to frequencies other than present in the parent tone to the total power in the circuit.

NON--MONOTONIC LOGIC : Is any system of logic in which the set of axioms and theorems is permitted to decrease in size as time goes on, and is one way to attack the frame problem.

NON--PROCEDURAL LANGUAGES : Languages in which the user tell the computer what to do rather than exactly to do the dask. Statement are more declarative of what is to happen than specific about the procedure to produce the desired results.

NON--PROGRAMMED DÈCISIONS : Decisions that are unstructed and for which an algorithm for solution cannot be specified.

NON--TERMINAL SYMBOL : In a grammer does not appear in the sentences of the language generated by the grammer, but rather represents a category or legal combination of symbols in the rewrite rules of the grammer, such as NP for "noun phrase" in English.

NON VOLATILE MEMORY : The memory component of a computer, of which all data gets erased when power supply to computer is cutoff. The RAM is of this type. See RANDOM ACCESS MEMORY.

NOR GATE : An OR gate to which has been added an inverter at its output. A combination of a NOT and an OR circuit. A binary circuit having two or more inputs and a single output, in which the output is off (0) if any one of the inputs is on (1), and is on (1) only if all inputs are off together.

NORMAL COUPLING : The coupling of circuits where the capacitance and inductive coupling effects act in the same direction. See ORTHODYNE CIRCUIT.

NORMAL RATING : The definition of the output which a machine, etc., is designed to give or the voltage, current, etc., which it is designed to deal with under ordinary working conditions. See CONTINUOUS RATING.

NORTON'S THEOREM : In a linear network, the current through an impedance connected across a pair of terminals A and B is the same as if the impedance were connected to an ideal current source, equal to the short circuit current between the terminals A and B, and, in parallel with an output impedance, equal to the network impedance between A and B with all voltage sources replaced by their internal impedances.

NOTCH FILTER : A filter network capable of rejecting a single frequency or frequency band within a range of possible input frequencies.

NOT CIRCUIT : A binary circuit having a single input and a single output in which the *output* is always the opposite of the *input*. When the input is on (1) the output is off (0) and vice versa. This circuit is also called an *inverter circuit*.

NOT CONNECTIVE : The property of negating a true function.

NOTE AMPLIFIER OR NOTE MAGNIFIER : An audio-frequency Amplifier applicable, strictly speaking, only to radio-telegraphy but sometimes used for any low-frequency amplifier, including a speech amplifier in radio-telephony.

NOTE TUNING : Tuning a circuit to respond to an audio-frequency.

NOVACHORD : An electrical musical instrument of the organ type in which a set of twelve oscillating circuits is employed to obtain the frequencies of the highest octave, those in the lower octaves being produced through Frequency Dividers.

N-P-N TRANSISTOR : A semiconductor device where the *collector* and *emitter* are made from *n*-type semiconductor material while the base is made from *p*-type material. See TRANSISTOR.

NRFD : Not Ready for Data; handshake bus line.

NRZ : Non Return to Zero.

NRZI : Non Return to Zero Inverted.

NULL : Empty or having no members. This is in contrast to a black or zero, which indicates the presence of no information. For example, in the number 540, zero contains needed information.

NULL MODEM : A device that mimics a modem, used for connecting two computers together.

NUMERIC : A reference to numerals as opposed to letters or other symbols.

NYQUIST DIAGRAM : A diagram that can be used as a criterion of stability in an amplifier. It shows the relation, in rectangular coordinates, between the amplification and the feedback of the device.

O

OAO : Orbiting Astronomical Observatory. (USA)

OBJECT LANGUAGE : The output of a translator, usually machine language.

OBJECT ORIENTED PROGRAMMING (OOP) : Is a metaphor for computation, often used in AI systems, that considers programs as collections of semiautonomous objects that get work done by sending messages back and forth, executing instructions received from other objects, and returning results.

OBJECT PROGRAM : A program that is in machine executable form. This means that the program is most likely to be in some sort of a binary pattern, ready to be used by the computer.

OCR : Optical Character Recognition

OCR (Optical Character Recognition) : The machine recognition of certain type styles and/or printed and handwritten characters.

OCTAL : A number system which has 8 distinct digits 0,1,2,3,4,5,6,7.

OCTAL NUMBER SYSTEM : A representation of values or quantities with octal numbers. The octal number system uses eight digits: 0,1,2,3,4,5,6, and 7, with each position in on octal numeral representing a power of 8. The octal system is used in computing as a simple means of expressing binary quantities. A number system having eight digits, 0 through 7.

OCTAVE : The interval between two frequencies having a ratio of 2:1.

OCTODE : A Thermionic Valve with eight electrodes, usually for combining the functions of two valves, e.g. a six-grid frequency changer in a superheterodyne receiver which combines an. oscillator with a Variable-Mu Pentode mixer valve.

OCTULE PHANTOM CIRCUIT : A superposed circuit each side of which consists of the sixteen conductors of a Quadruple Phantom Circuit in parallel.

ODOGRAPH : An instrument that produces a graph of an altermating voltage. It has a digital response and hence the graph of the waveform appears as a series of small steps rather than as a smooth curve.

OEM : Original Equipment Manufacturer

OERSTED : A name originally proposed for the unit of Reluctance, but now adopted for the unit of Magnetising or magnetic Force; named after the Danish Physicist, H.C. Oersted (1777-1851).

OFF-LINE : Describes any operations that is not directly controlled by the CPU. Refers to computer equipment which is not at the time directly connected to the computer.

OFFICE AUTOMATION : The use of technology like word processing, electronic mail, and similar systems to improve the productivity of knowledge workers.

OFFSET CURRENT : In a DC amplifier, such as an operational amplifier, the difference in the current flowing in the two input leads.

OHM : The Practical Unit of Resistance, named after the German physic-ist, G.S. Ohm (1787-1854). The true Ohm is 10 C.G.S. absolute el-ectromagnetic units of resistance. See also INTERNATIONAL OHM

OHMIC DROP : The potential drop between two points in a current carrying conductor, due to its resistance (IR Drop) as distinguished from drop due to other causes such as Inductance.

OHMMETER : An instrument that measures the electrical *resistance of conductors or insulators. The indicating scale is calibrated in ohms or suitable multiples or submultiples of ohms conductor under test and the potential drop across it. Used in conjunction with a magneto generator giving a few hundred volts for testing insulation resistance (see MEGGER) and in some forms of Resistance Thermometer. An instrument for measuring resistance by direct deflection, usually one in which the moving system is deflected by forces due to currents in two coils at right angles to one another, carrying currents proportional respectively to the current through the

OHM'S LAW : The law of the flow of unvarying currents, whereby the strength of the current varies directly as the electromotive force and inversely as the resistance of the circuit; usually expressed in symbols as $I=E/R$ where I is the current (in amperes), E is the electromotive force (in volts) and R is the resistance (in ohms).

OMNI-AERIAL : An aerial radiating at equal strength in all directions in the same angle with the horizontal.

OMNI DIRECTION RADIO BEACON : A radio beacon radiating its characteristic signals equally in all directions.

OMNIGRAPH : An apparatus used in teaching the Morse code, which consists of a disc with contacts so arranged that when rotated it produces Morse signals in a buzzer circuit to which it is connected.

OMR : Optical Mark Reader

ONDAMETER : See WAVE METER.

ONDOGRAP HOSPITALIER : An instrument for drawing alternating wave-form curves by a step-by-step method. A commutator driven through gearing by a synchronous motor connects a condenser momentarily to the circuit once in every hundred waves at successive points a little further along the wave each time. The condenser is discharged each time into a recording galvanometer which traces the required curve ona drum.

ONDOSCOPE : A glow-discharge tube that detects the presence of high-frequency radiation. A glow discharge tube used as an indicator of electric waves. See NEON TUBE.

ONE'S COMPLEMENT : A system of binary arithmetic in which a negative number is expressed as the complement of its postive equivalent. Arithmetic which provides a method of negating a binary number so that binary can be performed using addition techniques. To obtain the 1's complement of a binary number, all bits in that number must be complemented (i.e., is changed into 0s and vice versa).

ONESHOT : A monostable multivibrator.

ONE VALVE RECEIVER : See VALVE RECEIVER.

ONE WAVE RECTIFICATION : See HALF-WAVE RECTIFICATION.

ONE-LINE HELP : Help that you can request while using a program.

ON-LINE : Describes the condition of a piece of computer equipment which is directly connected to the computer.

ON-LINE SYSTEM : A system that has the capability to provide direct communication between the computer and remote terminals; files are updated immediately as data are entered. An electronic system when a printer is connected to a computer and turned on, or when a computer is connected to another computer through a modem and a telephone line to enter data or work on a program while it is connected to a computer that is running.

ON-LINE UPDATING : Pertaining to a system in which the data entered are used to update the files immediately.

OP : Operation

OPEN : To load a file and make it usable by program.

OPEN AERIAL : An aerial not forming a closed conducting circuit (as in a FrameAerial), but in which the circuit is completed by its distributed capacitance.

OPEN CIRCUIT : (1) A system of conductors, intended to carry a current, temporarily interrupted so that no current can flow. A machine, transformer, et.c, is said to be on open circuit when giving its normal voltage but no current to the circuit which it is arranged to supply.

OPEN-CIRCUIT IMPEDANCE : See network.

OPEN CIRCUIT SYSTEM : A system of electric signalling, fire alarms, etc.,in which the circuit is normally open when no signal is being made and the signals are made by passing currents when required. Cf. CLOSED CIRCUIT SYSTEM.

OPEN LOOP GAIN : The voltage gain of an amplifier connected to a load without any feedback path between its output and its input.

OPERAND : A quantity or data item involved in an operation. An operand is usually designated by the address portion of an instruction, but it may also be a result, a parameter, or an indication of the name or location of the text instruction to be executed. One of the numbers upon which the computer operates. The object or "target" of some program operation, usually a number or variable that is involved in some arithmetic operation.

OPERATING POINT : The point on the family of characteristic curves of an active electronic device, such as a transistor, that represents the magnitudes of voltage and current when designated operating conditions are applied to the device.

OPERATING SYSTEM : An organized group of computer instructions that manage the overall operation of the computer. A program that tells you computer how to work, Common operating systems include MS-DOS,PC DOS, and OS/2. A program that organizes the housekeeping details for the computer or a group of special programs that are used in various combinations to make a computers easier to use.

OPERATION CODE : Sometimes referred to as "Op-çode" this is the electronic binary pattern that directs the computer circuit to perform some particular operation. The Op-code for an ADD instruction is 11000101."

OPERATIONAL AMPLIFIER : An amplifier with an assumed infinite gain, infinite input impedance, and zero output impedance.

OPERATIONAL CONTROL DECISIONS : Day-to-day decisions concerned with the continuing operations of a company, such as inventory management.

OPERATIONALIZATION : Is the process of implementing in the performance element improvement discovered by the learning element in an AI system; it can sometimes be circumvented by the single-representation trick, which ensures that the knowledge used by the PE and LE is in a consistent, interchangeable format.

OPERATOR : A symbol indicating an operation and itself the subject of the operation. It indicates the process that is being performed. For example, + is addition, - is subtraction, x is multiplication, and / is division. Operators are procedures, modelled as mathematical functions, that can be applied to problem states in state-spaces to transform them into other legal states; solving a problem consists of applying a sequence of operators to transform the start into a goal state.

OPS5 : A most popular language for rule-based programming with production systems; it uses forward chaining control with a fast conflict-resolution algorithm.

OPTICAL AMMETER : An instrument that measures the current flowing through the filament of an incandescent lamp by comparing, photometrically, the illumination produced with that produced by a current of known magnitude in the same filament.

OPTICAL DISTORTION : A distortion that in an image is seen in electronic systems, such as cathode-ray tubes, television picture tubes, etc., and in facsimile transmission. It is due to errors in the electron-lens focusing systems.

OPTICAL PYROMETER : An instrument for ascertaining furnace temperatures, etc., by observing the glowing portion through a tube and filament heated electrically. The brightness of the filament is adjusted till a balance is obtained, and the temperature is deduced from the reading of an ammeter in the filament circuit. (Also called Disappearing Filament Pyrometer or Monochromatic Pyrometer.) See RADIATION PYROMEMTER.

OPTICAL STORAGE : Storage devices using laser optical disks, characterized by extremely high densities of data.

OPTICAL TELEPHONY : A form of telephony without line wires in which a beam of light (or ultra-violet or infra-red rays) is modulated by the transmitter, and falling on the distant receiver, affects a photoelectric cell so that the modulations are reproduced in the current in the receiving telephone.

OPTION : An add-on device that expands a system's capabilities.

OPTOPHONE : An instrument for enabling the blind to read ordinary print by means of a telephone receiver controlled by the variations in resistance of a Selenium Cell. The letters are intermittently illuminated in bands, each with a different frequency of illumination, and arrangements are made so that as black part of a letter passes through one of these bands as the "eye" of the instrument travels over it, the balance between the resistance of two selenuim cells is upset and a current of a frequency equal to that of the illumination of the band in question passes through the telephone receiver. Thus a definite signal consisting of a combination of consecutive and simultaneous sounds of different pitch (chords) is heard, from which, after a little training, the form of the latter is easily recognised.

OR CONNECTIVE : A condition in which any true input results in a true output.

OR GATE : A binary circuit having two or more inputs and a single output, in which the output is on (1) if any one of the inputs is on (1) and is off (0) only if all inputs are off (0) together.

ORDER : A sequence or an instruction or command.

ORDVAC : ORDnance Variable Automatic Computer

ORIGIN : The absolute address used as referred point for relative addresses.

ORIGIN DISTORTION : A distortion due to result of flattening of the waveform where it crosses the zero line in gas-focused cathode-ray tube employing electrostatic deflection. It occurs at low electing voltage when there is a nonlinear relationship between the angular deflection and the reflecting voltage.

ORSA : Operation Research Society of America.

ORTHICON CAMERA : An improved form of television camera similar in principle to the Emitron Camera but employing low-velocity electrons for scanning.

ORTHODYNE CIRCUIT : A circuit for a Thermionic oscillator in which Normal Coupling is employed between the *plate* and *grid* circuits. Cf. ANTIDYNE and RHEODYNE CIRCUITS.

OS : Operating System

OSCILLATING COMPONENT : A term used for an oscillating current when superposed upon a direct current as in the anode current in a radio receiver

OSCILLATING CURRENT (or voltage) : A current (or voltage) waveform that periodically increases and decreases in amplitude with respect to time, according to a particular mathematical function. Oscillating wave-forms can be *sinusoidal, sawtooth, square,* or many other shapes. A high frequency current, the frequency of which is entirely dependent on the constants of the circuit itself.

OSCILLATING ELECTRODE : The coiled wire intermediate electrode in a microray oscillator corresponding in construction but not in function to the grid in an ordinary triode. Cf. REFLECTING ELECTRODE.

OSCILLATING METER : A meter in which the moving system executes in oscillatory motion instead of the rotary motion usual in a motormeter;either the speed or the amplitude of the oscillation being proportional to the load

OSCILLATION : The setting up or existence of continuous Free Oscillations in an Oscillatory Circuits.

OSCILLATION CONSTANT (of a cymometer, also called resonance Constant) : The property of an oscillating circuit upon which its natural frequency depends, sometimes defined as the square root of the product of the *inductance* and the *capacitance,* and sometimes as the square of this quantity. This is read directly on the scale of cymometer. Also called "RESONANCE CONSTANT"

OSCILLATION VALVE : Any parrartus depending on electrolytic or ionic action which is conductive in one direction only, such as the original form of Thermionic Valve discovered by Fleming in 1887, consisting of a hot cathode in the form of a lamp filament from which a stream of electrons proceeds towards the anode in the form of a metallic cylinder or plate in the same bulb. Such an apparatus will conduct in the direction from the anode to the cathode only, and on can be used as a Detector in radio reception. See also LODGE VALVE, ELECTROLYTIC RECTIFIER.

OSCILLATOR : (1) A self-excited active circuit whose output voltage is a periodic function of time. (2) An amplifier with positive feedback. A system of conductors suitably arranged to have the properties of an oscillating circuit, usually in conjunction with some form of apparatus, such as a sparkgap, arc, or thermionic valve capable of exciting or maintaining such oscillations.

OSCILLATORY CIRCUIT : A circuit in which the relations of the inductance and capacitance are such that it has a natural period for which free electrical oscillations are possible, i.e. where $4L$ is greater than R^2C (where L is the inductance in henries, R is the resistance in ohms, and C is the capacitance in farads).

OSCILLATORY DISCHARGE : A sudden electric discharge, such as that of a condenser is not always unidirectional, but if the inductance of the circuit into which it is discharged is greater than $R^2 C/4$ (where R is the resistance in ohms, and C is the capacitance in Farads) the discharge executes rapid oscillations, i.e surges to and fro at a frequency of $1/2\pi \sqrt{(CL)}$, decreasing in amplitude until the whole energy of the discharge is dissipated by heat and radiation.

OSCILLATORY IMPEDANCE : The impedance of a conductor at high frequencies where, owing to the Skin Effect the resistance is higher than at low frequencies.

OSCILLION : A variety of Thermionic Valve similar to the Audion but used for the production of oscillations. See THERMIONIC OSCILLATOR.

OSCILLOGRAM : A record of wave form made photographically or otherwise by an oscillograph.

OSCILLOGRAM DUST : See DUST OSCILLOGRAM.

OSCILLOGRAPH : A recorder, generally using light-sensitive paper, that produces a plot of a time-varying waveform. An instrument for recording photographically or otherwise the wave form of alternating or other rapidly changing currents, voltages, etc., e.g. by the action of a beam reflected by a mirror attached to a moving system subjected to forces proportional to the instantaneous value of the quantity to be measured, and suffciently light and well damped to follow the rapid vibratory movement with accuracy. See also CATHODE RAY OSCILLOGRAPH.

OSCILLOSCOPE : An instrument using a cathode-ray tube to display a signal's variation with time. Also called a scope. An apparatus for rendering visible the wave form of electrical oscillations or alternating currents; e.g. the Gehrcke Oscilloscope, consisting of a tube containing nitrogen at a low pressure,with two aluminium electrodes nearly touching, between which a visible discharge takes place extending a distance up the electrodes proportional to the voltage. The wave-form is seen by viewing this in a rotating mirror. The name is also used for an instrument for indicating by the length of the negative glow in an auxiliary discharge tube, the amount of current passing in an"X" Ray Tube.

"O"- SECTION NETWORK : See QUADRIPOLE

OSISO : Trade name of a simple form of oscillograph for use as a Phonos-cope.

OSMO REGULATOR : An appliance used to regulate the vacuum in X-Ray tubes, consisting of a small mass of platinum, palladium, etc.,in a branch ofthe tube which, when heated, gives off occlued hydrogen and lowers the vaccum when it has become too high after continued use.

OSOPHONE : A telephone receiver designed for the use of the particulerly deaf by applying the vibrations direct to the bones of the head. Computer results, or data that has been processed.

OUTPUT : (1) The power (or Current) which is given out by a generator, motor,or other machine or apparatus; usually expressed in kilowatts or,in the case of a motor, in horse-power, and equal to the input minus the losses. (2) The information communication by an automatic computer, usually as a result of or representing the results of data processing. (3) The processing of doing this peres.

OUTPUT TRANSFORMER : A transformer used for coupling either the *plate* or *collector* circuit of an amplifier to a speaker or other load. A transformer used to couple a telephone receiver or loud speaker to a radio-receiver.

OUTPUT VALVE : A Thermionic Valve suitable for use in the last audio-frequency amplifying stage of a radio-receiver,i.e. for the immediate supply of the loud speaker either direct or through a transformer.

OVERALL EQUIVALENT (of a Telephone Circuit) : The actual attenuation as measured between terminals of a telephone circuit containing a repeater.

OVERCURRENT RELEASE (Also Known As Overload Release). A switch, circuit-breaker, or other tripping device that operates when the current in a circuit exceeds a predetermined value. A current that causes the release to operate is an overcurrent.

OVERFLOW : A quantity exceeding the maximum number a register can contain. In a counter or register, the production of a word of field which has more admissible marks than the capacity of the counter or register for example, in adding two numbers each 10 character,the result may be a sum of 11 characters long, which is beyond the capacity of the register. The character lost may be either the most of least significant of the number. This occurs when the calculations performed produce numbers that are too larger for the storage unit to handle. The result of performing an arithmetic operation within the computer which yields an answer that is too large tobe contained in the MPU.

OVERLAP : The ability of CPU to continue processing while input/output operations are under way. A term sometimes used in connection with radio-receiving sets for the extent to which a control dial, etc., has to be turned back from the point where its forward movement has just caused self-oscillation to a point where this just ceases without any other measures being taken. A segment of program read into storage on top of and hence obliterating other parts of the same program. The use of overlay is a technique for conversing on the use of storage, and usually is limited to rarely used parts of a program.

OVER-MODULATION In Radio Telephony : Excessive amplitude of the modulations in proportion to that of the carrier-wave, apt to produce blasting in receiving apparatus.

OVERSHOOT : See PULSE

OWEN BRIDGE : A four-arm bridge used for the measurement of inductance in terms of known resistance and capacitance.

OXIDATION ANODE : See ANODIC OXIDATION.

P

PACKING OF A MICROPHONE : The setting of the carbon granules into tightly packed groups, which do not easily respond to the vibrations of the diaphragm, but can be freed by tapping the instrument gently.

PAGEMAKER : A word processing cum page making software that provides all facilities for data entry, editing, styling and page formating of lengthy documents. Compare VENTURA PUBLISHER

PAGINATION : The numbering or ordering of pages.

PAGING : The segmentation of storage into small units that are moved automatically by hardware or software between primary and secondary storage to give the programmer a virtual memory that is larger than primary memory.

PAIRS : Large telephone cables are made up of the pairs of wires each belonging to one circuit, twisted together with the necessary insulation between them. Thus a cable of 100 pairs contains wires for 100 metallic circuits.

PALETTE : A selection of styles and colors available in graphics programs.

PALLO-PHOTOPHONE : A form of Photophone which has been used instead of a microphone in radio-telephony, employing a Photo-electric Cell.

PANEL : See PLUGBOARD.

PANEL AUTOMATIC TELEPHONE SYSTEM : A Rotary automatic telephone system use in the United States employing banks of multiple contacts arranged on flat upright panels over which the mechanically driven contact brushes move vertically; other features are the employment of a Call Indicator and a system of a Revertive Control. See also SENDER and SEQUENCE SWITCH.

PAPER CAPACITOR : A capacitor of medium loss and medium capacitance-stability that is used in high-voltage a.c. and d.c. applications. These capacitors are manufactured by winding together aluminium foils inter-leaved with layers of tissue paper. The moisture content the paper is removed by impregnation with a suitable oil or wax. Metallized paper capacitors use an *evaporated metal film* as the electrode instead of *aluminium*.

PAPER SIZE : Refers to actual size of paper. Standard Sizes include letter (8.5 x 11 inches), legal (8.5 x 14 inches), European A-4(8.27 x 11.69 inches) and European B5 (6.93 x 9.84 inches).

PAPER TRAY : A device that holds paper, feeds paper onto the printer, or holds paper after it emerges from the printer.

PARADOX : Paradox has been heralded as powerful, easy-to-use relational database that lets beginners create complicated databases programs without prior programming experience.

PARAGUTTA : An insulating material for submarine telephone cables composed of *balata* rubber and deproteinised rubber with a small quantity of wax.

PARALLEL : Of the presence of flow of information to any part of an automatic computer in multiple lines or channels simultaneously.

PARALLEL COUNTER : A syhchronous counting circuit.

PARALLEL D/A CONVERSION A D/A : Converter in which all bits are converted simultaneously according to their weights.

PARALLEL DATA : Data in which each bit is represented by single line; thus 16 bits require 16 lines.

PARALLEL DISTRIBUTED PROCESSING (PDP) : See connectionism.

PARALLEL OUTPUT : The method by which all bits of a binary word are transmitted simultaneously.

PARALLEL POLLING : A method of simultaneously checking status (e.g. request for service) on up to eight devices on the IEC bus at the same time.

PARALLEL PORT : A connector usually used to attach a printer to the computer. Often called a printer port or a parallel printer port for those who like redundancy.

PARALLEL TESTING : The testing of a new system at the same time an existing system is in operation. The results of a new system are compared.

PARALLEL TO SERIAL CONVERTER : A device that converts parallel binary data to serial data.

PARAMAGNETISM : An effect observed in certain materials that possess a permanent atomic or molecular magnetic moment.

PARAMETER : A variable that is given a value for a specific program or run. A definable characteristic of an item, device, or system. A quantity that is constant in a given case but has a particular value for each different case considered. Examples include the values of the resistances, capacitances, etc., that form an electrical network or the constants appearing in the eciuations connecting currents and voltages at the terminals of network. See also transistor parameters.

PARAPHASE AMPLIFIER : An amplifier having two output signals which are 180 degrees out of phase with each other, often used to drive a push-pull amplifier stage.

PRASITIC : An unwanted signal or oscillation.

PRASITIC ELEMENT : An antenna element that serves as part of a directional antenna system, but which has no direct connection to the receiver or transmitter. It reflects or re-radiaters electromagnetic energy that reaches it.

PARITY : An extra-bit of code that is used to detect data errors in memory by making the sum of the active but in a data word etiher an odd or an even number. Refers to two things that are similar in some respect-for example, two integers may both be even or both be odd.

PARITY BIT : A bit that is set to a 1 or 0 to make the total number of 1's in a binary word odd or even.

PARITY CHECK : A type of redundent check in which the *evenness* or *oddness* of the number of 1 bits is certified.

PARITY CHECKING : A method for detecting errors during file transmission through modems.

PARSE : Separation of an input string of symbols into its basic components.

PARSING : Is the lowest level of natural language processing, and is concerned with deciding whether and low a sentence is a member of a language by examining its synatctic structure.

PARTITION : An area on a fixed disk set aside for a specific purpose, such as a location for an operating system. A segement of a hard disk, usually used to divided a large hard disk into several smaller ones for faster access and organization.

PART-PROGRAMMING LANGUAGE : A group of symbols, codes,

format, and syntax (grammar) definitions which describe machining operations that are understandable to computers or control.

PARTY LINE : A telephone line serving a number of subscribers whose instruments are connected to parallel to it. Special arrangements are adopted for ringing up the individual stations on such a line.

PARTY LINE RINGING KEY : A key on a telephone exchange switchboard which gives automatically the series of current impulses required to call a particular subscriber on a party line.

PASCAL : A programming language designed for teaching programming named after the French philospher, Blaise Pascal. The SI unit of pressure, defined as the pressure that results when a force of one newton acts uniformly over an area of one square metre.

PASCHEN'S LAW : The breakdown voltage that initiates a discharge between electrodes in a gas is a function of the product of pressure and distance. For example, if the distance between the electrodes is doubled breakdown will only occur at the same potential difference if the gas pressure is halved. See also Hittorf's principle.

PASS : One reading of input data by computer, usually as part of a rum.

PASS BAND : The frequency range of a filter network where the *attenuation* is at a minimum. Also called bandpass, or filter passband.

PASSIVE : Denoting any component, device or circuit that does not inrtoduce gain or loss or does not have a directional function. As per this, only pure resistance, capacitance, inductance or any combination of them can be classified as passive Cf. ACTIVE.

PASSIVE MODE : In computer graphics, a mode of operation of a display device that does not allow an on-line user to alter or interact with a display image.

PASS-NETWC⁻⁷ : A pass network is an interconnection of a set of pass tra: istors to achieve a particular switching function.

PASS-TRANSISTOR : A transistor in an *MOS* or *pMOS* transistor used to block or conduct logic signals in MOS circuits.

PATCH : A section of coding inserted in a program in orderto rectify an error in the original coding or change the sequence of operation the basis for shading algorithms.

PATCH CARDS : The wires used on plugboards, as on punched card machines.

PATCHES : A Portions of solid-object surfaces, as calculated and displayed by three-dimensional graphics software. Patch definition often form the basis for shading algorithms in state-space search is a sequence of operators in search tree.

PATH : The route that the computer travels from the root directory to any subdirectories. The path also refers to the subdirectories the MS-DOS examines when you type a command.

PATTERN MATCHING : Takes a specific structure and a more general model structure (the pattern) and determines whether one is a specific instance of the other, and if so, how the variable elements in the pattern are instantiated in the example.

PATTERN RECOGNITION : Identification of visual images by classification into categories. Pattern recognition is usually considered a part of artificial intelligence.

p-BAND : The radio-frequency band from 225 to 390 MHz.

PC : Progarm Counter, Personal Computer.

P.C.B. : See PRINTED CIRCUIT BOARD.

PcFILE : Pcfile originally began as a shareware program but is now available commercially as well. Pcfile can use data stored in dBASE III Plus files and is easier to use.

PCP : Punched Card Punch.

PCR : Punched Card Reader.

PDE : Partial Defferential Equation.

PDP : Programmed Data Processor.

PEAK : The point where a load curve, wave-form curve, etc., reaches a maximum. See LIGHTING PEAK and POWER PEAK.

PEAKFACTOR : The ratio of the Peak Value to the R.M.S. Value, having a value of 2 for pure sine wave-forms. Syn. CREST FACTOR. The ratio of the peak value of a periodically varying quantity to the root-mean-square value. If the quantity varies sinusoidally the peak factor is 2.

PEAK INVERSE VOLTAGE : (1) The peak AC voltage that a semiconductor diode will withstand in the reverse direction. Also called inverse peak voltage. (2) Breakdown voltage.

PEAK LOAD : The magnitude of the load on a generating station or plant at the time of the day when it is a maximum, e.g. in a lighting station, just after dark.

PEAK METER : An additional meter started automatically when the demand exceeds a certain value in systems where a higher rate is charged for peak-load units.

PEAK VALUE : Syn. amplitude; crest value. The maximum positive or negative value of any alternating quantity, such as current or voltage, during a given time interval. The positive and negative values are not necessarily equal in magnitude.

PEAK VALUE (of alternating currents voltages etc.) : The value at the extreme crest or peak of the wave.

PEAK VOLTAGE : The *maximum* or *crest* value of an alternating voltage.

PEAK WAVE FORM : A wave-form curve of alternating current, e.m.f., etc., the maximum of which are more pointed than those of a true Siine Curve. Cf. FLAT-TOPPED WAVE-FORM.

PECN : Pacific Educational Computer Network.

PELTIER EFFECT : The raising or lowering of the temperature of a contact between two dissimilar metals by passsing a current through it by means of an external e.m.f., in the reverse and same direction, respectively, as their own thermoelectric e.m.f. when heated, owing to the evolution or absorption of energy.

PENETRATING RADIATION : See COSMIC RADIATION

PENETRATION : A thermionic value that contains five electordes and is effectively two triodes with a single common cathode, two anodes, and two control grids.

PEN PLOTTER : Device that draws lines on paper or other media to create drawings from a computer. (Also see electronic plotter)

PENTA-GRID CONVERTER : A Thermionic valve provided with five grids, for use in Superheterodyne Receivers to effect complete conversion from high to intermidiate frequency in one valve combining the functions of local oscillator, detector, and amplifier.

PENTATRON : A five electrode valve with two anodes, two grids, and one cathode, rendering the Push-Pull system of amplification possible with a single valve.

PENTODE : A vacuum tube that has five elements: cathode, plate,

control grid, screen grid, and suppressor grid used to prevent secondary electrons from the anode reaching the screen.
The suppressor grid must be of an open mesh design otherwise the passage of the primary electron beam would be
impeded by it. A Thermionic Valve with five electrodes, e.g.
with a cathode, anode, and three grids acting as Control Grid
Screen Grid, and a Suppressor Grid; used as a power amplifier of high amplification factor, especially where there is only
one stage of low frequency amplification. See also PENTA-
TRON.

PENTATRON : A thermionic valve containing *five* electrodes. It is
equivalent to a tetrode containing an additional electrode.
The suppressor grid between the screen and the anode. The
supressor gird is at a negative potential relative to both anode
and screen.

PEPR : Precision Encoding and Pattern Recognition is the process of
receiving continuous stimulation from the environment, in
forms like light for vision or sound for speech understanding,
and deriving its relevant semantic contents by constructing a
conceptual representation of it.

PERCEPTION : A robot's ability to sense its environment by sight,
tough, or some other means, and to understand it in terms of
a task: for example, the ability to recognize an obstruction or
to find a designated object in an arbitrary location.

PERFECT RECTIFIER : A rectifier which allows of no flow of current
in the reverse direction.

PERIDYNE RECEPTION : A system of radio-reception employing an
inductance enclosed in a metal case with a movable screen
which serves as a fine adjustment to the effective inductance
of the coil.

PERIKON DETECTOR : A detector of the rectifier class, consisting
usually of a contact between a point of copper pyrites (bor-
nite), and a mass of crystalline zinc oxide (zincite).

PERIQD : Usually used to signify the time occupied by one complete
Cycle, i.e. including two waves in opposite directions, but occasionally used in older works for half a cycle or one wave
only.

PERIODCITY : The unmber of periods of an alternating current, voltage, etc., making up one second. See also FREQUENCY.

PERIODMODE : The selected function of the counter/timer where it measures the time of one single period.

PERIPHERAL : An external input/output, or storage device.

PERIPHERAL : A device attached to the computer, such as a modem, disk drive, mouse or printer. Input/output equipment of auxiliary storage equipment of a computer system Input/output devices connected to a computer system. Pieces of equipment that are added to the computer to allow communication.

PERMALLOY : Originally an alloy composed of 78.5% nickel and 21.5% iron but now any of a variety of alloys made by the addition of copper, coalt, managnetic permeability at low values of magnetic flux density and by low hysteresis loss. An alloy which, after subjection to certain beat treatment, has a much greater permeability (about thirty times) and much lower magnetising forces; used in the Continuous Loading of submarine cables, and extending the possible use of them very considerably.

PERMANENT MAGNET : A magnetized sample of a ferromagnetic material, such as steel, that possesses high retentivity and is stable against reasonable handling. It requires a definite demagnetizing flux in order to destroy the residual magnetism. A simple magnet consists of a single bar, which can be horseshoe shoed, of the material. A compound magnet has several suitably shaped bars or laminations fastened together. See also *non-volatile memory; magnetic hysteresis*.

PERMATRON : A thermionic rectifier in which control is effected by a magnetic field instead of a variable voltage applied to a grid.

PERMEABILITY TUNING : Tuning of the high-frequency circuits of radio-receivers by moving the cores of the coils to change their inductance, without the employment of variable condensesrs. This is rendered possible by the use of Dust Cores.

PERMITTANCE : A name sometimes used for capacity under the same conditions as described under Permittivity.

PERMITTIVITY : The ratio of the electric displacement produced in a particular medium to the electric force ratio for a vacuum. Also called Dielectric Constant, Dielectric Coefficient, or Inductivity.

PERMUTATION : An ordered arrangement of a given number of different elements selected from a set.

PERSISTENCE : Syn. *afterglow* (1) The time interval after excitation during which a phosphor continues to emit light (see luminescence), particularly the phosphor on the screen of a cathoderay tube. The length of time an image produced on a display device by activated phosphors remains clear, bright, and sharp.

PERSONAL COMPUTER : See *Microcomputer*

PERSONAL INFORMATION MAGNGER (PIM) : A category of software that organizes information you would normally write in a calender scheldule organizer or network.

PERSPECTIVE PROJUCTION : Graphics displays simulating depth and distance by representing parallel lines merging at a vanishing point.

PERT : Program Evaluation and Review Technique.

PETRICK'S METHOD : A method of covering minterms by applying Boolean algebra to select prime implicants in tabular reduction.

PFS FIRST CHOICE PFS : First Choice remains popular for its low cost, simplicity and powerful features, including a *spell checker graphics* and *communication program.*

PFS FIRST PUBLISHER PFS : First Publisher is designed as a simple, low-end program for people who need to creat simple newsletter, brochures or flyers.

PHANOTRON : A metal-cased Mercury Vapour Rectifier with an independently heated catchode to supply the electron emission at starting.

PHANTOM CIRCUIT : An equivalent of an additional telephone circuit formed by the use, in parallel, of pairs of existing wires forming ordinary insulated circuits. The instruments belonging to the lines so used are connected to form a bridge accurately balanced so that the superimposed current of the "phantom" circuit does not affect then. *Cf.* PHYSICAL CIRCUIT.

PHANTOPHONE : A telephone system for speaking over lines used simultaneously for telegraphy, in which the receivers and transmitters are connected to the line through transformers with condensers in series with them, so that they are not effected by the telegraph signals.

PHASE : (1) The position of a particular point in the cycle of an alternating current e.m.f., etc., expressed as an angle (the whole cycle being taken as 360), i.e. the angle made by the vector position to the position corresponding to the moment in the cycle under consideration. (2) Used in practice also to signify one of different portion or Branches of a circuit in which currents are flowing which reach different phase angles at the same time. See SINGLE-PHASE, TWO-PHASE, THREE-PHASE, POLYPHASE, etc.

PHASE ANGLE : The angle between two vectors that represent two sinusoidally varying quantities of the same frequency (see phase). If the two quantities are nonsinusoidal but have the same fundamental frequency, the phase angle is the angle between the two vectors that represent the fundamental components. Waveforms that have a phase-angle of $\pi/2$ are said to be in quardrature. If the angle is π or 180, they are in opposition.

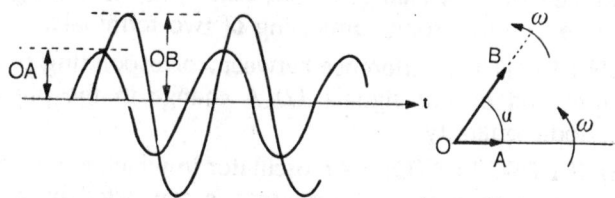

Phase angle between two quantities of the same frequency

PHASE CREEP : A progressive error in phase in the running of machine telegraph apparatus, etc., slightly out of synchronism.

PHASE DIFFERENCE : (1) The difference in phase between two sinusoidally varying quantities of the same frequency. It may be expressed as an angle the phase angle or as a time. (2) The angle between the reversed secondary vector of an instrument transformer and corresponding primary vector. The vectors represent current in a current transformer and voltage in a voltage transformer. The phase difference is positive if the reversed secondary vector leads the primary and negative if it lags. The terms phase error has been used in this application but this is deprecated.

PHASE DISTORATION : A distortion that occurs when the phase change introduced is not a linear function of frequency. Distortion of received signals due to unequal phase-shifting of components of different frequencies.

PHASE MODIFIER : An apparatus, such as a Sychronous Condenser or a Synchronous Reactor, which can advance or retard the phase of an alternating current with respect to the voltage as desired for regulation of power factor, line drop, etc.

PHASE MODULATED DIGITAL RECORDING : A system of magnetic recording in which 1s are represented as positive transitions in the middle of a bit and '0' are represented as negative transitions.

PHASE MODULATION : Modulation in which the angle of the sinewave carrier diviates from the original angle by an amount proportional to the modulating signal's amplitude.

PHASE RELATIONSHIP : In synchronous telegraph systems. The degree of divergence from synchrousims between the distributor brushes at the two stations.

PHASE REVERSAL : A change of one-half cycle, or 180 dgrees, in phase due to wrong connecting of two terminals.

PHASE SHIFT : (1) The difference between corresponding points on input and output signals. (2) A change in the phase of a periodic quantity.

PHASE SHIFT OSCILLATOR : An oscillator in which a network that has a phase shift of 180 degrees is connected between the output and input of an amplifier.

PHASING : A term used in Television for the adjustment required to bring the formation of the image into phase with the scanning of the object.

PHENONMONOLOGY : A phiosophical theory of intelligence, proposed as an alternative to the "rationality" represented by information processing, which stresses environmental factors and the contribution of social context and expectation to perception and cognitive Phenomonology acts.

PHON : The name sometimes used for the unit of loudness, on a logarthmic (decibel) scale, in sound measurements, defined differently in different countries.

PHONE : Poplar abbreviation of Telephone.

PHONIC DRUM MOTOR OR PHONIC WHEEL : A simple form of constant speed motor used for driving multiplex and printing telegraph apparatus, etc., in which iron projections on an unwound iron armature are attacted by field magnets energised periodically by current impulses controlled by a vibrating contact maker consisting of a tuni   rk or tuned reed receiver.

PHONOPLEX TLELGRAPH SYSTEM : A multiplex system of telegraphy in which a number of signals can be sent over the same line simultaneously by utilising interrupted currents of different frequencies for each signal, to each of which a particular receiver is tuned to respond.

PHONOPORE : A form of telephone receiver suitable for use on lines which are also used for telegraph working, in which the speaking current flows into a condenser of very low capacitance formed by two wires placed side by side.

PHONOSCOPE : An instrument for recording wave forms of audiofrequency. It is claimed that deaf persons can be trained to interpret such records of speech and music.

PHONOTELEMETRY : see RADUI-ACOUSTIC POSITION FINDING.

PHONOVISION : The use of a gramophone record made from the sounds produced in a telephone by the variations of current from a television transmitter for subsequent reproduction of the original image by an apparatus resembling a gramophone "pick-up" and television.

PHOSPHOR : A substance on the interior of a display screen capable of luminescence when excited by a energy source (e.g. electromagnetic waves, accelerated electrons, an electrical field), thus creating an image.

PHOSPHOR BRONZE : Bronze that contains at least 0.18% ot added phosphorus. The addition of the phosphorus enhances the tensile strength, ductility, and shock resistance of the alloy. Phosphorbronze in strip form has been widely used for galvanometer suspensions and other.

PHOSPHOR BURN-IN : What occurs when the same image is left on the screen for extended periods of time, burning itself in so the image can be seen even when the monitor is turned off.

PHOSPHORESCENCE : See LUMINESCENCE.

PHOSPHOR PERSISTENCE : Measure of the time required for a phosphor's brightness to drop to one-tenth of its initial value; the tendency of a phosphor to continue to emit light when no longer excited by an electron beam.

PHOTO-ACTOR : A device that provides that light source to activate a photocunductive photocell when the latter is used as a switch.

PHOTO CELL : see PHOTO-ELECTRIC CELL.

PHOTO CONDUCTIVITY : The property of certain crystals, especially those of highrefractive index, whereby they assume increased conductivity when light falls on them due to internal movment of electrons similar to the external movement in photo-electric cells of the emissive type.

PHOTO CURRENT : An electric current produced in a device by the effect of incident electromagnetic radiation. Any electronic device that detects (see detector) or responds to light energy. See photocell; photodiode; phototransistor. See also photocell;

PHOTO-ELECTRIC ALARM : An electronic alarm system that employs a photocell used as a switch. The most common form of operation is to arrange for the photocell to be subjected to a constant beam of light. It the light beam is interrupted the photocurrent ceases, causing the alarm to be automatically activated.

PHOTO-ELECTRIC CELL : Term indicative for all types of light sensitive cells like phot-conductive, photo-emissive, and photo-voltaic cells.

PHOTO-ELECTRIC CONSTANT : The ratio of the Planck constant, to the electron charge. The photoelectric constant can be determined by photoemission experiments and has been used to calculate the Planck constant'than a characteristic value-the photoelectric threshold of the material.

PHOTO-ELECTRIC EFFECT : An effect, first noticed by Heinrich Hertz, whereby electrons are liberate from matter when it is exposed to electromagnetic radiation of certain energies. In solids electrons are only liberated when the frequency of the exciting radiation is greater.

PHOTO IONISTAION : Ionisation of an atom or molecule that results from exposure to electromagnetic radiation. The mechanism is the same as that operating in the photoelectric effect but in . photoionization an electron is ejected from a single atom or rolecule, such as occur in the gaseons state.

PHOTOLITHOGRAPHY : A technique used during the manufacture of Integrated circuits, semiconductor components, thin-film circuits and printed circuits. Photolithography is used in order to produce a desired pattern from a photographic mask on a substrate material preparatory to a particular processing step.

PHOTOPIC VISIOIN : The eye-brain response to luminance levels sufficient to permit the full discrimination of colours. Also called daylight vision, as contrasted to twilight or scoptopic vision.

PHOTO RESIST : A photo sensative organic material used during photo lithography. Negative photoresists are materials that form polywers on exposure to light. positive photoresists are polymers that are depolymerized by the action of light. The polymerized material acts as a barrier during processing steps in the manifacture of solid-state devices.

PHOTO-SENSITIVITY : The property of responding to electromagnetic radiation, particularly in the ultraviolet, visible, or infrared porations of the electromagnetic spectrum. Various responses are observed which can be either physical or chemical. See photoconductivity; photoeletric effect; photoionization; photoresist;photovoltic effect.

PHOTO-TELEGRAPHY : The telegraphic reproduction of photographs, or other "Still Picture," by means of a series of signals, either in the rform of variations in the current in a circuit or waves radiated from a radio-transmitting station, controlled by photoelectric cells or otherwise, according to the density of that portion of the original mounted on a revolving drum passing a certain point. These signals control the photo falling upon a sensitive film on a synchronously revolving drum in the receiving apparatus, or otherwise cause marks to be made thereon. In the latest developments advantage is taken of the properties of recent forms of photoelectric cells, thermionic amplification, and carrier wave systems of telegraphy. See ALEXANDERSON SYSTEM, BELL SYSTEM, BEPHOTOTELEGRAPHY. IN SYSTEM, ELECTROGRAPH, FULTOGRAPH, KAROLUS SYSTEM, KORN SYSTEM and TELECTROGRAPH.

PHOTO-TELEPHONE : An instrument on similar principles to the Photophone for transmission of sounds or speech over considerable distances. The transmission of sounds or speech over considerable distances. The transmission of sounds or speech over considerable distances by means of variations in a beam of light.

PHOTO-TRANSISTOR : A photodetector that consists of a bipolar junction transistor operated with the base region floating. The potential of the base region is determined by the number of charge carriers stored in it. The electromagnetic (usually ultraviolet) radiation to be detected is applied to the base of the transistor and produces the base current; the transistor is operated essentially in common emitter conn :cu.'n.

PHOTO-TUBE : An electron tube that contains a photosensitive electrode, usually the cathode. A vaccum phototube is evacuated to a sufficiently low pressure that ionization of the residual gas in the tube does not affect the characteristics. A gas phototube is one that photoglow tube. The sensitivity of such a tube is increased by the presence of the glow discharge.

PHOTO-VOLTAIC EFFECT : An effect arising when a junction between two dissimilar materials, such as a metal and a semiconductor or two opposite polarity semiconductors, is exposed to electormagnetic radiation, ususlly in the range near-ultraviolet to infrared. A forward voltage appears across the illuminated junction and power can be delivered from it to an external circuit. Ther effect results from the depletion region and resulting potential difference in variably associated with an unbiased junction (See also *p-n* junction).

PHYSICAL BLOCK : The physical representation of data on the medium. It is used to prevent confusion between the industry usage of terms such as sector or record on disk, and a block or record on tape. If physical commands are issued, it's the responsibility of the host to ensure that the correct physical block address is used.

PHYSICAL CIRCUIT : An actual metallic telephone circuit as distinguished from a phantom circuit super imposed there on. See PHANTOM CIRCUIT.

PHYSICAL RECORD : One or more logical records read into or written from main storage as a unit.

PICK : Event triggered by an electronic device that reports identifying data for the detected display item and the segment containing it.

PICK-AND-PLACE : The simplest kind of material-handling applications, in which a rotbot picks up an object at one point and places it in another.

PICTURE QUALITY : See RESOLUTION

PIE-NETWORK : See QUADRIPOLE

PIE-SECTION : See QUADRIPOLE

PIERCE OSCILLATOR : A modification of a Colpitts oscillator in which a piezoelectric crystal replaces the inductor in the tank circuit between the collector (plate) and base (grid) of the transistor (vacuum tube) circuit.

PIEZO-ELECTRIC COUPLER : A Piezo-Electric Resonator with two pairs of electrodes by which energy can be transferred from one circuit to another at the frequency of resonance.

PIEZO-ELECTRIC CRYSTAL : A crystal that exhibits the piezoelectric effect. All ferroelectric crystals are piezoelectric as well as certain nonferroelectric crystals include quartz crystals, and some ceramics. The best-known examples of piezoelectric crystals include quartz crystal, Rochelle salt, and barium titanate, See also piezoelectric oscillator.

PIEZO-ELECTRIC EFFECT : The material causes it to contract or expand according to the sign of the electric field. An effect that occurs when certain materials are subjected to mechanical strees. An electrical polarizaton is set up in the crystal and the faces of the crystal became electrically charged. The polarity of the charges reverses if the applied across.

PIEZO-ELETRICITY : Electrical signals generated as a result of the piezoelectric effect.

PIEZO-ELECTRIC MICROPHONE : A microphone dependent upon piezoelectric forces.

PIEZO-ELECTRIC OSCILLA-TOR : A Piezo-Electric Resonator used in conjunction with a thermionic valve circuit to initiate oscillations of constant frequency. Used in some forms of Drive Oscillator for ensuring constancy and accuracy of frequency in wireless transmitting stàtions.

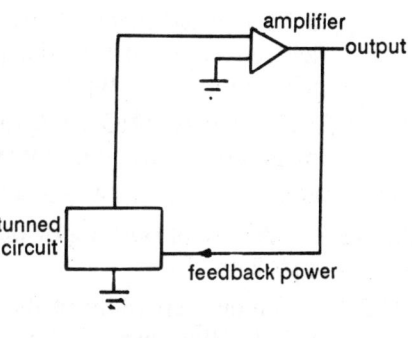

piezoelectric oscillator

PIEZO-ELECTRIC RESONATOR : Oscillators. An apparatus in which resonance is exhibited due to rapidly alternating Piezo-Electric and Converse Piezo-Electric Effects in a quartz or other crystal suitably mounted such resonators can be used as standards of frequency, and for the control of Drive.

PILON REGULATOR : A vacuum regulating device for X-Ray Tubes consisting of auxiliary electrodes of a material which gives off a small amount of gas when a current is passed through them by temporary connection to the cathode. Cf. OSMO-REGULATOR and MERCURY AIR VALVE REGULATOR.

PILOT : A simple programming language that is good for teaching and computer-assisted instruction.

PILOT RELAY : A relay controlling a Pilot Lamp on a telephone exchange switch-board.

PILOT SIGNAL IN TELEPHONY : Any signal by a lamp or otherwise, denoting a change in connections in one or more of group of circuits.

PILOT SPARK ARC RECTIFIER : A rectifier depending upon the unilateral conductivity of an arc which extinguishes itself after every half-wave in one direction and is restarted by a pilot spark at the commencement of the next half-wave in the same direction.

PILOT SYNCHRONISING : A method of controlling the frequency of the local oscillator in Suppressed Carrier and allied systems of radio-telephony by means of an auxiliary or pilot signal. PIN Connection point for logic elements and components in CD/CAM displays.

PINCH-OFF VOLTAGE : The voltage at which the current flow between the *source* and *drain* in a field effect transistor is blocked due to the deplated channel between these electrodes. For *n*-channel field effect transistors the pinchoff voltage is positive; for pchannel types it types it is negative.

PINCUSHION DISTORTION : A distortion that makes a displayed image appear to bulge inward on all four sides.

PI-NETWORK : See DELTA NETWORK.

PIN FEED : A way of pushing or pulling paper through a printer by sprockets.

PITCH : Rotation (especially of the "hand") in a vertical plane that includes the "arm", when the "arm" is extended horizontally. (See YAW, ROLL)

PIXEL : The acronym for picture element. A single cell in an image, usually represented as a bitmap of intensity values. A pixel is a single dot on a monitor that can be addressed by a single bit. A dot of light that appears on the computer screen. A collection of pixels forms characters and images on the screen. The smallest unit of resolution on a raster-scan display; on a computer system, a pixel is the smallest portion of the screen that can have its light characteristics converted to computer-readable expressions of electric current.

PL/I : Progamming Language One.

PLACE : The position of an admissible mark in a word or field.

PLAIN AERIAL TRANSMITTER : A transmitter on the Spark System of radio-telegraphy, in which the spark gap is in series with the aerial.

PLAIN COUPLER : A coupler for plain (non-threaded) conduit tube.

PLAIN TUNING ERROR : In case of direction finders, an error occuring when one loop is more nearly in tune with the in coming wave than the other.

PLANAR PROCESS : The most commonly used method of producing Junctions during the manufacture of semiconductor devices. A layer of silicon dioxide is thermally grown on the surface of a silicon substrate of the desired conductivity type. Photo lithography is used to etch holes in the oxide-layer, which then acts as a mask for the diffusion of suitable impurities into the substrate in order

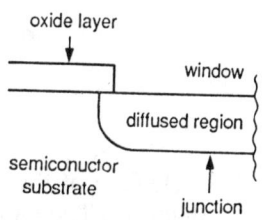

Planar Process

to produce a region. of opposite polarity. The junction between the two semiconductor types actually meets the surface of the substrate below the oxide, since the diffusion occurs in direction both normal to and parallel to the surface of silicon.

PLANNING : Is the process of preparing in advance of an action a procedure or set of guidelines for performaing it; search can be viewed as inefficient, primitive planning that does not allow for contigencies.

PLASMA PANEL : A type of CRT utilizing an array of neon bulbs, each individually addressable. The image is created by turning on points in a matrix (energized grid of wires) comprising the display surface. The image is steady, long-lasting, bright, and flicker-free; selective erasing is possible.

PLASMA PROCESSING PLASMA : Processing has become widespread in today's VLSI technology. Plasma processing comprises of plasma etching, plasma deposition and plasma ashing. Plasma etching uses reactive plasmas.

PLASMATRON : A gas-filled electron tube that contains a thermionic cathode and can pass large currents at only a few volts anode potential (see diagram). An auxiliary circuit is used to generate a plasma of electrons and positive ions in the tube before the main discharge occurs in order to provide a conducting path for it. Unlike the thyratron the plasmatron can operate continuously.

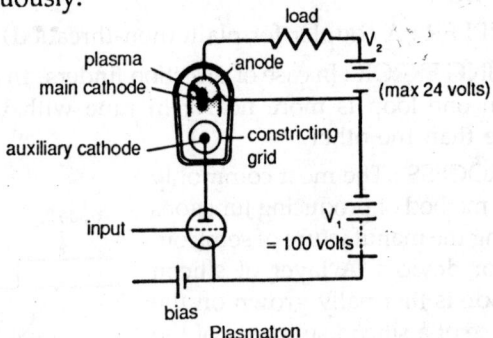

Plasmatron

PLATE : (1) One of the electrodes of a condenser or of an accumulator. (2) The name often used for the anode of a Thermionic Valve which receives the bombardment of electrons projected from the hot cathode: are often in the form of a cylinder than a flat plate.

PLATE CONDENSER : A simple form of condenser such as one for testing the specific inductive capacity of different materials, with the electrodes formed by two parallel plates, sometimes with adjustable distance between them.

PLATEN : A round rubber roller in a printer that holds paper in place.

PLATO : Programmed Logic for Automatic Teaching Operations.

PLIOTRON : The name given to the first practical three-electrode Thermionic Valve to employ a high enough vacuum to rely on a pure electron discharge. Designed by I. Langmuir for use as an amplifier (from Greek word, meaning more).

PLOTTER : A type of printer designed for printing color charts and graphs. A plotter uses a robotic arm for drawing an image. A graphic hard copy output device which can use any of a number of technologies to "plot" an image. Pen Plotters, electrostatic plotters, photo-plotters, ink-jet plotters, and laser plotters are some examples.

PLUG AND SOCKET : A device that enables electrical apparatus to be connected or disconnected from a source of supply or other equipment. It consists of two separable portions, the 'male' plug and the 'female' socket, with metal contacts that engage each other when connection is effected. A polarized plug is constructed so that engagement with the socket is only possible in one position.

PLUGBOARD : A wired panel, ofter removable, which is sometimes used to regulate and edit the handling of information done by machine.

PM : PERMANENT MAGNET or PHASE MODULATION.

PMOS : P-Channel Metal Oxide Semiconductor.

PMS : Processor-Memory-Switch.

PNEUMATIC LOGIC : See FLUID LOGIC.

P-N JUNCTION : A two- terminal, solidstate junction made from a semiconductor that has been treated to conduct current more readily in one direction than in the other. This treatment results in a p-type semiconductor at one end and an n-type semiconductor at the other end. The PN junction forms the basis of most solidstate devices, such as diodes, transistors, and silicon controlled rectifiers.

P-N-P TRANSISTOR : A semiconductor device where the collector and emitter are made from p-type semiconductor material while the base is made from n-type material. In normal operation the emitter is positive with respect to the base while the collector is negative with respect to the base. See TRANSISTOR.

POCKETING : Mass removal of material within a predetermined boundary by means of NC machining. The NC tool path is automatically generated on the system. Machining begins at an inner point of the pocket and continues to the outer boundary in ever-widening machining passes.

POINT BINARY : The equivalent of a decimal point in binary.

POINT CONTACT DIODE : A diode rectifier formed by making a point contact between a small metal wire with a sharp point and a semiconductor. The contact may be a simple mechanical contact or a small alloyed contact formed by electrical discharge, as in the gold-bonded.

PONIT CONTACT TRANSISTOR : Transistors, which are now obsolete. The point contact transistor consisted of a small crystal of semiconductor (usually germanium) with two rectifying point contacts attached in close proximity to each other and a single large area othmic contact at some distance from the point contacts.

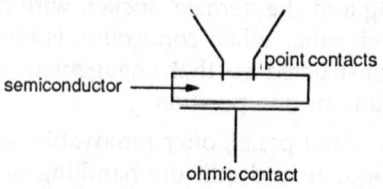

semiconductor

point contacts

ohmic contact

Point-contact transistor

POINTER : An arrow-shaped icon controlled by a mouse. Used to select items on the screen such as menu commands or text. Data that indicate the location of a variable or record of interest.

POINT OUT : Human language output produced by a computer, usually by means of a printer.

POL : Procedure- Oriented Language or Problem-Oriented Language.

POLAR : Denoting a component, such as an electrolytic capacitor, that can only operate normally in one direction of applied voltage or any system, such as a molecule, or any physical property that exhibits asymmetry.

POLAR COORDINATE SYSTEM : A mathematical coordinate system used to define positions on a plane using one linear and one circular axis. In this system, a position is defined by the number of degrees of rotation from the zero position and the distance from the centre of rotation.

POLARITY : (1) (magnetic) : The mainfestation of two types of regions in a magnet at which the inherent magnetism appears to be concentrated, (poles). There are two types of magnetic pole: north-seeking (N). (2) (electrical) The mainfestation of positive or negative parameters in an electrical circuit or device. The parameters include voltage, charge, current, and majority-carrier type. (3) The potential of a portion of the display signal representing a dark area relative to the potential representing a light area. Polarity is stated as "black positive."

POLARIZATION : (1) A phenomenon occuring in a simple electrolytic cell containing dissimilar electrodes. The current obtained from the cell decreases substantially soon after commencing the operation of the cell. This is due to the accumulation of bubbles of hydrogen gas, released from the electrolyte, around one of the electrodes. The bubbles partially cover the plate and thus increase the internal resistance of the cell and also create an e.m.f. of opposite polarity to cell e.m.f. Cells such as the Daniell cell and the Leclanche cell are designed to minimize ploarization.

POLARIZATION DIVERSITY : A diversity system that uses aerials that are arranged to receive oppositevely polarized waves.

POLE : (1) Each of the terminals or lines to a piece of electrical apparatus, a circuit, or network between which the circuit voltages are applied or produced. See also number of poles; quadripole (2) See magnetic pole. (3) An electrode of an electrolytic cell.

POLLAK VIRAGE WRITTING TELEGRAPH : A high speed system of telegraphy, in which currents in two electromagnets in the receiving instrument control the *vertical* and *horizontal* displacement of a mirror causing a spot of light to execute movements, approximating to the form of written letters when a suitable series of current impulses, controlled by a perforated strip passing through the transmitting instrument, are passed through the line. The received message is recorded photographically.

POLYCHROMATIC RADIATION : Electromagnetic radiation that contains more than one frequency. The term is also applied to particulate radiation when the particles are all the same type but of different energies. The description in homogeneous is usually preferred in this latter case Cf. MONOCHROMATIC RADIATION.

POLYGON FILL : Coloring or cross-hatching of a closed, multisided, program-defined surface.

POLYODE : A Thermionic Valve with more than three electrodes, e.g. with several additional grids, as in a pentagrid rectifier.

POLYPHASE SYSTEM : An electrical system or apparatus that has two or more alternating supply voltages displaced in phase relative to each other. In a symmetrical polyphase system each voltage is of the same magnitude and frequency and is displaced by an equal amount. If there are n sinusoidal voltages the mutual phase displacement is π/n radians and the system requires n lines at least. Thus in the three-phase system there is a phase difference of $2\pi/3$ between three voltage lines. An exception is the two-phase system that has a phase difference of $\pi/2$ between the two voltages. Cf. MONOCHROMATIC RADIATION.

POLYPHASE TRANSFORMER : A transformer that is used with a

ployphase system. The magnetic circuits required for each of the phase windings usually have portions in common with each other in order to retain the correct voltages.

POLY SILICON : Polycrystalline silicon. Silicon in polycrystalline form is most often used to form the gate electrodes in silicon-gate MOS integrated circuits and charge coupled devices. In this application the silicon is doped with a sufficiently high doping concentration so that it becomes degenerate and exhibits metallic properties.

POLYSTAGE AMPLIFIER : See MULTISTAGE AMPLIFIER.

POPCORN NOISE : Noise produced by erratic jumps of bias current at random intervals in operational amplifiers and other semiconductor devices.

PORT : The entry channel to and from the certral computer for connection of a communications line or other peripheral device. A circuit and connection that will allow external communication with the computer.

PORTABLE PROGRAMES : Software that is both host-computer and display device independent.

PORTABLE RECEIVER RADIO : A complete self-contained radio-receiving set, usually with a frame aerial and with the batteries and a loud speaker within the same case, which can easily be carried about.

POS : Product of Sums. Point-of Sale Terminal.

POSITIONAL WEIGHTING : A number system in which the value of a digit is determined by its position with repect to the input.

POSITIVE FEEDBACK : An output-to-input signal path where the out-put signal is fed back in phase with the input in order to increase amplification. Excessive possitive feedback causes *distortion, instability,* and when sufficiently high, *oscillation.* Also called regeneration,or regenerative feedback.

POSITIVE LOGIC : A logic system in which positive voltage represent 1's and negative values 0's.

POSITIVE MODULATION : (In television) : Modulation of the carrier wave in a manner whereby the peak value corresponds to the full white of the picture and 30 per cent modulation to full black, the remainder down to zero being available for synchronising signals. Cf. NEGATIVE MODULATION.

POSIVITE PHOTO-RESIST : See PHOTO-RESIST.

POSITIVE WIRE (In Telephony) : The wire forming that side of a telephone circuit within the exchange, normally connected to the
positive side of the battery.

POST-MARTUM ROUTINE : A routine that produces a record of
machine conditions and contents when the control sequence
of some other rountine is broken.

POST-OFFICE BOX : A box containing a number of resistance coils
connected to form three arms of a Wheatstone bridge, the resistance to be measured forming the fourth arm. Each resistance coil is connected between adjacent metal blocks so that
it may be shorted out of the circuit by the insertion of a metal
plug between the blocks. Resistances from 0.1 to 10^6 ohms
can be measured.

POST-PROCESSOR : A computer program that transforms cutter path
coordinate data into a form which a specific control system
can interpret correctly of the potential difference and the charge.
The potential gradient at a point is the potential difference
per unit length. See also electromotive force.

POTENTIAL DIFFERENCE (p.d.) : The difference in electric potential
between two points, equal to the line integral of the electric
field strength between the points. If a charge is moved from
one to the other of the points by any path, the work done is
equal to the product of the potential difference and the charge.
The potential gradient at a point is the potential difference
per unit length.

POTENTIAL FRONT : The steep front portion of a wave of e.m.f. in
a case like that of an inductive circuit to which an e.m.f. has
suddenly been applied and the current is taking an appreciable time to grow to its full value. Owing to the steep slope
of the curve of potential along the conductor there can be a
very considerable difference of potential between points comparatively near together, such as between the first few turns
in a winding.

POTENTIOMETER : (1) A form of potential divider that uses uniform wire as the resistive chain. A movable sliding contact is
used to tap off any potential difference less than that between
the ends of the wire. A typical use is for the measurement of
potential difference or e.m.f. by balancing the unknown e.m.f.
with that of a standard.

POWER : The functional area of a system that transforms an external power source into internal DC supply voltage. The ability to influence behavior. The power developed is equal to $VI \cos \phi$, where V and I are the root means-square values of voltage and current and ϕ is the phase angle between them. $Cos \phi$ is the power factor of the circuit or device and the apparent power is the product VI measured in volt-amperes. The product $VI \sin \phi$ is the reactive power. The rate at which energy is expended or work is done. In a direct-current circuit or device the power developed is equal to VI, where V is the potential difference in volts and I the current in amperes.

POWER AMPLIFIER : An AF or RF amplifier that delivers maximum output power to a load rather than providing a maximum voltage gain. An expression sometimes used in radio-reception for a low frequency amplifier for use as a final stage, particularly with a loud speaker of large size.

POWER DISSIPATION OF A LOGIC CIRCUIT : The supply power when a logic cirruit is operating with a 50% duty cycle (i.e., when it is the 0 state half of the time and in the 1 state the other half of the time).

POWER FACTOR : Is equal to the cosine of the phase angle between them. The power factor is also equal to the ratio of the resistance, R to the impedance, Z, and thus indicates the dissipation in an insulator, inductor, or capacitor. The radtio of the actual power in watts developed by an a.c. system, as measured by a wattmeter, to the apparent power in volt-amperes, indicated by voltmeter and ammeter readings. If the voltage and current are sinusoidal, the power factor is equal to P/VI.

POWER FREQUENCY : The frequenfy at which domestic and industrial mains electricity is supplied and distributed. In India and U.K. the standard value is 50 hertz; in the U.S. it is 60 hertz.

POWER RIINGING TELEPHONY : Ringing by alternating current from a power-driven generator.

POWER STATION : Also known as generating station. A complete assembly consisting of all necessary plant, equipment, and buildings at a suitable site for the conversion of energy of one type, such as nuclear energy, into electrical power.

POWER SUPPLY : A device that provides power to electronic equipment such as your computer. Power supplies are rated by wattage; the higher the wattage, the stronger the power supply. Any source of electrical power in a form suitable for operating eletronic circuits. Alternating-current power may be derived from the mains either directly or by means of a suitable transformer. Direct-current power may be supplied from batteries, suitable rectifier/filter circuits, or form a converter.

POWER TRANSISTOR : A transistor capable of handling high currents and power, generally greater than 1 watt. A transistor designed to operate at reatively high values of power or to produce a relatively high power gain. Power transistors are used for switching and amplifications. They usually require some form of temperature control since the power dissipation in them ranges from 1 watt to 100 watts.

POWER VALVE : (1) A valve for a Power Amplifier, or for the last stage of low frequency amplification, taking a higher filament current and anode voltage than other receiving valves, and capable of dealing with a considerable grid voltage swing while still working on the straight part of its characteristic. (2) A large Thermionic Valve used as a main oscillator in a radio transmitting station.

PP : Peripheral Processor

PPC : Parallel Poll Configure; addressed bus command.

PPL : Polymorphic Programming Language

PPR : Parallel Poll Response.

PPU : Paralel Poll Unconfigure; universal bus command.

PRAGMATICS : An area of linguistics that studies the interactions between sentences to understand ambiguities like pronoun reference and intended meanings; also, a level of dialogue analysis in natural language processing that seeks to solve the problems developed by such ambiguities.

PREAMP : See PREAMPLIFIER

PREAMPLIFIER : An amplifier intended to operate with low-level signals to provide gain and impedance matching to a level that can be handled by another amplifier. Also called PREAMP.

PRECISION : The dgree of definition. For example, the magniutre of a quantity might be expressed with a precision of ten significant places. The word length is a direct measure of the possible precision of an automatic computer.

PREFERENCE SEMANTICS : A system of semantic primitives that uses about 75 basic objects and relations in various combinatio s to represent more complex concepts.

PREMISE : An antecedent IF-part on the left-hand side of a production rule that lists the conditions under which the rule is applicable and whose satisfaction is necessary before the rule can be used.

PREPARE/DELIBERATE TRADEOFF : A fundamental constraint of knowledge-based systems, which states that the amount of search (deliberation) required to solve problems at a given level of performance will decline as the amount of knowledge (prepration) increases.

PREPROCESSOR : A computer program that takes a specific set of instructions and converts them into the form required to be run by the processor program.

PRESCALER : A frequency converter (plug-in) functioning on the divider principle.

PRESELECTOR (in Automatic Telephony) : A Selector to connect the calling subscriber's line to a vacant selector by which the connection to the required line is made through further selecting apparatus. (Also called Subsriber's Rotary Line Switch, or simply Line Switch)

PRESENTATION GRAPHICS : High-quality graphics intended to visually reinforce points made in the presentation of proposals, plans, or budgets, to top management.

PRESET : To cause a flip-flop to go the '1' state. When the memory or input divider of a digital instrument is set to a specific value.

PRESSURE AND PRESSURE GRADIENT MICROPHONES : Types of microphone actuated by pressure on the front and difference of pressure on the back and front respectively of the diaphragm.

PREVENTIVE RESISTANCE : (1) A resistance placed between two halves of the moving contact of a multiple contact switch, such as an accumulator regulating switch, to prevent short-circuiting individual cells, etc., when moving from one contact to the next. Cf. PREVENTIVE COIL, (2) A resistance in a preventive lead in an alternating current motor.

PRIMAL SKETCH : An intermediate representation in visual processing between the low-level bitmap and the high-level conceptual model of the image to be undetstood; it segregates the image into regions like lines and blobs, and indicates the two-dimensional orientation of each grouping.

PRIMARY COLOURS : A set of colors from which all other colors can be derived, but which cannot be produced from each other. The additive primaries (light) are *red, green,* and *blue.* The substantive primaries (colorant) are *yellow, magenta* (a deep pink), and *cyan* (a blue, green). The psychological primaries (perceived as basic and unmixed) are the pairs red-green, yellow-blue, and black-white.

PRIMARY MEMORY : The memory in which programs and data are stored and from which they are generally executed; main storage.

PRIME IMPLICANT : A tabular-reduced term incapable of being further reduced.

PRIMITIVE ATTRIBUTE : A general characteristic of a display primitive, such as color, intensity, linestyle, or linewidth.

PRIMITIVE ELEMENT : Graphic element such as a point or line segment which can be readily called up or extrapolated or cobined with other primitive elements to form more complex objects or images in two or three dimensions.

PRINT BUFFER or spooler RAM : Memory set side for temporarily strong data. Storing data in a print buffer lets the RAM send the data directly to the printer. Freeing the computer to do some-thing else.

PRINTED CIRCUIT : A physical realization of an electronic circuit design in which the connections between the terminals of individual components are fromed from copper conductors laminated onto a flat supporting sheet of an insulating material such as fiber glass. The conductor pattern is normally printed and etched onto the sheet and components are then attached to the copper "lands" by hand or dip soldering. The supporting sheet plus circuit is shown as a printed circuit board.

PRINTED CIRCUIT BOARD : Abbreviated PCB, an insulated board, generally made from epoxy or bakelite, on which a circuit has been printed and components mounted. A baseboard made of insulating material and an etched copper-foil circuit pattern, on which are mounted ICs and other components required to implement one or more electronic functions. The PC boards pug into a rack or subassembly of electronic equipment to provide the brains or logic to control the operation of a computer, a communications system, instrumentation, or other electronic systems. The name derives from the fact that the circuity is connected not by wires but by copper-foil lines, paths, or traces actually etched onto the board surface. CAD/ CAM is used extensively in PC board design, testing, and manufacture.

PRINTER : A device that prints data from the computer to paper.

PRINT HEAT : The part of the printer that strikes a ribbon against the paper.

PRINTING MULTIPLEX TELEGRAPH SYSTEM : A Multiplex system applied to a Type Printing Telegraph system, such as the Baudot system. Spoken of as double, treble, quadruple, quintuple, or sextuple, according as 2,3,4,5 or 6 ways are porvided for. Cf. MORSE MULTIPLEX SYSTEM, PRINTING TELEGRAPH BAUDOT CREED HUGHES MURRAY & STELJES See BAUDOT PRINTING TELEGRAPH, CREED PRINTING TELEGRAPH, etc.

PRINTOUT : A paper copy of data printed by printer. Also called *hard copy*. A hard copy or printed copy form of computer output.

PRINTQUALITY : The sharpness of print that a printer can produce. Print quality ranges from draft, near-letter quality (NLQ), letter quality, near typeset quality, and typeset quality modes.

PRIVATE BRANCH EXCHANGE : An intelligent, programmable switch that can route telephone calls or digital data from computers and terminals throughout a building or to an outside line for local or long-distance transmission.

PROBE : An electric lead that connects to a measuring or monitoring circuit, or contains such a circuit at its end or along its length, and that is used for testing purposes. The measuring or monitoring circuit may be formed from either active or passive components. A language specifically designed for one particular type of problem, such as civil-engineering computations.

PROBLEM ORIENTED LANGUAGES : A non machine language for programming.

PROBLEM PROGRAM : A user-written program that uses only nonprivileged instructions. It should be distinguished from a supervisory, or control, program, which may have privileged instruction.

PROBLEM SPACE : See STATE SPACE.

PROCEDURAL ATTACHMENT : An augmentation of a slot in a

frame with an entire procedure to computer its filler whenever its value is needèd, such as a function to claculate densities from masses and volumes.

PROCEDUUAL KNOWLEDGE : Is knowledge about how to perform various tasks, usually represented by procedures that do them.

PROCEDURAL LANGUAGE : A language designed to facilities the coding of algorithms to solve a problem, for example, COBOL.

PROCEDURE : A written (and often flow-charted or process charted) description of system.

PROCESS COLOURS : The transparent ink colors used in four-color process printing : *yellow, magenta, cyan,* and *black.* The white of the paper provides a fifth color essential to this type of color reproduction.

PROCESSOR : The portion of a computer that performs the arithmetic and provides system control. (1) In hardware, a data processor. (2) In software, a computer program that compiles, assembles, translates, and performs related functions for a specific programming language.

PRODUCTION RULE : An IF-THEN, premise- action association used to represent both declarative and procedural knowledge needed to solve problem.

PRODUCTION SYSTEM : Is an organized collection of production rules that work together to solve one or more problems under an appropriate inference algorithm.

PRODUCT OF SUMS (P.O.S.) : A Boolean expression formed by ANDing ORed term.

PROGRAM : A series of instructions or statements in a form acceptable to a computer, designed to cause the computer to execute a series of operations. Computer programs include softwares such as operating systems, assemblers, compilers, interpreters, data management systems, utility programs, sort-merge programs, and maintenance/diagnostic programs, as well as application programs such as payroll, inventory control, and engineering analysis program. A series of instruction that tells the computer what to do. Typical programs are word processors, spread-sheets, databases, and games. A set of instructions that directs the computer to perform a specific series of operations. A set of in-sequence, coded instruction for digital computer, sometimes known a routine or subroutine. A set of instructions that tells the computer the steps it must perform. It must be written in a language that the computer understands directly or for which you have a program to translate your code into machine code.

PROGRAM COUNTER : A register that contains that next address from which and instruction will be taken by the computer.

PROGRAM DATA : A programming instruction (e.g., range of frequency of an instrument used to remote-control functions).

PROGRAMMABLE CONTROLLER : A controller in which the operation is determined by user-programmed codes or instructions.

PROGRAMMABLE READ ONLY MEMORY : A read only memory with contents capable of being permanently changed by the designer.

PROGRAMMABLE UNIJUNCTION TRANSISTOR : Abbreviated PUT, a three-terminal, solid-state device having an *anode* a *cathode*, and a *gate*. Its action is similar to that of a silicon controlled switch in that for a constant gate voltage the device remains non-conductive until the anode voltage is greater than the gate voltage by a predetermined amount, after which it conducts.

PROGRAMME CIRCUIT : A Telephone Circuit specially balanced to be able to deal with a larger range of frequencies than that required for ordinary speech, so as to be used for the transmission of musical programmes over long distances for broadcasting purposes.

PROGRAMMED DECISIONS : Generally, decisions that can be made automatically by following certain rules and procedures.

PROGRAMMER : A person who writes programms for a computer to use.

PROGRAMMING : Preparing Program.

PROGRAMMING LANGUAGE : A set of rules for writing instruction for the computer. Popular programming languages include BASIC, C, Pascal, Modula-2 and Ada.

PROGRAM SYNTHESIS/TRANSFORMATION : See AUTOMATIC PROGRAMMING.

PROGRESSIVE SCANNING (In Television) : A system of Scanning in which contiguous lines of the field are covered successively in order. Also called Sequential Scanning. Cf. INTERLACED SCANNING.

PROLOG : Programming in Logic is the most popular logic programming language used for AI applications, especially natural language processing with its built-in definite clause grammer facility and expert systems with its backward-chaining thereon prover.

PROLONGATION : The continuance of a telegraph signal current after the source of e.m.f. has been disconnected from the line by the gradual discharge of the capacitance of the line, or the continuance of a current in an inductive circuit after the e.m.f. has ceased.

PROM : Programmable read only memory. A type of memory that is programmed after manufacture.

PROMPT : A character or series of character that person on the screen to request input from the user. A symbol that the computer displays on the screen when it needs input from the keyboard. A message or menu on the display surface calling for operator action.

PROPAGATION DELAY : The time delay between the application of a signal to the input of a logic circuit and the change of state at the output.

PROPAGATION TIME : The time for a level change on the input to result in a response at the output.

PROPORTIONAL SPACING : Printing or displaying characters so the left and right margins align to an arbitarary setting. Books and magazines print examples or proportional spacing with left and right margins artificially enforced to enchance the appearance of the text.

PROSTHETIC ROBOT : A controlled mechanical device connected to the human body which provides a substitute for human arms or legs when their function is lost.

PROTECTION : Maintenance of the integrity of information in storage by preventing unauthorised changes.

PROTOCOL : A set or rules procedures for devices to communicate with each other. The rules under which computers exchange information, including the organization of the units of data to be transferred.

PROTO-TYPE : A mode of a system or a version without all the final features desired used to provide early feedback to users.

PRT : Program Reference Table.

PSEUDO COLOURS : Also called false colors. Colors atbitrarily assigned to an image to represent data values, tather than natural likeness. Often used in satellite imagery.

PSEUDO INSTRUCTION *Syn.* : *Pseudo Code* : Command(s) or instructions used by a programmer in writing program which have the form of a regular commands or instructions, but have the function of fefining operands, specifying origins or otherwise directing the translation process. A human or machine translation of interpretation is required before a program incorporation pseudocodes can be used by an automatic computer, some pseudocodes are actually macrocodes of macrocommand.

PSEUDO NMOS LOGIC : Pseudo-*n*MOS logic consists of an N-net passing of 0'f of the the function and a P-net with a single conducting PMOS transistor passing the 1's of the function. The main problem with this gate is the static power dissipation that occurs whenever the pull down chain in turned on. But a gate so implemented may have a density advantage over other CMOS configurations.

PSW : Program Status Word.

PTR : Printer, more rarely, Paper Tape Reader.

PUBLIC DOMAIN : Material that is not copyright, and thus available for public use, free of charge.

PUCK : Hand-held device with crosshairs, used to input coordinate data through data tablet.

PUFFT : Purdue (University) Fast Fortaran.

PUISSANCEGRAPHE : A variety of the Ondograph for tracing vurves of Instantaneous Power in an alternating current circuit in which a frrom a watt meter replaces the ammeter or voltmeter.

PULL DOWN MENU : A menu display that appears at the top of the screen, organizing specific program commands into categories. Choosing a specific category displays a menu of actual progarm commands to select.

PULL TRACTOR FEED : A deyice in printer that pulls paper through using sprockets and special computer paper with holes on both sides.

PULSATANCE : An expression sometimes used for the angular velocity of the vector expressing an alternating current voltage, etc.

Equal (in radians per second) to 2 times the frequency. Usual symbol (Greek "omega"). Also called Pulsation.

PULSATION : In the case of a varying frequency, or speed, the ratio of the difference between the maximum and minimum values to the average value.

PULSE : A sharp brief difference between the normal level of an electric current and some different level of the some electric current. A momentary, sharp change in a current, voltage, or other quantity that is normally constant characterized by a rise and fall of finite duration. A pulse consists of a voltage or current that increases from zero (or a constant value) to a maximum and then decreases to zero (or the constant value), both in comparatively short time. The zero or constant value in the absence of the pulse is termed the *base level*. A pulse is described according to its geometrical shape when the instantaneous value is plotted as a function of time : it can be rectangular, square, traingular, etc. Unless otherwise specified a pulse is assumed to be *rectangular*.

In practice a perfect geometrical shape is never achieved and a practical rectangular pulse is shown in figure (a). The portion of the pulse that first increases in amplitude is the leading edge. The time interval during which the *leading edge* increases between specified limits, usually between 10% and 90% of the pulse height, is termed the *rise time*. The pulse decays back to the base level with a finite decay time, usually taken between the same limits as the rise time. The major portion of the decay time is termed the *trailing edge* of the pulse.

A *spike* is an unwanted pulse of relatively short duration superimposed on the main pulse; *ripple* is unwanted small periodic variations in amplitude. A practical pulse frequently rises to a value above the pulse height and then decays to it with damped oscillations. These phenomena are known as *over-*

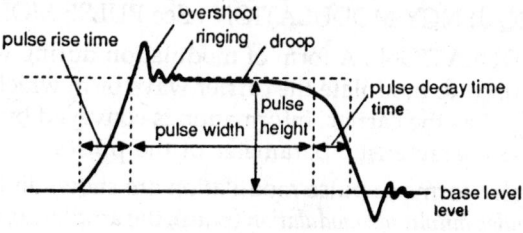

(a) Practical rectangular pulse

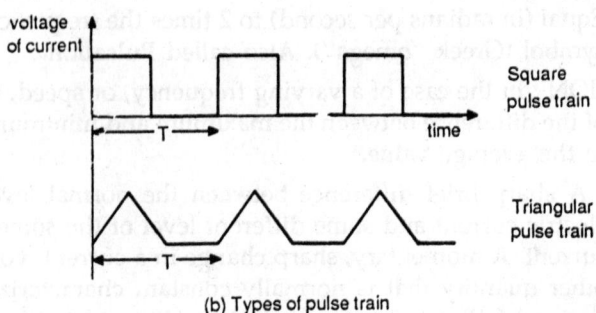

(b) Types of pulse train

shoot and *ringing*. A circuits may frequently incorporate a *smearer*, which is a circuit designed to minimize overshoot.

A group of regularly recurring pulses of similar characteristics is called a pulse train; it is usually identified by the type of pulses in the train, e.g. square waves or sawtooth waves (Fig b). The time interval between corresponding portions of the pulses in the train, e.g. the pulse rise times, is the *pulse spacing* (or pulse-repetition period), T. The *pulse-repetition frequency* (or pulse rate) is the reiprocal of the period and is the rate at which pulses are transmitted in the pulse train; it is measured in hertz.

PLUSE (Pair) RESOLUTION : The minimum time in a counter between the two pulses of pulse pair which the counter recoginizes as separate pulses.

PULES AMPLITUDE MODULATION : Abbreviated PAM, amplitude modulation of a pulse train used as a carrier signal. : See PULSE MODULATION

PULSE CODE MODULATION : Abbreviated PCM, modulation in which the signal is periodically sampled and each sample is then grouped and transmitted as a digital binary code. See PULSE MODULATION.

PULSE DURATION MODULATION : See PULSE MODULATION.

PULSE FREQUENCY MODULATION : See PULSE MODULATION.

PULSE MODULATION : A form of modulation during which pulses are used to modulate the carrier wave or in which pulse train is used as the carrier. Information is conveyed by modulating some characteristic parameter of the pulses.

Different forms of pulse modulation are shown in the diagram. In *pulse-amplitude modulation* (pam), the amplitude of the pulses is modulated by the corresponding samples of the modulat-

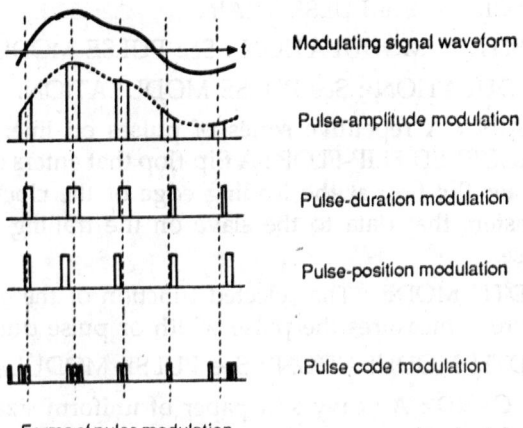

Modulating signal waveform

Pulse-amplitude modulation

Pulse-duration modulation

Pulse-position modulation

Pulse code modulation

Forms of pulse modulation

ing wave. In *pulse-time modulation* (ptm). the samples are used to vary the time of occurrence of some parameter of the pulses. Particular forms of pulse-time modulation are pulse-duration modulation (pdm), in which the time of occurrence of the leading edge or trailing edge is varied from its unmidulated position, *pulse-frequency modulation* (pfm)m in which the repetition frequency (see pulse) of the carrier pulses is varied from its unmodulated value, and *pulse-position modulation* (ppm), in which the time of occurrence of a pulse is modulated from its unmodulated time of occurrence, i.e. the pulse repetition period is varied. All these types of pulse modulation are examples of uncoded modulation.

In pulse code modulation (pcm) only certain discrete values are allowed for the modulating signals. The modulating signal is sampled, as in other forms of pulse modulation, but any sample falling within a specified range of values is assigned a discrete value. Each value is assigned a pattern of pulses and the signal transmitted by means of this code. The electronic circuit or device that porduces the coded pulse train from the modulating waveform is termed a coder (or pulse coder). A suitable decoder must be used at the receiver in order to extract the original information from the transmitted pulse train. Morse code is a very well known example of a pulse code.

Pulse modulation is commonly used for time-division multiplexing.

PULSE REPETITION RATE : The number of electric pulses per unit time. See PULSE

PULSE SPACING : See PULSE TRAIN.

PULSE POSITION MODULATION : See PULSE MODULATION.

PULSE MODULATION : See PULSE MODULATION.

PULSE TRAIN : A repetitive series of pulses on line. See PULSE TRIGGERED FLIP-FLOP : A flip-flop that enters data into the master flip-flop at the leading edge of the clock pulse and transfers that data to the slave on the trailing edge of tha pulse.

PULSE WIDTH MODE : The selected function of the couter/timer where it measures the pulse width or pulse duration.

PULSE WIDTH MODULATION : See PULSE MODULATION.

PUNCHED CARD : A heavy stiff paper of uniform size and shape suitable for being punched with a pattern of holes to represent data for mechanical handling. A card used to enter data by means of punched holes.

PUNCHED TAPE : A tape on which a pattern of holes is used to represent data.

PUPIN'S LAW : Two telephone lines, one with inductance evenly distributed and the other with Load-equivalent for transimssion purposes if half the angle obtained by reckoning the wave length as 180 coils per wave length is so small that its sine is approximately equal to its value in circular measure.

PURE CONTINUOUS WAVES : Electric waves of unvarying amplitude and pure sine wave form without harmonics.

PURE PROCEDURE : A program in which no part of the code modifies itself. Because a re-entrant program is not modified execution, it can be shared by many users.

PURITY : An objective term that denotes a measurement which can be visualized on a chromaticity chart as a position between the equal energy mixture of all colors (white) to the dominant wavelength of the color. (Also see *chorma* and *stauration.*)

PURPLE BOUNDARY : Straight line drawn between the ends of the spectrum locus on the CIE chromaticity diagram.

PUSH-PULL AMPLIFIER : A balanced amplifier that uses two similar vacuum tubes or transistors working in phase opposition. An amplifier that uses two similar vacuum tubes or transistors with their grid or base leads connected in phase opposition while the plate or collector leads are connected in parallel to common load. It is used primarily as a frequency multiplier to emphasize evenorder harmonics.

PUSH-PULL MICROPHONE : See DIFFERENTIAL MICROPHONE.

PUSH-PULL SYSTEM OF AMPLIFICATION : An arrangement of the last stage of low frequency amplifier in which two valves are used with both their grid and plate circuits connected as in a three-wire system, in opposite directions to tappings on the input and output transformers respectively.

PUSH TRACTOR FEED : A device in printers that pulls paper throuth using sprockets and special computer paper with holes on both sides.

PUT : Abbreviation for Programmable Unijunction Transistor.

PYRO ELECTRICITY : The development of opposite electric charges at the ends of the polar axes in certain crystals when subjected to a temperature gradient. Tourmaline and lithium sulphate are two of the crystals that exhibit pyro-electricity.

Q-BAND : A band of microwave frequencies ranging from 36 to 46 gigahertz. See FREQUENCY BAND. The radio-frequency band from 36 to 46 GHz.

Q-CHANNEL : The 0.5 MHz- wide band used in the American NTSC color television system for transmitting green magenta color information.

Q-FACTOR : (L/C) : Where R is the resistance of the circuit, L the inductance, and C the capacitance. Also known as quality factor. A factor that is associated with a resonant circuit and describes both the ability of the circuit to produce a large output at the resonant frequency and the selectivity of the circuit. The factor is given by $Q=(A/R) \sqrt{L/C}$ where R is the resistance of the circuit, L the inductance and C the copacitance. See QUALITY FACTOR.

Q-METER : An instrument for measuring the Q of a circuit or circuit element. Also called a *quality-factor meter.*

Q-MULTIPLIER : A filter circuit which gives either a very sharp peak or rejection at a particular frequency, which is equivalent to increasing the Q of a tuned circuit at that frequency.

Q-POINT : See QUIESCENT POINT.

Q-SAM : QUEUED SEQUENTIAL ACCESS METHOD.

Q-SIGNAL : The narrow band signal produced by mouldating the color subcarrier at a phase removed from the burst reference phase. Also called "quadrature" signal. The signal reproduces a range of colors from purple to yellow-green.

QUAD CABLE : A telephone cable which contains a number of Quad Cores. Compare MULTIPLE TWIN CABLE.

QUADRANT ELECTROMETER : An electrometer that consists of a flat cylindrical metal box formed form individually isolated quardrantal segments and containinig a light foil-covered van that is supported by a quartz fibre and is free to move.

QUADRAPHONIC BROADCASTING : See QUARDRAPHONY.

QUADRAPHONY : A sound reproduction system that is an extension of stereophonic sound reproduction and uses for channels, feeding four sepatate loudspeakers. The sound may be recorded either using four coincident directional microphones at right angles to each other and placed at the position of the listener, or using separated microphones whose outputs are divided between the channels and combined in the correct proportions to achieve the desired effect.

Placed at the position of the listener, or using separated microphones whose outputs are divided between the channels and conbined in the correct proportions to achieve the desired effect.

QUADRATURE : Two periodic quantities that have the same frequency and waveform are in quadrature when the phase difference between them is $\pi/2=(90°)$. They therefore differ by one quarter of a period, one wave reaching its peak value when the other passes through zero.

QUADRATURE AMPLIFIER : An amplifier that shifts the phase of a signal by 90 degrees.

QUADRATURE MODULATION : The modulation of two carrier signals 90 degrees apart by separate modulating functions.

QUADRIPOLE : A network that has only four terminals, i.e. a pair of input terminals and a pair of output terminals. The behaviour of q quadrepole is usually described by the impedances presented at its terminals at specified frequencies. Also known as four pole network.

QUADRUPLE PHANTOM CIRCUIT : A superposed circuit each side of which consists of the eight conductors of a Double Phantom Circuit in parallel.

QUADRUPLEX TELEGRAPH SYSTEM : A telegraph system in which Duplex and Diplex working are so combined that four messages, two in each direction, can be sent simultenously over a single line. Cf. MULTIPLEX.

QUALITATIVE REASONING : Is understanding and reasoning about real-world physical processes, such as those operating in a digital computer or nuclear reactor, in non-numeric terms at every stage during manufacture.

QUALITY CONTROL : A method of inspection used during the mass production of elctronic components, circuits, devices, or other apparatus in order to ensure that the finished product conforms to specifications. It usually involves exhaustive checking of random sample at every stage during manufacture.

QUALITY FACTOR : Abbreviated as Q-factor, a dimensionless figure of measurement or selectivity for a resonant circuit or system. The Q of an inductor at a given frequency is the ratio of its reactance to its series resistance, or $2 \ fL/R$. The Q of simple LC resonent circuit is $Q_r/(Q_1+Q_c)$. The Q of a bandpass, or notch filter, network is the ratio of its resonant, or center, frequency to its band width, f_r/BW.

QUANTIZATION : Keep the loss to minimum; this leads to a certain amount of noise, or quantization distoration. A method of producing a set of discrete or quantized values that represents continous waveform. The waveform is divided into a finite number of subranges each of which is represented by an assigned value within the subrange. This technique is used whenever a set of discrete values is required, for example to produce data suitable for a digital computer or in pules modulation. Some loss of information of the original waveform is inherent in this process, although suitable choice of the subranges can.

QUANTIZATION ERROR : The voltage deviation form a linearity line of D/A converter caused by the inability to express voltages in other than step functions.

QUANTUM : See quantum theory.

QUANTUM EFFICIENCY : See quantum yield.

QUANTUM MECHANICS : Insulators, and semiconductors. See Quantum Theory. A mathematical physical theory that grew out of the quantum theory proposed by Planck Quantum mechanics in the form of wave mechanics is used to solve the energy states of tatoms and molecules and hence explains the electrical behaviour of metals.

QUANTUM NUMBERS : A set of numbers that are used to label the various possible values of certain physical properties. The property is restricted to certain discrete values (quantized) as a result of applying the principles of quantum mechanics to a physical system. A set of quantum numbers is usually in the form of a set of integers and half-integers.

QUANTUM THEORY : A theory introduced by Max Planck in 1960 that departed readically from classical Newtonian mechanics. It was found that certain physical phenomena could not be satisfactorily explained by classical mechanics. Quantum theory successfully described the photoelectric effect, atomic structure and the Compton effect. Despite its success in the above applications it gave misleading results in many other problems and has been further refined and developed into the system known a Quantum mechanics.

QUANTUM EFFICIENCY : The number of reactions of a particular type induced per photon of absorbed electromagnetic radiation.

QUARK X PRESS : Quark X-press is extremely popular with many Macintosh magazines such as MacWeek. It is considerably more powerful than PageMaker, but also more complex and expensive.

QUARTER PHASE SYSTEM : See POLYPHASE SYSTEM.

QUARTER WAVE ANTENNA : An antenna whose electrical lenght is one-fourth the wavelength of the transmitted or received signal.

QUARTER WAVE LENGTH AREIAL : An aerial having an effective height equal to a quarter of the wave length to be radiated. Cf. HALF WAVE-LENGTH AERIAL.

QUARTER WAVE LENGTH LINE : A transmission line of length equal to one quarter of the wave-length of the fundamental frequency. It is used for impedance matching, particularly in systems designed to operate at high radio-frequencies, for the suppression of even-order harmonice in filter networks, and for coupling and feeders used with aerials. Also known as Quarter wavelength transformer.

QUARTER WAVE TRANSMISSION : Long distance alternating current transmission in which line losses and pressure drop are lessened by arranging the natural period of oscillation of the line to be equal to four times the frequency, so that resonance is produced. Cf. HALF-WAVE TRANSMISSION.

QUARTER WAVE LENGTH TRANSFORMER : See quarter-wavelength line.

QUARTZ : Naturally occurring crystalline silicon dioxide exhibits marked piezoeletric properties and is frequently used as the piezoelectric crystal in piezoelectric oscillators. It also exhibits marked dielectric strength. Quartz may be readily drawn into extermely fine uniform filaments that are very strong, elastic, and physically and chemically stable. Such quartz fibres are frequently used as torsion threads in delicate measuring instruments, such as electrometers.

QUARTZ CRYSTAL OSCILLATOR : See PIEZOELECTRIC OSCILLATOR.

QUARTZ LAMP : A Mecury Vapour Lamp in a container or tube of quartz instead of glass. This enables increased current densities to be employed. It is also more transparent to ultra-violet rays.

QUARTZ OSCILLATOR : A Quartz (or Piezo-Electric) Resonator connected between the grid and one of the other terminals of a thermionic valve, and producing continuous oscillations when the circuit constants are suitable. Used for frequency control of transmitting stations for standardising wavemeters.

QUATTRO PRO : Quattro Pro is a director competitor to Lotus 1-2-3. Unlike Lotus 1-2-3, Quattro Pro can run on a standard IBM PC-XT.

QUENCH : A capacitor, resistor, or combination of the two, that is used in parallel with a contact to an inductive circuit and that inhibits spark discharge across the contact when the current ceases. A quench is commonly employed across the make and break.

QUERY : In data communications, the process by which a master station asks a slave station to indentify itself and to give its status.

QUERY LANGUAGE : A language used to provide access to data stored in a database or file.

QUEUE : A list of messages or files waiting to reach a particular destination. such as printer or another computer. A priority-ranked collection of tasks waiting to be performed on the system.

QUEUED : Describes a unit's ability to accept multiple operation commands from the host and execute them in a sequence, according to unit-defined or host-defined algorithms.

QUICK-BASIC : See QUICK C.

QUICK BREAK SWITCH : A manually operated switch that is designed to operate at a speed independent of that at which the operating handle is moved. The speed of separation of the contacts is usually controlled by a spring.

QUICK C (IBM) : Microsoft BASIC, C and Macro Assembler have set the standard for language compilers. For novice programmers, Microsoft offers Quick BASIC and Quick C which provide many features of Microsoft's professional compilers but at a much lower cast.

QUIESCENT CARRIER TELEPHONY : A system of radio-telephony in which the carrier wave is suppressed when transmission is not actually taking place. See TELEPHONY.

QUIESCENT COMPONENT : A component of an electronic circuit that, at a specific instant will become operative.

QUIESCENT CURRENT : The current that flows in any circuit under specified normal operating conditions but in the absence of an applied signal.

QUIESCENT PERIOD : The period between transmissions in a pulse transmission system.

QUIESCENT POINT : The region on the characteristic curve of an active device, such as a transistor, during which the device is not operating.

QUIESENT PUSH PULL : An arrangement of a final amplification stage in a radio-reciver with the two output valves arranged on one transformer in Push-Pull connection and so connected that practically to current flows when a signal is not being received. The two sets of eletrodes are sometimes placed in one bulb.

QUIET AUTOMATIC VOLUME CONTROL : See DELAYED AUTO-MATIC VOLUME CONTROL.

QUIET TUNING : Automatic reduction of or "Muting" of output of radio-receiver at all times except when accurately tuned to an incoming carrierwave.

QUIKTRAN : Quick Fortran

QUINE-MCLUSKEY ALGOTRITHM : A method of minimization using tables instead of maps.

QWERTY KEYBOARD : The standard typewriter keyboard layout. The name is derived from the first six letter on the top alphabetic row of a typewriter.

R

RACELOGIC : Problem intorduced into a system by propagation delay through two possible paths to the destination.

RADAR : A cronym for Radio Detecting and Ranging , a system that uses beamed and reflected RF energy for the detection and location of objects, as well as for measuring distances or altutides. A system that locates distant objects using reflected radiowaves of microwave frequencies. Modern radar systmes are highly sophisticated and can produce detailed information about both stationary and moving object and can be used for navigation and guidance of ship, aircraft and other vehicles and system. Also known as RADIO LOCATION.

RADIAL : Refers to the bit-significant address of units or devices.

RADIAN : The natural unit of measure of the angle between two intersecting half-lines on the angles from one half-line to another intersecting half-line. It is the angle subtended by an arc of a circle equal in length to the radius of the circle. As the circumference of a circle is equal to 2 times its radius, the number of radians in an angle of 360° or in a complete turns is 2.

RADIANCE : The electormagnetic flux radiated per unit solid angle (steradian) per unit of projected area of the source. Measured as watts per steradian per square meter.

RADIANT ENERGY : Energy transmitted in the form of electromagnetic radiation, such as heat, light, or radio-waves. Energy transferred by an electromagnetic wave; in colour, this refers to the visible part of the electromagnetic spectrum.

RADIANT HEAT THERAPY : Treatment of disease by radiation of a wavelength between 4000 and 7000 A°.

RADIATING CIRCUIT : An oscillating circuit such as the aerial of a radio transmitting station, from which energy is being radiated in the form of Electric Waves.

RADIATION : Any form of energy that is propagated aswaves or streams of charged particles. Radiation take place suddenly, releasing successive photons. The transfer of energy from a system by the sending out of electrical distrubance through the *"Ether"* in the surrounding space, including the low frequency radiation or Electric Waves used in radio-communication, the higher frequency waves or rays known as RADIANT HEAT or INFRA-RED RAYS. Visible radiation or light rays of still higher frequency, visible but chemically active called Ultraviolet Rays, and the extreme cases of "X" or *Rontgen Rays* and *Gamma Rays.* Recent researches point to such radiation being of a discrete corpuscles end owed with a property similar to mass and known as *photons,* rather than being purely wave motion and point to the view that the energy of radiation from atoms is supplied by the slipping of electrons into orbits of less diameter, and that these changes takes place suddenly, releasing successive photons. See QUANTUM THEORY, FLUOURSCENCE, PHOSPHORESCENCE, and LUMINESCENCE.

RADIATION ANGLE : The angle between the surface of the earth and the center of the beam of energy radiated upward into the sky from a transmitting antenna. Also called angle of radiation.

RADIATION CONSTANT : The product of the Radiation Height of an aerial system and the Aerial Current.

RADIATION PYROMETER : A pyrometer such as the Fery ɔryometer in which the temperature of part of a furnace is measured by focusing the rays of radiant heat from the body onto a thermo-junction and finding its temperature rise from the e.m.f. produced.

RADIATION RESISTANCE : The ratio of the total power radiated by an antenna divided by the square of the effective antenna current as measured at the point where power is supplied to the antenna.

RADIATIVE CIRCUIT : An Oscillating Circuit which is capable of radiating energy.

RADIATOR : (1) In an antenna, the part that radiates eletromagnetic waves either directly into space or against a reflector for its direction. (2) A body or surface that emits radiant energy.

RADIO : (1) Indicative of the transmission of eletromagnetic waves through space. (2) A receiver apparatus. (3) A prefix that denotes the use of rsdiant energy in the form of radio waves. Radio also the process of transmitting or receiving such signals. (4) Short for radiofrequency. Denoting electromagnetic radiation in the radiofrequency range or any device, component, or other apparatus used to transmit or receive information at frequency. The use of electromagnetic radiation of frequency within the radiofrquency portion of the electromagnetic spectrum (See frequency band) for the transmission and reception of electrical impulses or signals without connecting wires or waveguides. It is RADIO used in connection with the giving off of other kinds of rays. (5) An expression now used, chiefly to refer, either as a complete to refer, either as a complete word or as a prefix, to all matters connected with communication by means of electromagnetic waves, otherwise called wireless telegraphy or Telephony, but also radio Facsimile (6) A slang term for radio receiver. See RADIO-ACTIVITY, RADIO-GRAPHY.

RADIO ACOUSTIC POSITION FINDING : A subanqueous method of should ranging in which a charge exploded under water is caused to close a radio circuit, and the distance from the observing station is calculated from the time between the receipt of the sound of the explosion transmitted through the water. For application of a similar priciple to sea sounding see ECHO-DEPTH SOUNDER.

RADIO ACTIVITY : The phenomena of the giving out of energy by certain substances, such as radium, due to the detachment of electrons from and consequent successive internal re-arrangement of the atoms, causing the formation of a series of disintegration products. See METABOLON.

RADIO ASTRONOMY : The study of astronomicall bodies and events by means of the radio signals associated with them.

RADIO BUTTON : A round button that appears in a dialog box, offering a choice. Only one radio button can be selected within a group of radio button.

RADIO COMMUNICATION : The transmission of signals by means of electric waves without conductive circuits between the sending and receiving stations. See RADIO-TELEGRAPHY, RADIO TELEPHONY and TELEVISION.

RADIO COMPASS : The navigational aid carried on board aircraft and ships. It consists of a radio receiver together with a directive aerial and isessentially a direction finder. The aerial is rotated in order to find the direction of a specific radio transmitter relative to the craft; the information is presented as the heading of the craft telative to the transmitter.

RADIO ENGINEERING : The practical application of Radio-Communication on a large scale.

RADIO FREQUENCY : Abbreviated RF, electromagnetic radiation having any frequency in the region from 10 kHz to 300 GHz. Also called radio spectrum. This range is divided into eight regions as in the table above. Any frequency of electromagnetic radiation or alternating currents in the range three kilohertz to 300 gigahertz (see frequency band). An electronic that operates in this range is known as a radiofrequency device or radio device; similarly any associated radio frequency. A frequency such as used for oscillations employed by the carrier wave in radio-communication. Cf. AUDIO-FREQUENCY.

RADIO FREQUENCY AMPLIFICATION : Amplification of radio signals at the frequency of the received carrier wave before they are rectified by the detector, and not at the frequency of the modulations.

RADIO FREQUENCY (or RF) AMPLIFIER : An amplifier that amplifies radio frequency signals. In a receiver, it is sometimes referred to as the front end.

RADIO FREQUENCY HEATING : Dielectric heating or induction heating that is carried out using an alternating field of frequency greater than about 25 kilohertz.

RADIO FREQUENCY (or RF) INTERFERENCE : Abbreviated RFI, interference from sources of radio-frequency energy located outside a system.

RADIO FREQUENCY (or RF) OSCILLATOR : An oscillator that generates AC signals at radio frequencies.

RADIO GONISCOPE : An automatic Radio-goniometer with visual indicator.

RADIOGRAM : (1) Popular term for a message received by Radio Telegraphy. (2) See "X" RAY.

RADIO GRAMOPHONE : A combined electric gramophone and radio-receiving set.

RADIO GRAPH : A photograph taken by "X" Rays. See X-RAY PHOTO GRAPH

RADIO GRAPHER : An expert in X-Ray photography usually working under a Radiologist. Photography by means of "X" Rays or rays from Radioactive materials.

RADIO INDUCTIVE INTERFERENCE : Interference with radio-reception due to inductive effects of neighbouring power or communication circuits.

RADIO INTERFEROMETER : See RADIO-TELESCOPE.

RADIO LOCATION : Obsolete name for radar.

RADIOLOGIST : One who practises the medical or industrial applications of "X" Rays.

RADIO METER : Any instrument for the measurement of radiation, originally limited to the Crookes radiometer for the frequencies of radiant heat and light, consisting of a pivoted system carrying light vanes enclosed in a high vacuum. One side of each vane is blackened and its greater absorption causes a difference of mechanical pressure between the two sides of the vanes, resulting in rotation at a speed dependent upon the amount of radiation fallingupon it.

RADIO NOISE : Any unwanted sound or distortion appearing at the loudspeaker of a radio receiver.

RADIO-PHONOGRAM : A written message taken down from a received radio telephone message, for further delivery.

RADIO-PHOTO : A form of the Telephote employing radio-communication.

RADIO PILL : A small capsule that contains a miniature radio transmitter and may be swallowed. Miniature transducers respond to conditions within the body and the signals produced are transmitted to a suitable receiver outside the body. The pill may be recovered after passing through the alimentary tract.

RADIO RECEIVER : Syns. radio; wireless; radio set. A device that converts transmitted radiowaves into audible signals. A simple receiver containsan aerial, a tuner that can be adjusted to the desired carrier frequency, a preamplifier, detector, audiofrequency amplifier, that are associated with the audiofrequency amplifier and are used to restore the lower audiofrequency signals; the treble compensation acts on the higher audiofrequencies. Stereophonic radio receivers contain suitable detecting circuit that the demodulate stereophonic radio transmissions and produce two outputs, each of which is separatley amplified and output to a loudspeaker or frequency-modulated signals. A complete equipment for reproducing in intelligible form signals radiated by radio-transmission. They are described as A/M receivers or F/M receivers, respectively. A receiver that has the facility to detect both types of signal is an AM/FM reciver. Highfidelity (Hi-Fi) radio receivers usually contain extra circuits and a loudspeaker. Modern radio recivers commonly employ super heterodyne reception in order to improve the signal-to-nosie ratio of the output. Radio receivers can detect either amplitude-modulated signals.

RADIO SCOPE : An appparatus for projecting home sound films by controlling the synchronisation of a home cinema with a gramophone at a broad casting station by special signals.

RADIO SONDE : A small radio transmitter together with suitable transducers that is carried by a balloon or kite into the upper atmosphere and transmits meteorological and other scientific data to the ground.

RADIO SPECTRUM : The entire range of frequencies, from approximately 10 kHz to 300 GHz, in which radio waves can be produced. Also called radio frequencies. VLF 10-30 kHz LF 30-300 kHz, MF 300 kHz, -3 MHz, HF 3-30 MHz, VHF 30-300 MHz, UHF 300 MHz-3 GHz, SHF 3-30 GHZ, EHF.

RADIO TELEGRAPHY : The transmission and reception of code signals by means of key-controled electric waves.

RADIO TELEPHONE : Apparatus for sending or receiving messages by Radio telephony. The transmission and reception of spoken words or other sounds by electric waves.

RADIO TELESCOPE : A telescope used in radioastronomy that detects extraterrestrial radio signals. There are two basic forms of radiotelescope. The first consists of a large steerable parabolic reflector; or dish, that can be directed at a desired region of the sky.

RADIO TRANSMISSION : The sending out of electric waves for radio communication. An installation for sending signals by electric waves in radio-communication.

RADIOTRON : A name sometimes given to Thermionic Valve, paticularly when used as an oscillator.

RADIOVISOR : A name adopted for a relay worked by Light Sensitive Cells for such purposes as lighting and extinguishing lamps at dusk and dawn, automatically, burglar alarms and other pupposes.

RADIO WAVE : An electromagnetic wave that has a frequency laying in the radiofrequency range.

RADIX : A number that is arbitrarily made the fundamental number of a system of numbers; a base. Thus, 10 is the *radix*, or *base*, of the common system of logarithms, and also of the decimal system of enumeration. The base of a number system; the number of symbols in the number system. The fundamental number of a number system e.g. 10 in the decimal system. 8 in the octal system, and 2 in the binary positive integer by which the weight of the digit place is multiplied to obtain the weight of the next higher digit place for example in the decimal-numeration system, the radix of each digit place is 10; in a binary code, the radix of each positon is 2.

RADIX POINT : An indicator that separates integral and fractional digits. e.g. the decimal point into the decimal system.

RAM : Acronym for random-access memory. Memory that can be written onto and read from called "random" because it gets or stores information quickly using vertical and horizontal components rather than sequential ones. See RANDOM ACCESS MEMORY.

RAMAC : Random Access Method of Accounting and Control.

RAM-CACHE : Memory set aside for temporarily holding data stored on a disk. By letting the computer access data direct from its memory (the RAM cache) instead of the disk, a RAM cache can make a computer run faster.

RAMDIS : Memory set aside to act like a floppy disk where you can store files. RAM disks are used to make programs run faster, RAM cahes, but a RAM disk generally uses more RAM.

RAMP CONVERSION : A method of A/D conversion by counting the time it takes for reference voltage to equal the input voltage.

RAMDOM ACCESS : The ability to retrieve records without serially searching a file. Access in a storage device under conditions such that the next address from which information is to be obtained is choosen at random.

RANDOM ACCESS MEMORY : The computer system's high-speed work area that provides access to memory storage location by using a system of vertical and horizontal coordinates. The computer can write information into or read information from the random access memory. The part of the computer's memory where data, instructions and results are stored temporarily. RAM is measured in kilobytes and megabytes. Generally the more RAM a computer has, the more capable it is. RAM is a volaltile memory in the sence that it looses all data, information or program entered into it when electric power supply gets cut off. RAM is of two types Dynamic RAM (DRAM) and static RAM (SRAM). RAM is also known as RANDOM-ACCESS STORAGE. Dynamic RAM (DRAM) require refreshing because of the cell circuit design (although it is still volatile, loosing data on removal of power). SRAM is more expensive than DRAM, but is faster and requires less control circuity. An auxiliary memory device on which the programmer can directly access each separate data area without having to search throgh the whole data file.

RANDOM-ACCESS STORAGE : A storage device such as magnetic core, magnetic disk, or magnetic drum in which each record has a specific predetermined address that may be reached directly; access time in this type of storage is effectively independent of the location of the data.

RANK (Automatic Telephony) : Selectors, etc., are said to be of the first second, etc., "rank" according as they perform the first second, etc., stage of selection.

RASTER : The horizontal scanning pattern of the electron gun in television sets and computer monotors that use CRT (cathode-ray tube) display. A raster display device stores and displays data as horizontal rows of uniform grid or picture cells (pixels).

RASTER PLOTTER : Plotter that reproduces displays in dot-matrix patterns.

RASTER SCAN : The generating of image on a screen by focussing an electronic beam on phosphor-coated screen.

RASTER UNIT : On a graphic display screen, a raster unit is the horizontal or vertical distance between two adjacent addressable points on the screen. The distance between two adjacent addressable points located horizontally or vertically on a CRT display.

RATE MULTIPLIER : A divide by n counter where n is programmable.

RATIONALISTATION OF UNITS : The alternation of the "unrationalised" systems of electrical and magnetic units, so that unit magnetic pole and electric charge respectively should emit unit magnetic and electric flus instead of 4 units.

RATIONAL NUMBER : A real number that is the quotient of an integer divided by an integer other than zero.

RAY LOCKING DEVICE IN CATHODE RAY OSCILLOGRAPHS : Electrostatic or other arrangements contained in a special chamber to deflect the electron jet a way from the aperture in the diaphragm to prevent in affecting the photographic film before and after the events to be recorded.

RAYS : A term used in general for transfer to energy throuth space, either by true Radiation or by projection of particles, but used by some writers to denote only the path of such radiation, etc.

R-C FILTER : Any passive filter network made up of only resistors and capacitors.

R-C NETWORK : Any L shaped circuit made up of resistors and copacitors arranged to perform a specific function.

RCTL : Resistor-capacitor-transistor logic; a variant of resistor transistor logic in which a copacitor is connected across the series resistor to permit faster switching.

REACH : Defines the robot's arm movement or work envelope.

REACTANCE CHART : A chart that is presented in a form that enables the user to read directly the reactance of any given capacitor or inductor at a specified frequencey, and conversely to deduce the capacitance or inductance of a given reactance at a particular frequency.

REACTANCE DIMMER : A Dimmer used for A.C. stage lighting circuits in which the reactance of a choking coil in series with the lamps is varied by altering the saturation current is sometimes controlled through a thermionic valve by manipulation of a small rheostat in the grid circuit, so that a compact control board can be used in a large number of circuits.

REACTIVE CURRENT : Syns. *reactive component*, wattless component, idle component, quadrature component of the current. The component of an alternating current vector that is in quadrature with the voltage vector.

REACTIVE FACTOR : The ratio of the ractive volt-amperes of any load, circuit, or device to the total volt-amperes.

REACTIVE LOAD : A load in which the current and voltage at the terminals are out of phase with each other. Compare NONREACTIVE LOAD

REACTIVE VOLT-AMPERES : Syns. *reactive component, wattless component, idle component, quadrature component of the voltage.* The component of an alternating voltage vector that is in quadrature with file current vector.

REACTIVE VOLTAGE : Also known as *reactive component wattless component*, idle *component, quadrature component of the voltage.* The component of an alternating voltage vector that is in quadrature with the current vector.

READ : Retrieve data from memory or graphic input device; process halts graphics oprations. To sense, or obtain the state of data either in memory, or from an input device.

READ AROUND RATIO : Roughly, the number of times a bit stored in an electrostatic storage can be read or written before the surrounding bits must be regenerated.

READ IN : To place words of fields in storage at specified addresses.

READ ONLY MEMORY : Abbreviated ROM. A type of memory that contains permanent data or instructions. The computer can read from but not write to the read-only memory. A memory element capable of being read by incapable of being altered. Memory that cannot be written under program control; used to store microinstructions. The part of the computer's memory that contains a permenent store of instructions for the computer.

READ OUT : To copy words or fields from specified addressed in storage into a acternal storage device, or to output words by copying them from specified addresses in storage.

READTIME : Of processing data about an operation first enough so that the processed data can be used to control the operation before it has gone to completion, as for example in controlling a moving friggate. A memory which cannot be modified or reporgrammed. Typically used for control and program execution.

REAL NUMBER : An ordinary number, either rational or irrational; a number in which there is no imaginary part, a number generated from the single unit, 1; any point in a continum of natural numbers filled in with all rationals and all irrationals and extended indefinitely, both positive and negative.

REAL TIME : Solving a problem by computer in the time the activity is taking place. The flight of rocket ships is an example of real-time solution. You can't wait for the answer until after the plane has crashed time logged on the robot. The time in which it is or could be performing work. The actual time required to solve a problem. (2) The process of solving a problem during the actual time that a related physical process takes place so that results can be used to guide the physical process.

REAL TIME IMAGE GENERATION : Performance of the computations, necessary to update the image is completed within the refresh rate, so the sequence appears correctly to the viewer. An example is flight simulation, in which thousands of computations must be performed to present an animated image, all within the 30 to 60 cycles per second rate at which the frames change.

REASON MAINTENANCE SYSTEM (RMS) : Is an extended truth maintenance system capable of handling inferences in nonmontonic logic.

REASONABLENESS CHECKS : General range checks on data to be sure that values are within reason.

REASONING UNDER UNCERTAINYT : Is the problem of making inference and solving problems when some of the input data or stored knowledge is not known to be true; it often involves assigning certainties or probabilities to pieces of information and developing procedures that can calculate the cartainty of their conclusions from the certainties of their inputs.

RECEIVER :The part of a telecommunication system that converts transmitted waves into a desired from of output. A complete apparatus for the reception of incoming electrically transmitted meassages, signals, etc. Used in telephony for the actual apparatrs which is held to the ear.

RECEIVING AERIAL : An aerial used for the reception of radio signals.

RECEIVIING CIRCUIT : The apparatus and connections of the equipment of a radio-telegraph or telephone station used exclusively for reception of messages. Cf. TRANSMITTING CIRCUIT.

RECEIVING EARTH IN A SUBMARINE CABLE : A Sea Earth used for receiving purpose only usually at a point out at sea where the depth is over 500 ft. Cf. TRANSMITTING EARTH.

RECEIVING PERFORATOR : An apparatus such as that used in the Creed Printing Telegraph system which automatically punches a paper strip according to the signals received, thus reproducing the strip which is being passed through the transmitting apparatus.

RECEIVING STATION : (1) A station at the far end of a long power transmission line where the energy is utilised or whence it is distributed in a convenient form. (2) A station for the reception of radio or other electrical signals messages.

RECIPROCAL COUNTER : A special purpose counter which measures the period of a periodic signal and converts the result of the measurement with an internal arithmetic unit into frequency, which is displayed.

RECIPROCAL THEOREM : If an electromotive force, E applied at a point in a networks produces a current, I, at a second point, then if the same e.m.f. is applied at the second point it will produce the same current at the original point.

RECORD : Used by database to denote a unit of information, such as someone's name, address, and phone number. A series of adjacent characters with a storage medium. A group of words or fields on a related subject e.g. the memorandum an employee's employment history.

RECORD LOGICAL : A collection of related data items.

RECORD PHYSICAL : One or more logical records combined to increase input/output speeds and to reduce space required for storage.

RECORDING : The ability to prevent two or more people from modifying a single database record. Used with Local Area Networks (LANs).

RECORDING ATTACHMENT TO A GRAMOPHONE : An applicance similar to a Gramophone Pick-up, but with a recording instead of a reproducing needle, mounted on an arm with mechanism giving an axial feed motion.

RECORDING DENSITY : The closeness with which data is stored on magnetic tape. The most common densities are 200, 556, 800 and 1600 characters per inch.

RECTANGULAR COORDINATES : Spatial coordinates defined by three distances. Some robots are programmed in rectangular coordinates.

RECTANGULAR PULES : See PULSE.

RETIFICATION : The conversion of an alternating current into a unidirectional current. See RECTIFIER and RECTIFIED CURRENT.

RECTIFIED CURRENT : A current that as beem converted from, an alternating current into a unidirectional current, either by suppressing alternate waves, or by reversing their direction; with or without special means being taken to smooth out the pulsations.

RECTIFIER : An apparatus which converts an alternating current into a unidi-rectional current, without the use of dynamo electric machinery, including apparatus depending upon synchronously moving commutators and electrolytic and therminonic applicances, mercury vapour.

RECTIFIER PHOTO ELECTRIC CELL : The class of Photo-electric cells of the Photo-voltaic type in which electrons pass from a semiconducting to a conducting layer under influence of radiation. Also called Barrier-Layer Cell.

RECTIFYING DETECTOR : A detector of electric waves, such as a crystal or thermionic valve, which acts by rectifying the received oscillations so that they given an audiable effect in a telephone receiver, due to the successive impulses from each Carrier Wave having a cumulative effect on the diaphragm instead of cancelling out.

RECTIFYING VALVE : An electrolytic or therminonic valve used as a Rectifier.

RECTIGON RECTIFIER : A rectifier of the thermionic valve used as a Rectifier. A rectifier of the Thermionic Valve type, with tungsten filamentacathode and graphite anode in an argon-filled bulb; used for battery charging from A.C. circuits.

RECUPERATIOR : A form of Phase Advancer proposed by Lebalnce, in which a copperdisk with suitable connections to the sliprings of an inductionmotor is allowed to vibrate freely in a direct current field. Cf. KAPP VIBRATOR.

RECURRENT CIRCUIT : A circuit consisting of successive sections, each containing the same arragement of resistance, inductance, and capacitance as used for Artificial Lines and some kinds of Filters. (Also called Lattice Circuit.)

RECURSION : Occus when a procedure causes itself to be executed, eiher by calling it directly or by calling another procedure that in trun calls the original, before it has completed its computations; it is contrasted with iteration, when a procedure loops within RECURSION itself.

RECURSIVE TRANSITION NETWORK (RTN) : A device that paresc sentences with context-free gra   rs by representing the set of rules for rewriting each nonterminal symbol as a separate finite automaton that can make recursive jumps to other automata as the condition for branching to a new state.

RED TAPE : Computer operations which do not directly contribute to useful data handling extract one of fields, verifying the identification of the input etc. in general, the set-up (including initialization) and clean up operations in a program.

REDIFFUSION RADIO : See AUDIO-FREQUENCY

REDIFFUSION REDUCTION (of Boolean Expressions) : The process of determining the minimum expression capable of satisfying a logic problem.

REDUNDANT CHECK : A check which uses extra bits or characters in words, but not complete diplication to help detect errors.

REED RELAY : A relay that has contacts mounted on magnetic reeds sealed inside a small glass tube. An actuating coil is wound either around the tube or on a ferrite core to provide the magnetic field to open and close the reeds. The contacts may be either dry or mercury-wetted. Also called a magnetic reed relay. See also MAGNETIC REED RELAY.

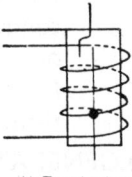

i b Reed relay

REED TYPE LOUD SPEAKER OR TELEPHONE : A Loud Speaker telephone receiver in which the variable pull of the magnetic system is not exerted directly upon the diaphragm, but acts upon a spring-mounted "reed" connected mechanically to a large diaphragm of non-magnetic material.

REENTRANT PROGRAM : Synonymous with pure procedure.

REFERENCE DIPOLE : A straight, halfwave, dipole antenna designed to operate at a specific frequency and used as a comparison for antenna measurements.

REFLECTANCE MODEL : Function which describes light on a surface by making assumptions concerning light sources, angles, surface texture, and other factors. Also called illumination model.

REFLECTED WAVE : A wave reflected from a surface, discontinuity, or junciton of two different media, such as the wave that travels back to the source end of a mismatched transmission line.

REFLECTOR : A single rod, system of rods, metal screen or sheet used behind an antenna to increase its directivity.

REFLIX : Reflex can display data in several different views that let you see relationships between data. To help you, Reflex includes graphing capabilites plus copatibility with both dBASE III Plus and Lotus 1-2-3 files.

REFRESH : The act of periodically restoring degraded digital data to their original value.

REFRESH (or vector refresh) : A CAD display technology that involves frequent renewing of an image displayed on the CRT to keep it bright, crisp, and clear. Refresh permits a high resolution. Selective erase or editing is possible at any time without erasing and repainting the entire image. Although substantial amounts of high-speed memory are required, large, complex images may flicker.

REGENERATION : Positive feeback also called regenerative feed back. (1) Retroaction in wireless receiving appparatus. (2) The reconditioning of exharsted apparatus as in a Regenerative Cell, or in the case of a Thermionic Valve by heating the filament for some hours without an anode voltage to bring the thorium or other active material to the surface.

REGENERATIVE DETECTOR : A vacuum tube (or transistor) detector circuit in a radio receiver in which RF energy is fed back from the plate (collector) circuit to the grid (base) circuit to give positive feedback at the carrier frequency, which thereby increases amplification and sensitivity of the detector REGENERATIVE FEEDBACK See POSITIVE FEEDBACK.

REGENERATIVE RECEIVER : A radio receiver that uses a regenerative detector. A repeater for cable signals for converting signals of distortedwave form into sharp signals with flat wave form of the correct lenght, one variety of which consists of a form of commutator rotating synchronously with the signal elements. See RETROACTIVE CIRCUIT.

REGIATRATION : The degree of accuracy in the positioning of one layer or overlay in a CAD display or artwork, relative to another layer, as reflected by the clarity and sharpness of the resulting image.

REGISTER : A memory element capable of containing one binary word. The hardware for storing a word temporarily, usually in the arithmatic and logic unit, not in the storage unit. A device capable of temporarily storing a specified amount of data, usually one word, while or until it is used in an operation. An electronic circuit within the microprocessing unit that is capable of storing one or more bytes of information. The most common register is the accumulator register, in which all of the arithmetic operations are performed. Memory locations in the ALU and control unit. A certain type of temporary storage unit for digital information. In general, storage locations capable of holding data. In particular, index, registers that can be used to modify instruction addresses, or artihmetic registers that perform calculations.

REGULAR EXPRESSION : (RE) A simple formalism for representing regular languages.

REGULAR GRAMMER : (RG) is a restricted context-free grammer, allowing only rules with a single mon-terminal symbol on the right-hand side, that is just powerful enough to represent regular languages.

REGULAR LANGUAGE : Is the simplest type of language, and includes all finite languages as well as those infinite languages which follow simple linear patterns; natural languages are most definitely not regular.

REINARTZ CIRCUITS : High or low frequency amplifying stages. An arragement of radio-receiving sets, particularly for short waves, comprising an untuned aerial circuit coupled to a tuned secondary circuit connected to the grid of the detector valve, and containing a retroaction coil. This can be combined with high or low frequency amplifying stages.

REJECTOR CIRCUIT : A circuit containing a wave-filter, consisting of a combination in parallel for priventing the passage of currents of a certian frequency. Used in radio telephony for tuning out stations not required, and in power circuit for removing harmonics liable to cause inductive interference. (Also called Stopper Circuit) Cf. ACCECPTION CIRCUIT. See WAVE TRAP.

RELATIONAL DATABASE : A database program that can retrieve and manipulate data stored in two or more files. Common relational databases include dBASE III Plus, Oracle, and Paradox.

RELATIVE ADDRESS : A symbolic address that includes specific indication of the number of absolute address it is separate from some origin.

RELATIVE LUMINOSITY : The ratio value of the luminosity at a particular wavelenght to the value at the wavelenght of maximum luminosity (555 nanometers for photopic vision).

RELAXATION : A method of constraint satisfaction and computation in connectionist networks that repeteatedly updates and contents of a network until it reaches quiescence, indicating that it has stabilized in a solution that satisfies all the local constraints.

RELAXATION OSCILLATIONS : Oscillations in which the damping conditions of the circuit are such as to cause them to be interrupted and re-established again repeatedly.

RELAY : An electrical device in which one electrical phenomenon (current, voltage, etc.) controls the switching on or off of an independent electrical phenomenon. A switch operated by electricity. Unlike a transistor, a relay has moving parts. There are many types of relay, most of which are either electromagnetic or solid-state relays. See MAGNETIC DIAPHARAGM RELAY, MAGNETIC REED RELAY, SOLID STATE RELAY, THERMIONIC RELAY.

RELAY (Automatic Telephone System) : A development of the Betulander system of working in which electromechanical switches are entirely replaced by Relays.

RELTIVE COORDINATE SYSTEM : A coordinate system whose original moves relative to world or fixed coordinates.

REMRAND : An Abbreviation for Remington Rand a division of Sperry-Radn corp.

REMOTE : A term used to refer to devices that are located at sites away from the certral computer.

REMOTE BATCH SYSTEM : A type of computer system in which batch jobs are entered into the computer from a remote location and the output is returned to that location.

REMOTE CENTER COMPLIANCE (RCC) : Lateral and ratational float and greatly earses robot or other menchanical assembly in the presence of errors in parts, *jigs, pallets* and *robots*. It is especially useful in performing very-close-clearance or interference insertions. A complaint divece used to interface a robot or other mechanical workhead toits tool or working medium. The RCC allows a gripped part to rotate about its tip or to translate without rotating when pushed laterally at its up. The RCC thus provides general.

REMOTE CONTROL : Heavy current or high voltages up to the operating positions of a switchboard. The control of switchgear, etc., by electromagnetic, electropneumatic, mechanical or other means from a point or points at a distance from the switches, etc., themselves. Partly for purposes of convenience and partly to avoid bringing connections for.

REMOTE CUTOFF TUBE : An electron tube, used principally in RF amplifiers, whose control grid wires are further apart at the center than at the ends. Consequently, the tube's amplification is not directly proportional to the bias. Also called a variablemu or extended-cutoff tube.

REMOTE TERMINAL : Computer terminal which is cabled to a larger computer. A remote terminal may or may not have local processing capability.

REN : Remote enable; general interface-management bus line.

RHEODYNE : A circuit for a Thermionic oscillator in which resistance coupling is used between the plate and grid circuits.

REPAINT : Refresh a display surface with an updated display.

REPEATABILITY : (1) Closeness of agreement of position movements, repeat under the same conditions during a long time interval, to the same location. (2) Ability of a hard copy or soft copy graphics device to retrace the same lines without deviation. In robots, the ability to repeat a motion and arrive at the same endpoint. A robot with a repeatability of 0.005 inch can always put the next part within that distance from where it put the last one. (See "Accuracy.")

REPEATER : (1) Any apparatus by means of which the variations of current in one circuit are reproduced in another circuit in amplified form, including a relay used in long telegraph circuits for the purpose of increasing the working speed by subdividing the line into independent sections, the working speed of each of which is considerably better than that of the whole line. (2) A telephone Repeating Coil. (3) A Thermionic Value used as a telephone Relay.

REPEATER STATION (Telephone etc.) : A station where incoming modulations are amplified by thermionicvalves for transmission to a further length of land line. Extensively used both in turnk line telephony and in simultaneous broadcasting.

REPEATING STATION : A telegraph station at an intermediate point in a longline where repeaters are used.

REPLICATE : To generate an exact copy of a design element and locate it on the CRT at any point(s), and in any size or scale desired Report Output, usually in human language form.

REPORT GENERATOR : A program that reads information concerning items to be retrieved from a file and their required output format, processes the file to retrieve the desired records, and produces a report according to the specified format.

REPORT PROGRAM GENERATOR (RPG) : A class of languages used to prepare programs to print reports quickly from a set of files. Can also be used to program complete applications.

RESEARCH : A company formed in 1972 by Seymour Cray at which a series of high-speed super computers have been developed including Cray-1 and Cray-2.

RESET : When a device or an instrument is returned to zero or to another specified condition. To restore an electrical or electornic device or apparatus to its original state following operation of the equipment. See also clear and hence become relatively remote from the plates. Only the charge near to the plates is removed by rapid discharge.

RESIDUAL CHARGE : The portion of the charge stored in a capacitor that is retained when the capacitor is discharged rapidly and may be withdrawn from it subsequently. It results from viscous movement of the dielectric under charge causing some of the charge to penetrate device.

RESIDUAL CURRENT : Current that flows for a short time in the external circuit of an active electronic device after the power supply to the device has been swithced off. The residual current resulted from the finite velocity of the charge carriers passing through the device.

RESIDUAL INDUCTION : See MAGNETIC HYSTERESIS.

RESIDUAL MAGNETISM : Syn. for REMANENCE. See MAGNETIC HYSTERESIS.

RESIDUAL RESISTANCE : The inherent resistance of a conductor that is independent of temperature variations. It is usually ascribed to irregularities in the molecular structure of the material.

RESISTANCE BRAZING : Electrical brazing that utilizes the heating effect of a current to provide the heat source.

RESISTANCE COUPLED AMPLIFIER : A multivalve amplifier with resistance, or resistance capacity coupling, between the stages.

RESISTANCE GAUGE : A guage that is used to measure high fluid pressures by measuring the change in electrical resistance of a sample of manganin or mercury when the sample is subjected to the pressure. The gauge must be calibrated against known pressures before use.

RESISTANCE LAMP : An electric lamp that is used to limit the current flowing in a circuit.

RESISTANCE SPEECH, THERMIONIC, TUNED CIRCUIT AND VALVE RESISTOR : A component which resists the passage of electricity.

RESISTANCE THERMOMETER : An electrical thermometer that utilizes the change in electrical resistance with temperature of a wire to measure the temperature of its surroundings. It consists of a small coil of wire (usually platinum but other metals may be used at low temperatures) would on a mica former and enclosed in a sheath of silica or porcelain (see diagram). The change in resistance is determined by placing the coil in one arm of a Wheat stone bridge. Compensating leads are usually added to the other arm of the bridge to compensate for termperature variations in the leads since the coil is usually remote from the measuring instrument. Resistance theremomters can be used over a wide range of temperatures from 200°C to over 1200°C.

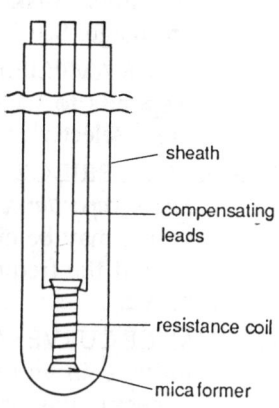

sheath

compensating leads

resistance coil

mica former

Resistance thermometer

RESISTANCE WELDING : Electrical welding in which the heating effect of a current is utilized to provide the heat source. An electric current is passed across the contact surface between the components to be welded and local melting of the metals occurs the surface in control.

RESISTANCE WIRE : Wire constructed from a material, such as *nichrome* or *constantan*, that has a high resistivity and low temperature coefficient of resistance. It is used for accurate wire-wound resistors.

RESOLUTION : The sharpness of an image in print or on the screen, resolution is usually measured in dots per inch (dpi) for printers and pixels for screens. The higher the *resolution*, the sharper the image. The voltage increment of D/A or A/D converter that results from an change of one least significant bit. In practice, the value of the least significant digit (LSD). A measure of ability to delineate detail, or to distinguish between nearly equal values of a quantity. Aso called *resolving power*. (1) The least interval between two adjacent discrete details which can be differentiated one from the other. The smallest increment of distance that can be read and acted upon by a numerical-control system. (2) Number of pixels per unit of area. A display with a finer grid contains more pixels and thus has a higher resolution, making it capable of reproducing more detail in an image. See VERTICAL RESOLUTION.

RESONANCE AMPLIFIER : An amplifier with a tuned circuit which responds to the frequency in question, thus obtaining very great selectivity and sensitivity.

RESONANCE BRIDGE : A four-arm AC bridge that measures *inductance, capacitance, or frequency*. The inductor and capacitor, which may be either in series or parallel, are tuned to resonate at the frequency of the source before the bridge is balanced.

RESONANCE CURVE : A curve showing the way in which the current induced in an oscillating circuit varies according to the frequency of the oscillations inducing it (other factors being equal); showing a sharp peak at the natural or resonating frequency of the circuit.

RESPONDER : An early form of detector used in the de Forest radio system, consisting of a tube containing Lead Trees which are broken up by the received oscillations, causing a sufficient alternation of resistance to produce a click in a telephone receivor.

RESPONSE TIME : The time from submission of a request until the computer responds. The time span between the end of a request for information or action and the beginning of receipt of a reply.

RESULT : To change the contents of a register or address to some arbitrary chosen word, usually zeros, or to the original contents.

RETARDATION METHOD : (1) Of measuring losses : A method of measuring the friction, windage, and iron losses in a machine, by dirving it at a known speed, throwing off the power suddenly and taking observations of the rate at which the speed diminishes as it gradually comes to rest, with the brushes up and down and the field excited and non-excited. The moment of inertia of the armature must be known. (2) Of simultaneous telegraphy and telephony : The class of system in which a high inductance or *Retardation Coil* is relied on to prevent telephone currents passing through the telegraph instruments.

RETORACTION : The effect of coupling, intentional or otherwise, between the anode and grid circuits of a thermionic valve, or the input and output circuits of a group of valves tending to magnification or reduction of the amplitude of the output, but if carried too far set up disturbing self-oscillations unless special precautions are taken. Also called Reaction and Feed-Back.

RETROACTIVE CIRCUIT- : An arrangement of a radio receiver to take advantage of Retroaction to obtain increased sensitivity. Also called a Regenerative or Reaction circuit.

RETUN TO ZERO : A method of magnetic recording in which each bit it recorded individually with the magnetization being natural (zero) between each bit in a sequences of bits.

REVERSE GRID CURRENT : A Grid Current in a direction such that it flows out of the valve at the grid.

REVERSE VIDEO : A display of characters on a background, opposite of the usual display.

REVERSING KEY : A simple appliance for reversing small currents, as in telegraphy and experimental work.

REVERTIVE CONTROL : Control of selectors, etc., in automatic telephony by impulses originating in the selector itself when once started, counted by the control apparatus and automatically stopped when the movement is finished.

RE-WRITE RULE : A rule in a grammer that allows the replacement of a combination of symbols by another combination of symbols so as either to generated sentences of the language (top-down) or reduce a given sentence to the root.

RGB : Acronym for Red-Green-Blue, refers to a type of color monitor. Commonly used to refer to the color space, mixing system, or monitor in color computer graphics. In RGB, a colour is defined as percentages of red, green, and blue, with 0,0,0 equivalent to blank and 1000, 1000, 1000, equivalent to white. See RGB MONITOR.

RGB MONITOR : A red/green/blue monitor, the type of CRT screen that produces colour images when a trio of red, green, and blue electron guns focus on phosphor triade RGB have equal percents of red, green and blue between 0 and 100 i.e. 25,25,25 is a dark gray.

RHEO : A prefix used in several practically obsolete electrical terms meaning anything flowing, i.e. a current. Thus the term "Rheostat" (which is the only one with this prefix in common use now) signifies an apparatus for keeping a current constant, i.e. a variable resistance. The prefix "Rheo" has thus become falsely indentified with resistance, although originally it was only used to mean current. Examples of both uses are given below.

REODYNE CIRCUIT : A circuit for a Thermionic Oscillator in which resistance coupling is used between the plate and grid circuits. Cf. ORTHODYNE CIRCUITS.

RHEO-MICROPHONE : A word sometimes used for a microphone depending on variation of resistance.

RHEOSTAT : Alternatively a number of small resistors may be used with a rotary switch that selects the appropriate value. The term is usually applied to physically large devices. Small rheostats are usually called potentiometers. A variable resistor that may be connected, in series, into a circuit and used to alter the current flowing through the circuit. A common arrangement is a linear or circular wire-wound resistor with a sliding contact whose position can be altered. An adjustable resistance for current whose position can be altered. An adjustable resistance for current regulation; originally applied to forms in which the resistance wire was wound on the outside of a cylinder arranged to be rotated to bring any point a fixed contact. Now applied to any form of variable resistance.

RHOMBIC ANTENNA : A horizontal, directional antenna that has four sides forming a diamond or rhombus. It is fed at one apex and terminated with a resistor or impedance at the opposite apex. Also called a *Bruce antenna.*

RHUMBATRON : See CAVITY REASONATOR.

RIBBON CABLE : A flat cable that consists of parallel strips of wires. A group of attached parallel wires.

RIBBON MICROPHONE : A microphone that consists of a very thin ribbon of aluminium alloy a few millimeters wide loosely fixed in a strong megnetic flux density parallel to the plane of the strip. A form of Moving Conductor Microphone in which the moving conductor is a thin fluted aluminium ribbon also forming the diaphragm.

RIGHI OSCILLATOR : An oscillator in which the spark gap is between two metal spheres placed between two smaller spheres.

RILM : Repertoire International dela Litterature Musicale.

RING : (of a Telephone plug) : The metal contact roing on a telephone plug between the TIP and the sleeve.

RING COUNTER : A counter formed by circulating a 1 in a shift register whose serial output has been connected to its serial input. See shift register detector. A circuit in which four diodes are connected in series to form a ring around which current flows readily in only one direction. The input and output connections are made to the four modes of the ring. Used as a balanced modulator, demodulator, or phase.

RINGING : See PULSE.

RINGING GENERATOR : A generator in a telephone exchange for supplying low frequency alternating current for actrating bells throughout the system.

RING NETWORK : A system in which all stations are linked to form a continuous loop or circle.

RING TONE (in automatic telephony) : A signal heard in the receiver after dialing is completed consisting of an intermittent low note with alternate long and short intervals signifying that connection has been made with the required line which is being rung.

RING WINDING : Gramme winding. A method of fabricating a coil, particularly for a winding in an electrical machine, in which the coil is wound on an annular core. One side of each turn of the coil is threaded through the ring to form a toroid.

RING WIRE IN TELEPHONY : See "R" WIRE.

RIPPLE : (1) An alternating-current component superimposed on a direct-current component resulting in variations in the instataneous value of a unidirectional current or voltage. The term is particularly applied to the output of a rectifier. See PULSE also.

RIPPLE COUNER : A counter formed by connecting each flip-flop output to the T-lead of its next stage.

RISC : (reduced instruction set computer) : A computer in which the CPU has a streamlined set of instructions for the operations most often executed.

RISE TIME : The time required for the leading edge of pulse to rise from one-tenth of its final value to nine-tenths of its final value. See PULSE.

RJE : Remote Job Entry

RMS : Abbreviation for Root Mean Square.

ROBOT : A programmable machine intended to perform a job originally conceived in terms of a human operator. A programmable, multifunctional manipulator. (See"Manipulator"). The word had been derived from the Czech word robota, meaning forced labour. Machine which can perform some of the movement of a person or animal, or a machine which moves and uses tools according to programmed instructions.

ROBOTICS : A research program that studies the construction of artifical autonomous agents, fully intelligent and able to perceive their environment and move about in it just like a human being.

ROCKY POINT EFFECT : A name used for an effect in large thermionic valves, somtimes causing considerable disturbance, due to a sudden increase in emission caused probably by irregularities in the cathode structure.

ROLL : A loss of vertical synchronization which causes the displayed image to move up or down.

ROLL-OFF : A gradual decrease in signal level.

ROM : Acronym for Read Only Memory.

ROM BIOS : Acronym for Read Only Memory, Basic Input/Output system. A special chip used to provide instructions to the computer when you turn the computer on.

RONTEGENOGRAM : See X-RAY PHOTOGRAPH.

ROOT (NODE) : A node at the top of a tree, from which all other nodes in the tree terminate. In search trees it represents the *start* state, in game trees it represents the *starting* position, and in backward chaining it represents the *initial* fact to be proved.

ROOT-MEAN-SQUARE (R.M.S.) VALUE : Syns. *effective value; virtual value.* The square root of the mean value of the squares of the instantaneous values of a periodically varying quantity the r.m.s. value is equal to the peak value divided by the square root of 2. In case of a true sine wave, $1/\sqrt{2}$ of the maximum value. The value of the square root of the mean value of the square of an alternating current or voltage or other periodic current, being the value that a direct current would have to be equivalent in heating effect, etc. Also called Effective or Virtual Value.

ROS : Read-Only Storage.

ROSSTALK : Interference due to cross coupling between adjacent circuits or to intermodulation (see MODULATION) of two or more carrier channel, producing an unwanted signal in one circuit when a signal is present in the other. It is common in telephone, radio.

ROTARY AUTOMATIC TELEPHONE SYSTEM : An automatic telephone system, such as that of Lorimer and one form of Western Electric system, in which the moving members of the selectors have a continuous rotary motion instead of a step by step action. Cf. STEP BY STEP AUTOMATIC TELEPHONE SYSTEM.

ROTARY HARMONIC ABSOBBER : A Harmonic Absorber in the form of a *Synchronbus Motor* running at a speed corresponding to synchronism at the frequency of the harmonic to be absorbed.

ROTARY PHASE ADVANCER : A *Phase Advancer* with a rotating armature, e.g. the *Walker* and *Scherbius Phase Advancers*. Cf. VIBRATORY PHASE ADVANCER.

ROTATING LOOP RADIO TRANSMITTER : The name usually given to a particular form of Rotating Radio-Beacon, with a rotating loop aerial, used for aircraft to obtain their own bearings by taking the time that elapses between the reception of a non-directional signal coinciding with the North position of the "Beam" and the directional signal of the beam it-self.

ROTATING RADIO BEACON : A radio transmitter which emits a concentrated beam which rotates in a horizontal plane at a constant speed, giving out different signals in different directions, so that ships and aircraft can as certain their bearing, without directional receiving apparat. For another system see ROTATING LOOP RADIO-TRANSMITTER.

ROTATIONAL DELAY : On rotating secondary memory devices, the time required for particular record to arrive under the read-write head.

ROTE LEARNING : It is learning by memorization, free or any real under-standing of what is being learned or why.

ROUND OFF : To change a more precise number to less precise one, usually by choosing the nearest precise number of a given length.

ROUTER : A PC or IC design application program which automatically determines the optimal interconnection of signals, based on design parameters established by the user.

ROUTING : A program or part of program. Set of instructions to the computer to perform a certain function. For example, an inking routine translates the graphic input form a light pen into visible continuous lines, giving the user the impression of sketching with the stylus. A group of program instructions which accomplishes a particular task that needs to be done frequently. A group of instructions can be used over and over again, simply by referring to the largest of five numbers.

ROW DOMINANCE : A technique in tabular minimization of eliminating an implicant if it covers fewer of the same minterms than another implicant.

RPG : Report Program Generator.

RS FLIP FLOP : A flip-flop whose inputs are designated R and S. Logical $1s$ should not be allowed to appear on the inputs together. A flip-flop with two inputs a *set* input and a *reset* input. If the set input is enabled (high), the flip-flop goes to the 1 state. If the reset input is enabled (high), the flip-flop goes to the 0 state.

RS-232 : A standard communication interface between a modem and terminal devices that complies with EIA Standard RS-232. Another term for a serial port, a common standard for serial data communication between pieces of computer equipment it permits linkage of host computer and graphics terminal or other peripherals over long distances. See also SERIAL PORT.

RST FLIP FLOP : Has three inputs designated R, S and T. The R and S inputs produce outputs as described above. Application of a plause to the T input causes the device to change state.

RTL : Resistor-transitor logic; a family of IC logic. A system of transistor logic in which a resistor is included in series with the base of each transistor in order to reduce differences in transistor currents.

RUBBER-BANDING : A capability that allows a component to be tracked (dragged) across the CRT screen, by means of an electronic pen, to desired location, while simultaneously stretching all related interconnections to maintain signal continuity.

RULE : See PRODUCTION RULE.

RULE BASED : Programming/System describes a methodology that views programs declaratively, as production systems or sets of logical aximos, rather than as collections of interactive procedures; it embodies the knowledge based approach used in applied AI.

RUN : (1) To operate an automatic computer (2) The time necessary for the computer to execute a program. (3) A specified part of the program.

RUN-LENGTH ENCODING : A data-compression technique for reducing the amount of information in a digitized binary image. It removes the redundancy that arises from the fact that such images contain large regions of adjacent pixels that are either all white or all back. RW Read/Write.

'R' WIRE : The wire in a telephone circuit within the exchange connected to the *Ring* of the plug and to the "*B*" *wire* of the line. Also called RING WIRE.

SAGE : Semi Automatic Ground Environment.

SAINT : Symbolic Automatic Integrator.

SAMPLE : Query a graphic device for coordinate data or operating status.

SAMPLING : A technique in which only some portions of an electrical signal are measured and are used to produce a set of discrete values that is representative of the information contained in the whole. In order that the output values represent the input signal without significant loss of information the rate of sampling of a periodic quantity must be at least twice the frequency of the signal.

SAMPLING RATE : The rate at which measurement are made. In, for example the frequency mode, the sampling rate is the inverse of the sum of gate time and display time. Frequency at which points are recorded in digitizing an image. Sampling errors can cause aliasing effects.

SAP : Symbolic Assembly Program.

SAPATIAL DATA : Represented spatially, i.e. on a map.

SATELLITE : An artificial body that is projected from earth to orbit either the earth or another body of the solar system. There are two mainclasses: information satellites and communications satellites.

SATELLITE COMMUNICATIONS : The use of orbiting satellites to receive, amplify, and retransmit data to earth stations.

SATIRA·TOPM (magnetic) : The maximum possible degree of magnetization of a material; it is independent of the strength of the magnetic flux density applied to the material. All the domains in the material at saturation are assumed to be fully oriented with respect to .

SATURATION : A transistor is saturated when a further increase of base current causes no further increase in collector current. (1) A condition where the output current of an electronic device is substantially constant and independent of voltage. In the case of a device such as a field-effect transistor or thermionic valves saturation is an inherent to the device. In the case of a bipolar junction transistor, saturation occurs because the output from the collector eletrode is limited by the circuit elements of the external circuit and changing these alters the magnitude of the saturation current drawn from the device. (2) A subjective term which usually refers to the difference of a hue from a gray of the same value. Colours can be desaturated by adding white, adding black, adding gray, or adding the complementary colour. In a subtractive system, adding the complement will the magnetic flux density (see ferromagnetism). Make the colour darker. In an additive system, adding the complement will make the colour lighter. This creates confusion since value, as well as saturation, is changed. (See "Chroma" and "Purity")

SATURATION CURRENT : The current shown on the portion of the static characteristic of an electronic device where it is substantially constant and independent of voltage. Very little further increase of current with voltage occurs until breakdown is reached. The value of the saturation current is a function of the device and the external circuit.

SAW TOOTH WAVE : A periodic waveform whose amplitude varies approximately lineraly between two values, the time taken in one direction, the active interval, being very much greater than the time taken in the other. The shorter period is termed the *flyback*. See PULSE.

S-AXIS : The horizontal axis on the screen of a cathode-ray tube.

S-BAND : A band or microwave frequencies ranging from 1.55 to 5.20 giga-hertz. See FREQUENCY BAND.

SCALE : To change the units of measurement of a quantity, or to bring its representation to within the capacity of a register or of an addrees. For example, $5 might be scaled as 5000 mils.(2) To enlarge or diminish the size of a displayed entity without changing its shape, i.e., to bring it into a user-specified ratio to its original dimension. Scaling can be done automatically by a CAD system. Used as a noun, scale denotes the coordinate system for representing an object.

SCALING : (1) The same as dividing. (2) The scaling mode of a counter is the special function where the frequency of an input signal is divided by a specified power of 10 in the time-base chain.

SCAN : To digitize an image or text from paper to a computer.

SCAN CONVERSION : Process of putting data into grid format for display on a raster device.

SCANNER : A device that converts images, such as photographs or printed text, into a format that a computer can use. A device that examines a spatial pattern one part after another, and generates analog or digital signals corresponding to the pattern. Scanners are often used in mark sensing, pattern recognition, or character recognition. Causing one complete horizontal or vertical traverse of the spot of light on the screen of a cathode-ray tube (CRT), i.e. one sweep of the screen, in response to a voltage generated by a time base circuit. Process of reading data in regular horizontal sweeps to cover the entire image or screen. Ther process adopted in Television transmitters to explore optically all parts of the object in successiooon, and in television receivers to build up the image by directing the rays illuminating it to all parts of the screen in succession, in synchronism with the scanning of the object by the transmitter.

SCANNING LINE : The series of signal values throuthout one transfer across the picture during Scanning in Television. The total number of lines scanned in one direction during a frame interval, including those which are blanked during the vertical retrace. Calculated by dividing the line fequency by the frame frequency.

SCATTER PLOT : Also called scatter diagam or dot chart. Shows a two-variable frequency distribution by plotting a dot or symbol at each data point. Sometimes a line or curve is added to show the correlation (if there is one) between the variables represented on the two axes.

SCENE ANALYSIS/UNDERSTANDING : See IMAGE ANALYSIS/UNDERSTANDING.

SCHERING BRIDGE : A four arm bridge for the measurement of capacitance.

SCHMITT TRIGGER : A device with electronic hysteresis that is used for "squating up" pulses.

SCHOTTKY BARRIER : See SCHOTTKY DIODE.

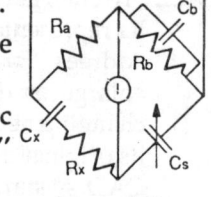

Schering bridge

SCHOTTKY DIODE : A semiconductor diode formed by contact between a semiconductor layer and a metal coating, which has a nonliner rectification characteristic. Also called a Sckottky barrier, or Hot-carrier diode.

SCIENTIFIC NOTATION : A mathematical technique for writing very large or very small numbers which makes use of 10 raised to a positive or negative power.

SCISSORING : The automatic erasing of all portions of a design on the CRT which lie outside user specified boundaries. (See "clipping.")

SCOTOPIC VISION : The eye-brain response to luminance levels below that required for the full discrimination of colours. Also called *twilight* or *night* vision, as contrasted to *photopic* or *daylight* vision.

SCRAMBLER : A circuit or device that is used in communication systems to produce an unintelligible version of the signal to be transmitted, in a predetermined manner. The received signal is rendered intelligible by an unscrambling circuit used at the receiver, in sympathy with the scrambler.

SCRATCH FILE : A file on which data are stored temporarily and which is not saved.

SCRATCH FILTER : A Wave Filter used in conjunction with a Gramophone Attachment or Pick-up to silence "needle scratch" by suppressing oscillations of the corresponding audio frequency.

SCRATCH PAD MEMORY : A high-speed memory used to temporarily store small amounts of information that can be fetched when needed. Fast-access temporary storage registers.

SCREEN : The glass part of a computer monitor that displays information to the user. Used in Cathode Ray Oscillographs and Television Reveivers. (1) Any partition or shield which can isolate apparatus from undesired effects of external electric or magnetic fields. (2) (in a Cathode Ray Tube). A chemically treated surface which shows luminosity at the point where the electron beam impinges upon it.

SCREENED AERIAL : An aerial provided between it and the earth, acting as a counter poise and to intercent the lines of force from the aerial to earth and to carry the return current, so as to eliminate a large fraction of the earth losses. An aerial provided with a screen of parallel wires between it and the earth, acting as a counterpoise and to intercept the lines of force from the aerial to earth and to carry the return current, so as to eliminate a large fraction of the earth losses.

SCREENED VALVE : A Thermionic Valve with an additional grid, called a Screen Grid, maintained at a positive potential less than that of the anode, which has the effect of almost completely neutralising the capacitance between the control grid and the plate in order to avoid self-oscillation. A Thermionic Valve with an additional grid, called a Screen Grid, maintained at a positive potential less than that of other connected to its middle joint for the three-phase side) is called the *Teaser Transformer*.

SCREEN GRID : A vacuum tube element positioned between the control grid and the plate. It is kept at a fixed potential in order to reduce the electrostatic effect of the plate element. Also called the second grid or G2. See SCREENED VALVE.

SCREEN GRID VALVE : See SCREENED VALVE.

SCREEN-SAVER : A special program that automatically blanks out your screen after a specified period of time when you do not touch the key-board. See also PHOSPHOR BURN-IN.

SCRIPT : A structure that represents stories and plans as sequences of primitive actions represented in conceptual dependency theory; together with information about time. location, etc. It can be used in language understanding for filling in missing details, answering questions about narrative texts, and other purposes.

SCROLL : To move text up/down or right/left on a computer screen. An arrow displayed at the ends of each scroll bar. Clicking or holding the mouse pointer on the scroll arrow causes the document in the window to move.

SCROLL BAR : Rectangular bars that apperas on the right and bottom of a window. Clicking in the scroll bar moves the document displayed in the window.

SCROLL BOX : A box that slides in a scroll bar to indicate the relative position of a displayed document.

SCROLLING : Translating text strings or graphics vertically.

SCS : Society for Computer Simulation.

SCSI : *(Small Computer system Interface)* : A connector that permits communication between computers and computer peripherals such as a hard disk. SCSI devices tend to work faster than ordinary devices connected to *serial* or *parallel* ports.

SCULPURED SURFACE : A mathematically described surface consisting of a composite on inter-connected, bounded, parametric surface patches, with each patch representing image of a unit square in parametric space.

SDC : Selective Device Clear; addressed bus command.

SDI : Selective Dissemination of Information.

SEAC : Standards Eastern Atomatic Computer.

SEARCH : A process of exploring alternative courses of action in order to solve problems or make plans for their later solution. A technique for finding a particular item in an ordered set of items by repeatedly dividing in half the portion of the ordered set containing the sought for item until only the sought for item remains.

SEARCH FUNCTION : A robot system can adjust the position of data points within an existing cycle, based on changes in external equipment and workpieces. One use of the search function is in stacking operation, especially when the stacked items are fragile or have irregualr thicknesses. The time delay inherent in deceleration from the input signal activation will permit some movement beyond the robot's receipt of the signal; so, if the signal originates through a limit switch that is closed upon contact with the stack, some compliancy must be built into the robot girpper. A fragile workpiece would also require a slow velocity during the search segment.

SEARCH SPACE : see state space.

SEARCH TREE : Is a model of state-space search processes in which the start states is the root node, the children of a node are those states which can be reached from it by the application of a single operator, and the goal states are leaves in the fully const~ucted tree.

SECOND : The SI unit of time defined as the duration of 9 192 631 770 periods of the radiation corresponding to the transition between two hyperfine levels of the ground state of the Casesium-133 atom. It was formerly defined as 1/86 400 of the mean solar day.

SECONDARY ELECTRON : See SECONDARY ELECTRONS.

SECONDARY ELECTRON MULTIPLIER : An electron tube in which the electron stream is focused on to a succession of tartgets, each of which adds its secondary electrons, thus producing a considerable amplifying effect, applicable to the amplification of the effect of photo-electric tubes and other purposes.

SECONDARY ELECTRONS : Electrons emitted by the anode, or other target of an Electron Tube when struck by the cathode stream of electron when they are projected with sufficient velocity.

SECONDARY EMISSION : The emission of electrons from the surface of a material, usually a metal, as the result of bombardment by high-velocity electrons or positive ions. The probability of secondary emission occurring when an ion approaches the surface is given by the Massey formula.

SECONDARY EMISSION VALVE : A valve in which use is made of secondary electrons to incresase the current through it.

SECONDARY MEMORY : Random-access devices such as disks and drums; programs are not executed from secondary memory devices but must be loaded into primary memory.

SECONDARY VOLTAGE : The voltage developed across the secondary (output) winding of a transformer.

SECONDARY 'X'-RAYS : Rays of similar character to ordinary "X" Rays, which are given off by solid and other bodies when ordinary "X" Rays fall upon them, at a frequency or frequencies characteristic of the substance.

SECTOR DISK : A section of a disk. A disk is divided into sectors by radial lines.

SEEBECK EFFECT : The fact discovered by Seebeck in 1821 that heating the point of contact of two dissimilar metals can produce an e.m.f.

SEEK TIME : For movable-arm disks, the time required for the reading mechanism to position itself over the track desired.

SEGMENT : A portion of a program, often used as an overlay consisting of one or more complete routines.

SEGMENTATION : Division of a display into parts that can be recalled or transformed individually.

SEGMENT ATTRIBUTE : A general characteristic of a retained segment, such as visibilty, highlighting, detectability, and image transformation.

SEIS-MICROPHONE : A special form of microphone for placing in contact with the ground to detect sounds coming through the earth; e.g. for locating enemy tunnelling operations in war.

SEL : Systems Engineering Laboratories.

SELECTANCE : The property of radio-receiving systems whereby they can be tuned to discriminate between simultaneous signals of not greatly differing wave lengths.

SELECTION BOX : A box drawn by dragging the pointer to enclose one or more graphic icons on the screen.

SELECTIVE ERASE : Deletion of specified portions of a display without affecting other portions.

SELECTIVE RADIATION : Radiation by a hot body or by luminescent gases, etc., of rays of a limited range of frequencies only, i.e. giving a discontinous spectrum.

SELECTIVE RECEIVER : A radio receiver which responds only to the particular wave length for which it is adjusted.

SELECTIVE RESONANCE : Resonance to one frequency only.

SELECTIVITY : The degree to which the response of a radio-receiving apparatus of a radio receiving apparatus is limited to one wave length, otherwise called *Sharpness of Tuning.*

SELECTOR : In general : An apparatus for responding to a prearranged series of current implses or wave trains only, or an automatic or other apparatus for making connection to any one of a number of circuits at will. In automatic telephony: An automatically actuaed multiple contact with any desired circuit, either controlled by a *step by step* method depending on a varying number of current impulses, or arranged to select a disengaged line by "hunting" over a number contacts and stopping at one when conditions are realised which bring a relay into action.

SELECTOR REPEATER (in automatic telephony) : A Selector which, after acting as a Group Selector, repeats all further impulses.

SELENIUM : A semiconductor element, atomic number 34. In the form of its grey allotrope selenium is markedly light-sensitive and is extensively used in photoconductive photocells.

SELENIUM AMPLIFIER : See SELENIUM MAGNIFIER.

SELENIUM BRIDGE : See SELENIUM CELL.

SELENIUM CELL : A small block or "cell" containing between suitable selenium, a substance the resistance of which becomes less when light falls upon it. Used, on account of this property, for various purposes but superseded to a considerable extent by other forms of Light Sensitive Cell. Cf. THALOFIDE CELL and PHOTO-ELECTRIC CELL, PHOTO CONDUCITIVITY.

SELENIUM MAGNIFIER OR "AMPLIFIER" : An apparatus for incresing the strength of signals received over submarine cables in which a beam of light is deflected by a mirror on the moving coil, causing variation of relative resistances of two selenium cells in opposite arms of a Wheatstone's Bridge, upon which it normally falls equally.

SELENIUM RECTIFIER : A Schottky diode that consists of a selenium iron junction and is used as a retifier. It is usual to construct a stack of such junctions in series.

SELENOPHONE : An early sound film system.

SELF CHECK : A redundant symbolization; for example, a number incorporating a check digit.

SELF COMPLEMENTING : The property of a BCD code where by each binary code is the complement of its 9's complement decimal number.

SELF-INDUCTANCE : The property of a circuit by which self induction occurs, measured by the rate of change of flux linkage caused by unit rate of change of current in the circuit itself. Cf. MUTUAL INDUCTANCE, and see also INDUCTANCE.

SELF OSCILLATION : Oscillation produced in a valve circuit of a radio-receiver, either caused accidentally by excessive retroaction, or purposely for Endoyne Reception, etc.

SELF-OSCILLATION TRANSMITTER : A Thermionic Valve transmitter in which the oscillations are produced by the effect of Retroaction between the anode and grid circuits (except in the case of magnetron and electron oscillators.)

SELF REGULATING : Machinery or apparatus which itself automatically regulates current, voltages, speed, or other function without the necessity of hand or automatic operation of a separate regulating apparatus.

SEMANTIC GRAMMER : A type of grammer used in natural language interfaces in which the semantic analysis of input sentences is collapsed into the syntactic parsing process.

SEMANTIC NETWORK : A knowledge representation formalism that presents each concepts as a node in a graphy and the relations between concepts as labelled arcs between nodes; the meaning of a concept is derived from its relationships to all other concepts in the network.

SEMANTIC PRIMITIVES : Are elementary units of meaning out of which all other concepts are composed, and are used to provide a base level of representation which, if it can be reasoned with soundly, allows us to manipulated all other concepts that can be represented.

SEMANTICS : A level of analysis in natural language processing that is concerned with the meaning of sentence, which is derived from the meanings of component words and phrases and from the surrounding context. The meaning of a programming language statement or group of statements.

SEMI-AUTOMATIC TELEPHONE SYSTEM : A telephone system in which personal operators are employed who ascertain the calls varbally but actuate automatic apparatus to obtain the required connections. (Also called Automanual System) See TRAFFIC DISTRIBUTOR.

SEMI CONDUCTOR: A small component having an electrical conductivity between the high conductivity of metals and the low conductivity of insulators. A material having a resistivity in the range between conductors and insulators and having a negative temperature coefficient of resistance. The conductivity increases not only with temperature but is also affected very considerably by the presence of impurities in the crystal lattice. Semiconductors are used in a wide variety of solid-state devices including transistors, integrated circuits, diodes, photodiodes, and light-emitting diodes. An intrinsic semiconductor is a perfect crystal semiconductor in which the energy gap, eg, between the conduction band and the valence band (see energy bands) is comparable to thermal energies. The simplified energy bands of an intrinsic semiconductor are shown in Figure.

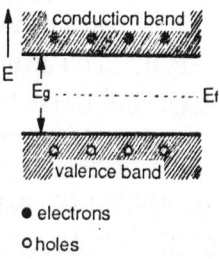

- electrons
- holes

a Intrinsic semiconductor

SEMI-CONDUCTOR MEMORY : Memory consisting of transistor devices; generally faster than core storage.

SEMI-CUSTOM LOGIC : Chips whose logic functions can be determined in the final stages of manufacturing, such as gate arrays.

SENDER : A piece of apparatus in the Panel Automatic Telephone System which receives the calling impulses and translates them into suitable form for operating the various selectors.

SENSE : To determine the state of something, as the position of a switch.

SENSE LINE : A line within a memory that senses whether the cell is a 1 or a 0.

SENSITIVITY : (1) In general, the change produced in the output of a physical device per unit change in the input. (2) The magnitude of the change in the indicated value or deflection of a measured quantity. It is usually quoted as the magnitude of the measured quantity required to produce full scale deflection. (3) The displacement in mm. of the luminous spot on the screen per volt applied to the deflecting plates (electric and sensitivity) or per ampere applied to the deflecting coil (magnetic sensitivity).

SENSOR : A transducer or other device whose input is a physical phenomenon and whose output is a quantitative measure of that physical phenomenon.

SENTINEL : A symbol indentifying or indicating the beginning or end or a word, field, record, block, or file

SEPARATE HETERODYNE : Hetrodyne reception in which an independent oscillator is used to produce the auxilitary oscillations.

SEPOL : Settlement Problem Oriented Language.

SEQUENCE : (1) A single operation. A sequence has one output (coil), and is identified by a unique (address). (2) A series of states that follow each other in a definite order to accomplish a function.

SEQUENTIAL : Following each other according to a certain order.

SEQUENTIAL ACCESS MEMORY : An auxiliary memory device which lacks any addressable data areas. A specified piece of data can only be found by means of a sequential search throuth the file.

SERCUS : Society for the Exchange of Raytheon Computer Users Software.

SERIAL : The handling of data in a sequential fashion, so as to transfer or store data in a digit-by-digit time sequence or to process a sequence of instructions one at a time. Of the presence or flow or information through any part of an automatic computer using only one line or channel at a time.

SERIAL ACCESS : A sequentially organized file from which information can be retrieved only by processing through the file in order.

SERIAL DATA : Data in which bits are transmitted over after another over the same line.

SERIAL OPERATION : (1) Relating to the sequential or consecutive execution of two or more operations in a single device, such as an arithmetic or logical unit. (2) Information flow throuth a computer in time sequence, usually by bit but occasionally by character.

SERIAL OUTPUT : Sending only one bit at a time to and from interconnected devices.

SERIAL POLLING : A method of sequentially determing which device connected to the IEC bus has requested service. Only one instrument is checked at a time, while its status byte is ready by the controller.

SERIAL PORT : An interface, designed to connected a computer to items like a modem, mouse, or printer. Also called an RS-232 port. See RS-232. Cf. Parallel Port.

SERIAL TO PARALLEL CONVERTER : A circuit that receives serial data and provides parallel outputs; usually a shift register.

SERIAL TRANSMISSION : Moving data in sequence one character at a time, as opposed to parallel transmission.

SERIES CAPACITOR : A bank of oil-immersed condensers in series with the line for power factor improvement, usually with an automatic by-pass device for short-circuit conditions.

SERIES MODE SIGNAL : The unwanted superimposed as signal on the dc voltage to be measured.

SERIES MODULATION : A system of modulation employed in large radio-transmitters in which the modulated amplifier valve and the modulated amplifier valve and the high-tension supply are all in series. The variations in impedance of the modulator anode circuit caused by the modulations of the grid voltage directly affect the voltage applied to the amplifier valve.

SERRODYNE : A frequency converter or mixer in which the output of the local oscillator is a sawtooth waveform, which in turn is used to phase modulate the input.

SERVER : The main computer that controls a Local Area Network (LAN).

SERVO-MECHANISM : An automatic control system which incorporates feedback that controls the physical position of an element by changing either the values of the coordinates or the values of their time derivatives.

SET INPUT : When the set input to a filp-flop is enabled, the flip-flop goes to the 1 state. See FLIP-FLOP and S-FLIP-FLOP.

SETTLING TIME : The time required for an error signal in a feedback control system to decrease to a given percentage of its peak value. Typically this percentage can range from one to five percent. The time necessary for all logic elements to reach their final logic states.

SETUP : The ratio of the difference between black level and blanking level to the difference between reference *white* level and *blanking* level, expressed in percent or IRE units.

SEVEN SEGMENT DISPLAY : A liquid crystal or light-emitting diode display composed of seven individual segments arranged in figure-eight pattern to form any number from 0 to 9 when the individual seqments are lit. The individual segments are lettered a through in a clockwise fashion.

SEXADECIMAL : Petaining to a selection, choice, or condition that has sixteen possible different values or states.

SHADES OF GRAY : A division of the gray scale from black to white into a series discrete luminance shades with a square-root-or-2 difference between successive shades.

SHADING : An unintentional large-area brightness gradient in a display. Also used to describe graphics software algorithms that establish the appearance of solid-object surfaces.

SHADOW MASK : Perforated metal plate positioned behind the color raster display surface so that electrons from a foucused triad gun strike only assigned phosphors.

SHAKE : Vibration of a robot's "arm" and "hand" during or at the end of a movement. Lack of shake is one of the hallmarks of a quality robot.

· SHAPE FACTOR : A dimensionless quantity used to quantify the selectivity of a broadband filter or tuned amplifier stage. It is equal to the ratio of the 60-dB bandwidth to the 6dB bandwidth.

SHAPE FILL : The automatic paintiing in of an area on an IC or PC board layout, defined by user specified boundaries; for example, the area to be filled by copper when the PC board is manufactured.

SHAREWARE : Programs you can legally copy and give away. If you like the program and use it, you are legally bound to send in a registration fee. In return for this fee you get printed manual, the latest version of the program, and telephone support.

SHARPNESS (of directivity of a directive aerial) : The degree of directional property as measured by the change of signal strength corresponding to a small angular displacement.

SHARP TUNING : The necessity for very fine adjustment of the tuning apparatus of a radio-receiver to obtain response to a particualr wave-length.

SHEARER TUBE : An "X" ray tube with an envelope of metal with procelain insulation instead of glass.

SEATH : (1) See CABLE SHETHING. (2) Anode (Plate) of a therminonic valve when it surrounds the Cathode. (3) In a discharge tube. The regions near the electrodes or near the wall of the tube where the presence of space charges causes unequal positive and negative electron concentration Cf. PLASMA.

SHEET FEEDER : A tray that holds individual sheets of paper and slides each page in to a printer one at a time.

SHELL : A program that lets you choose operation system commands by choosing from a menu. Shell progarms try to make the computer easier to use.

SHELL TYPE TRANSFORMER : A tranformer in which most of the windings are enclosed by the core (see diagram). The core is made from laminations and usually the windings are assembled and them the laminated core built up around them. Compare CORE-TYPE TRANSFORMER.

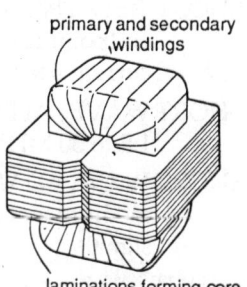

primary and secondary windings

laminations forming core

Shell-type transformer
(Single phase)

SHF : Abbreviation for super high frequency.

SHIELD : Define an opaque viewport or window for menu, title, or message display.

SHIELDED : "X" RAY TUBE. An "X" Ray Tube enclosed, except for a small window, by an earthed metallic container. Cf. RAY PROTECTED "X" RAY TUBE.

SHIFT : To move the admissible marks in a word or field one or more places to the left or right. In the case of a number, this is equivalent to multiplaying or dividing by a power of the radix.

SHIFT-CLICK : Holding down the shift key while clicking the mouse buttom.

SHIFT REGISTER : A series of flip-flops capable of shifting a binary number to the right or left. A storage device consisting of a chain of flip-flop in which the contents can be shifted one or more positions. In a shift to the right, the right-most bits on the far right of the number stored are lost; in a shift to the left, the bits on the far left are lost. In a circulating shift register no data are lost because the data leaving the register at one end are reinserted at the other. A Program, entered by the user into memory of a programmable controller, in which the information data (usually single bits) is shifted one or more position on a continual basis. There are two types of shift registers: asynchronous and synchronous.

SHOCKLEY DIODE : A four-layer, controlled semiconductor rectifier diode without a base connection, used as a trigger or switching diode.

SHORT CIRCUIT : A connection, accidental or otherwise, between two points having a difference of potential between them, of sufficiently low resistance to allow a very much larger current to flow than is normally the case; so large, sometimes, as to produce considerable damage.

SHORT RELATIVE VECTOR (SRV) : Vector of a limited length with endpoint identified in terms of x, y, and z distance from the current beam position.

SHORT WAVE : Abbreviated SW, radio frequencies in the range from 1.5 to 30 MHz.

SHORT WAVE ADAPTER : An apparatus which can be connected to a medium wave radio-receiver to encable it to receive abort wave signals.

SHUNT : (1) In general, syn. for parallel (2) Syn. instrument shunt. A resistor, usually of relatively low value, that is connected in parallel with a measuring instrument, such as a galvanometer. Only a fraction of the current in the main circuit passes through the instrument so that the shunt increase the range of the instrument and also protects it form possible damage caused by current surges.

SHUNTED CONDENSER : A condenser in parallel with a high resistance used in telegraph circuits etc., to compensate for the retarding effect of inductance and to improve the speed of working.

SIAM : Society for Industrial and Applied Mathematics.

SIC : Special Interest Committee (of ACM).

SIDE BAND : The frequency components of a modulated carrier above and/or below the carrier frequency. The lower sideband is the difference of the carrier and modulating components while the upper sideband is the sum of the carrier and modulating components. A term used in Radio and Carrier Wave Telephony, etc., for a group or Band of Waves of frequencies formed by the interaction of the carrier wave and the modulations composed of frequencies of the sum and difference of the modulating and carrying frequencies.

SIDE CIRCUIT : See PHANTOM CIRCUIT.

SIDE TONE REDUCTION WIRING : A method of wiring Central Battery subscribers' telephone sets with the microphone between the switch-hook and the induction coil to eliminate the effect of the condenser produced in Side-Tone Wiring.

SIDE TONE WIRING : The method of wiring Central Battery subscribers telephone sets with the microphone between the line and the switch hook in which the charging and discharging of the condenser, with the variation of the primary current, intensifies the Side Tone effect.

SIEMEN'S AUTOMATIC TELEPHONE SYSTEM : A system of automatic telephone exchange working in which a Pre-selector, actuated when a subscriber lifts his receiver, first connects a free Line Selector to his line, which automatically connects to the line corresponding to the series of current impulse sent by the manipulation of the subscriber's dial switch. The preselector, which is the chief distinguishing feature of the system is driven by an alternating current from a special generator.

SIEMEN'S ELECTRO DYNAMOMETER : An electordynmometer that may be calibarted as an *ammter*, *voltmeter*, or *wattmeter*. The signal to be measured produces an electromagnetic torque on the movable coil that in balanced against the torque of a spiral spring connected to it by adjusting a calibarated torsion head attached to the spring. At the balance position the deflection of the movable coil is zero, and the value of the measured parameter is given by the setting of the torsion head.

SIG-GRAPH : (Special Interest Group on Computer Graphics) : A group within the Association for Computing Machinery (ACM) that was the catalyst for subsequent committees on standards for computer graphics.

SIGHT CHECK : To verify visually the sorting or punching or punched cards by examining the petterns of punched holes.

SIGN : A bit or character used to designate whether a quantity is positive (plus) or negative (minus).

SIGNAL : A variable electrical parameter, such a current or voltage, that is used to convey information through an electronic circuit or systems. A visible, or other indication that conveys information.

SIGNAL GENERATOR : A generator that Produces pulse wave forms in normally referred to as a pulse generator, the term signal generator being reserved for a continous wave generator, particularly of a sinusoidal wave from any electronic circuit or device that produces a variable and controllable electrical parameter. The term is most commonly applied to a device that supplies a specified voltage of known variable amplitude.

SIGNALLING KEY : A key used in Line or Radio-Telegraphy to control the sequence of the current impulses forming the signals.

SIGNAL STRENGTH IN RADIO COMMUNICATION : A term used by

radio engineers to signify the strength of a recevied signal produced by a transmitter at a particular distance, as measured by the component of e.m.f. induced in the aerial by the modulations, expressed in millivolts per metre of effective.

SIGNAL TO NOISE RATIO : Abbreviated S/N ratio, the ratio, expressed in decibels, of the amplitude of the desired singal at any point to the amplitude of the noise signal at the same point. The peak value is used for pulse noise while the RMS value is used for random noise.

SIGN BIT : The bit of binary word that represents the arithmetic sign.

SIGNIFICANCE : Relative importance, as of digits or bits occupying various place positions in the representation of a number. By conversion, the left most non zero digit or bit is the most significant, and the right most the least significant, whether or not a zero.

SILENCER : (1) An enclosing vessel to diminish the noise made by the spark in a radio-transmitting apparatus.

SILICA : An extremely abundant compound occurring in several different natural forms, the best known of which are probably quartz and common sand (in which the silica is discoloured by ferric oxides). Silica is important as a source of silicon for the manufacture of electronic components, devices, and integrated circuits, as a grown oxide for the passivation of such equipment, ans as natural quartz, which has marked piezoelectric properties.

SILICA GEL : Deliquescent crystals consiting mainly of silica (SiO_2) that are used as a drying agent, particulary during dispatch and delivery of electronic and electrical equipment.

SILICA VALVE : A Thermionic Valve in a silica bulb as in large transmitting valves.

SILICON : A semiconductor element, atomic number 14. It is very abundant in nature in the form of silicon dioxide (silica) and is the most widely used semiconductor in solid-state electronics. It is cheap and extremely versatile and rapidly replaced germanium.

SILICON COMPILER : A software package that takes over chips creation from design to mask production.

SILICON CONTROLLED RECTIFIER (SCR) : Syn. reverse blocking triode thyristor. A *pnpn* device in which the forward anode-cathode current is controlled by means a of a signal applied to third electrode, called the *gate*. A solid-state eletronic control device by which small electrical currents can be utilized to control high electrical loads. Abbreviated SCR, a four-layer, three-terminal solid state device that exhibits an open circuit between *anode* and *cathode*

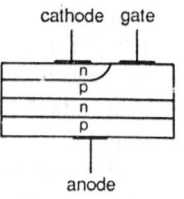

a Silicon-controlled rectifier

until a significantly large current flows between the *gate* and the *cathode*. It remains as a short circuit between anode and cathode until the anode-to-cathode voltage decreases to zero or goes negative.

SILICON DETECTOR : A rectifier of electrical oscillations consisting of a piece of silicon in a particular crystalline state against a metal contact, which can be used in series with a telephone receiver as a wireless detector. See CRYSTAL DETECTOR.

SILICON DIODE : A diode made using silicon as the semiconductor.

SILICON RECTIFIER : (1) A silicon diode used as a rectifier. (2) A Schottky diode, in which the semiconductor is silicon, used as a rectifier.

SILICON SOLAR CELL : A solar cell constructed as alternate layers of p and n type silicon.

SIMM (Signle in line memory module) : A specialtype of RAM chip that plugs into a socket on the computer's motherboard.

SIMPLE RECTIFICATION : See ANODE BEND RECTIFICATION.

SIMPLEX : Data transmission mode that permits communication in one direction at a time.

SIMPLEX DIALLING : A system of Dialling using both conductors in parallel with earth return. Cf. DUPLEX DIALLING.

SIMPLEXED CIRCUIT : A telegraph circuit consisting of the two wires of a telephone circuit in operation in parallel with earth return.

SIMPLEX TELEGRAPH SYSTEM : A system in which single messages in one direction only at a time are sent. Cf.DUPLEX, and MULTIPLEX SYSTEMS.

SIMPLEX WORKING (in Radio Telegraphy) : The use of the same aerial alternately for transmission and reception.

SIMULA : Simulation Language.

SIMULATION : (1) The modeling of some process that often involves the use of a computer program and probability distributions. (2) A software program used to execute programs written for one machine on another. An imitative type of data processing in which an automatic computer is used to implement an information model or some entity, as, of a chemical process. Information enters the computer to represent the factors entering or affecting the real process. The Computer produces information that represents the results of the process. The processing done by the computer represent the process itself.

SIMULATOR : A device, such as an analog computer, that simulates the behaviour of an actual physical system and can therefore be used to solve complex problems associated with the operation of the system. A simulator is usually fabricated from components that are easier, cheaper, or more convenient to manufactrue than the system itself.

SIMULTANEOUS CONTRAST : Changes in the appearance of a color relative to its background or adjacent colors.

SIN : Symbolic Integrator.

SINE CURVE : A curve in which the sine of an angle is plotted vertically and the angle itself horizontally.

SINE-WAVE : A fundamental waveform whose amplitude varies as the sine of a linear function of its argument, expressed in either radians, degrees, or time.

SING : See SINGING.

SINGAL STRENGTH IN RADIO COMMUNICATION : Height. Cf. FIELD STRENGTH.

SINGING :A thermionic Telephone Repeater is said to "Sing" or sometimes to "Howl," when, owing to unsuitable adjustment of its circuits, it acts as an oscillator, producing sustained oscillations of frequency which causes an audible note to be produced in the telephones connceted thereto.

SINGLE CURRENT KEY : A telegraph signalling key sending currents in one direction only. Cf. DOUBLE CURRENT KEY.

SINGLE CURRENT TELEGRAPH SYSTEM : A system employing currents in one direction only. Cf. DOUBLE CURRENT SYSTEM.

SINGLE DENSITY : The standard recording density of a diskette. Single-density diskettes can store approximately 3400 bits pre inch (dpi).

SINGLE NEEDLE : (Telegraph system) : A system in which the signals are made by the deflection, to one side or the other, of a vertical needle in the receiving instrument worked by a simple form of galvanometer. The signals can be read audibly to a certain extent from the sound of the needle hitting against its stops. See DOUBLE NEEDLE and FIVE NEEDLE SYSTEMS and IN-DUCED NEEDLE.

SINGLE OCTETE MODE : A mode of information transfer that uses Bus A in a unidirectional manner of transfer 8 bits of information from host to unit, and Bus B in a unidirectional manner to transfer 8 bits of information from unit to host.

SINGLE PHASE : A term characterising current etc., or apparatus relating to the system of alternating current working, where a single alternating current in one pair of wires is employed in the ordinary way. Cf. TWO-PHASE, THREE-PHASE, and POL-YPHASE.

SINGLE PRECISION VALUE : The number of words or storage positions used to denote a number in a computer. Single precision atithmetic in the use of one word per number, double-precision arithmetic is the use of two words per number, and so on. Foor variable word-length computers, precision is the number of digits used to denote a number. The higher the precision, the greater the number of decimal places that can be carried.

SINGLE REPRESENTATION TRICK : see under operationalization.

SINGLE SHOT : A monostable multivibrator.

SINGLE SIDE BAND TELEPHONY : A from of Suppressed Carrier Wave line in Radio-Telephony in which one Side Band is filtered out as well as the carrier frequency to minimise the range of frequencies occupied.

SINGLE SIDED : A term used to describe a diskette that contains data on one side only

SINGLE WAVE LENGTH WORKING : The working of two separate broad-casting stations, e.g. Relay stations at the same wave length. This requires very accurate frequencey control and results in a rather limited service area uound each station surrounuded by a "Mush" area where interference spoils the signals of both; the service area is greater if the same pro-gramme is being broadcast.

SINGLE WAVE RECTIFICATION : See HALF-WAVE RECTIFICA-TION.

SINGLE WAY SWITCH : A switch arranged to open and close one set of contacts only. Cf. TWO-WAY SWITCH, etc.

SINUSOIDAL OR SINUSOIDAL CURRENT VOLTAGE WAVES ETC.: An alternating current, etc., with a wave form following a true Sine Curve.

SINUSOIDAL : Denoting a periodic quantity that has a wave form graphically indentical in shape to a sine function; wave forms represented by the functions sinx and coxs would both be described as sinusoidal or as sine waves.

SITE ERROR IN DIRECTION FINDERS : Any error due to irregulatrity on the site of the reciving apparatus.

S.I. UNITS : The internationally agreed system of units intended for all scientific and technical purposes. The system is based on the MKS system and replaces the CGS and Imperial systems of units. There are three types of units in the SI system: base units. Derived units, and supplementary units. All are absolute units. The base units are an arbitrarily defined set of dimensionally independent physical quantities. In any purely mechancial system of units only three base units of *mass, length,* and *time* are required. In a consistent electric and magnetic system four base units are needed. In the SI system there are seven base units: the metre, kilogram, second, ampere, kelvin, candela (the standard of luminous intensity), and mole (the standard of amount of substance).

The derived units are formed by combining, by multiplication and/or division, two or more base units. Thus the coulomb, which is the derived unit of charge, is formed form a combination of one ampere times one second. There are at present two supplementary units: the radian and steradian, which are the units of plane and solid angle, respectively.

When considering electric and magnetic quantities a fourth term is required in addition to the fundamental units of mass, length, and time, for their complete definition. In the MKS system the fourth quantity is the permeability of free spaces which is defined as 4×10^7 henry per metre.

A set of 14 prefixes, including micro and kilo, are used with the SI units to form multiples and submultiples of the units.

SJCC : Spring Joint Computer Conference.

SKIAGRAM : See X-RAY PHOTOGRAPH.

SKIN EFFECT : A non-uniform distribuion of current over the cross section of a conductor when carrying alternating current, with the greater current, with the greater current density located at the surface (or 'skin') of the conductor. the *skin effect* is caused by electromagnetic induction in the wire and increases in magnitude with increasing frequency. At sufficiently high frequencies the current is almost entirely confined to the surface of the conductor and results in a greater IR loss than when the current is uniformly distributed. The effective resistance of the conductor is therefore greater than the d.c. or ohmic resistance, when carrying alternating current, and for high frequency applications the high frequency resistance of a conductor can be substantially greater than the nominal d.c. value.

SKIP DISTANCE : The portion of the range of a radio-transmitting station where signals are weak as it is too great to reach by Direct Rays and not far enough for Indirect Rays, which would have to strike theionosphere at too acute an angle to be totally reflected.

SKYWAVE : (1) Expressed in volts per microseconds, a radio wave that travels upward into space and may or may not be returned to earth by reflection from the ionosphere. (2) Ionospheric wave.

SLABY ARCO SYSTEM OF RADIO TELEGRAPHY : An early form of the Quenched Spark system.

SLEEVE : (1) of a Telephone Plug. The metal sleeve forming one of the contacts round the shank of the plug. Cf. TIP and RING. (2) In conduct wiring. See PLAIN COUPLER.

SLEW RATE : The maximum velocity at which a mainpulator joint can move; a rate imposed by saturation somewhere in the servo loop controlling that joint (e.g., by a value reaching its maximum open setting). The maximum speed at which the tool tip can move in an inertial Cartesian frame. Expressed in volts per microseconds, a measure of an operational amplifier's switching speed, defined as the maximum time rate of change of the output voltage when sujected to a square wave input signal when the closed-loop gain is unity.

SLIDE WIRE : A wire of uniform resistance provided with a sliding contact that can make a connection at any desired point along the length. Slide wires are used to provide a variable resistance, as a pontentiometer, or to provide a desired resistance ratio (see Wheatstone bridge). An overall length of one metre is connonly chosen.

SLOPE IN A THERMIONIC VALVE : The "slope" of that portion of the characteristic of a thermionic valve over which it works, i.e. the ratio of the change of anode current to the change of grid voltage causing it.

SLOT : An element in a frame that represents a property or feature of the object, whose value is contained in the filler for the slot.

SMALL SIZE :Of an automatic computer with a combined purchased price of not more than about $100,000. (Year 1990).

SMALL TALK : An object-oriented programming language and graphics-based development environement that is often used for AI applications.

SMART SENSOR : A sensor whose output depends on internal data or on input from another part of the system.

SMIS : Society for Management Information Systems.

SMOOTHING : Fitting together curves and surfaces so that a smooth, continuous geometry results.

SMOOTHING CIRCUIT : Syns. RIPPLE FILTER, RECTIFIER FILTER. A circuit that is designed to reduce the amount of ripple present in an essentially unidirectional current or voltage. A typical smoothing circuit consists of a low-pass filter (see diagram) but a single inductance may be used.

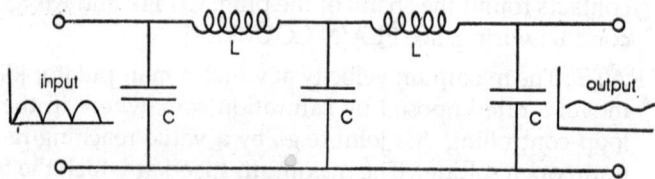

Simple low-pass filter as smoothing circuit

SNEAK CURRENT : A term used in telephone engineering for leakage currents from power circuits, etc., through telephone circuits which, althouth too weak to do immediate damage, may produce harmful heating effects if allowed to continue for an extended period; guarded against by Heat Coils.

SNOBOL : String Oriented Symbolic Language.

SNOOK RECTIFIER : A revolving four arm rectifying switch driven by a synchronous motor for obtaining a high undirectional voltage from an A.C. transformer for working "X" Ray Tubes.

SNOW : A display condition produced by random noise on a display signal, usually indicative of a weak signal.

SNR : Signal to Noise Ratio.

SN-RATIO : Signal-to-noise ratio; the ratio of the amount of signal carrying information to the amount of signal not carrying information. The S/N ratio is mostly expressed in *decibels* :

SOAP : A Symbolic Optimum Assembly Program used with the IBM-650.

SOFT-COPY : Storable video-tape copy of display.

SOFT-ERROR : Alteration of the information in a memory cell resulting from an alpha particle striking the cell.

SOFT-RADIATION : A term sometimes used for radiation of wave length intermediate between the ordinary range of "X" Rays and ultraviolet "light."

SOFT-VALVE : A thermionic valve with only a moderately high vacuum, even as low as 1/1,000 atmosphere, and depending upon gaseous ions rather than free electrons.

SOFTWARE : Sets of instructions called programs, used to give directions to the physical parts of the computer. Program, languages, procedures, rules, and associated documentation used in the operation of a data-processing system. A computer program stored on a floppy or hard disk that makes your computer do something useful, such as woed processing or playing chess. A string of instructions that, when executed, direct the computer to perform certain functions. Instructions that control the physical hardware of the computer system. Computer user aids that are not hardware; for example, manuals desrcibing and explaining how a computer operates, dump rountines, and equipment maintenance service.

SOFT "X" RAYS : "X" rays of poor penetrating power, i.e. comparatively long wavelength. Cf. HARD "X" RAYS.

SOLAR CELL : A deviec that utilizes the photovoltaic effect in order to convert radiation from the sun directly inot electrical energy. Solar cells are tha most important ling-duration power supply for satellites and space vehicles. They have been fabricated from a range of different semiconductors including silicon, gallium arsenide, selenium-cadmium sulphide, and thin film cadmium sulphide.

SOLENOID : A coil of wire that has a long axial length relative to its diameter. The coil is usually tubular in form and is used to produce a know magnetic flux density along its axis.

SOLID CONDUCTOR : A conductor that is composed of a single wire or uniform thin metal rather than being stranded or otherwise divided.

SOLID STATE : Pertaining to a circuit, device, or system that depends on a combination of *electrical, magnetic,* or *optical* phenomena within a solid that is usually a crystalline semiconductor material.

SOLID STATE DEVICE : An electronic component or device that is composed chiefly or exclusively of solid materials, usually semiconducting, and that depends for its operation on the movement of charge carries within it. A solid-state device has no moving parts.

SOLID STATE RELAY : A relay that has all its components made from solid-state device and involves no mechanical movement. Isolation between input and output terminals is provided using a light-emitting diode (LED) in conjunction with a photodetector. The switching is acheved using a silicon-controlled rectifier or more commonly two SCRs (a triac). This type of relay is compatible with digital circuitry and has a wide variety of uses with such circuits. The relay connot normally be formed on a single chip since the LED is usually formed in gallium arsenide and the photoodetectoor in silicon. Isolation may also be achieved by transformer-coupling on the input. Again a single chip may not be used. Examples of solid-state relays are shown in Figs. d and e.

Solid-state relays have advantages over electromechanical realys because of increased lifetime, particularly at a high rate of switching, decreased electrical noise, compatibility with digital circuitry, and ability to be used in explosive environments since there are no contacts across which arcs can form; the lack of physical contacts and moving elements also gives increased resistance to corrosion. No mechanical noise is associated with them. This is particularly important for certain applications where noise could be an annoyance, as in hospitals. Disadvantages include the substantial amount of heat generated at a current above several amperes, necessitating some form of cooling, and greatly increased production costs for multipole devices compared to single pole devices in certain applications a physical disconnection may be required for safety purpose and this is not available in solid-state relays.

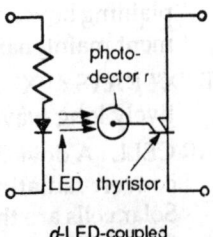

d-LED-coupled
solid-state relay

SOLOMON'S UNIT : A unit of "X" Rays quantity used in france, having a value of about 2.29 Rontgen.

SOMEBRERO FILTERING : A filtering process for images that applies a function shaped like a Mexican hat to each neighbourhood in the bitmap and has the effect of highlighting edges by transforming them into zero-corssings.

SONIC FREQUENCY : See AUDIO FREQUENCY.

SOP : SUM OF PRODUCTS.

SOROBAN : A Japanese abacus with one 5-bead and four 1-beads per digit.

SORT : To place records in some order on the basis of the key and the collation sequence.

SOUNDER : A telegraph receiving instrument in which the attraction of an amature by an electromagnet makes an audible sound as it hits against its stops at the beginning and end of each current impulse, and permits of the message being read in the Morse or other code by ear.

SOURCE CODE : The actual instruction of a progarm, written in a computer language such as BASIC, C., or Pascal.

SOURCE DOCUMENT : A document, often a form used in data acquisition, carrying input information in either a human or machine language.

SOURCE LANGUAGE : A Symbolic language comprising statements and formulas used in computer processing. It is translated into object language (object code) by an assembler or compiler for execution by a computer the input language to a translation process.

SOURCE PROGRAM : In a language, a program that is an input to a given translation process. A program that is to be translated, as for example, a program written in a symbolic language.

SP : Structured Programming.

SPA : Systems ans Procedures Association.

SPACE CHARGE : The electric charge in the space between the electrodes in a thermionic valve due to the presence of free ions or electrons.

SPACE CHARGE GRID : An extra inner grid in a thermionic valve, manintained at a fixed potential, having the effect of reducing the anode resistance of the valve by counter acting the space charge near the cathode.

SPACE CURRENT : The current through a Thermionic Valve due to the passage of the projected electrons towards the anode.

SPACED AERIAL : An aerial system used in receiving stations for short waves, in which a number of separate aerials at considerable distances apart coupled to the same receiver.

SPACE DIVERSITY : A Diversity system that employs several receiving aerial spaced several wavelenghts from each other.

SPACED LOOP DIRECTION FINDER : A Direction Finder employing two separate aerials at a short distance apart coupled the same receiver. A Spaced Aerial Direction Finder employing Loop Aerials.

SPACE TELEGRAPH AND TELEPHONY : Another name for what is commonly called *Radio-Telegraphy* and *Telephony*.

SPACING CURRENT : The current which flows between the impulses forming the signals in the Moorse or similar code, in systems where the signalling is effected by some change in the magnitude or direction of the current and not by its intrruption. Cf. MARKING CURRENT

SPACING WAVES : The waves radiated between these forming the signals of the Morse or similar code, in systems where the signalling is effected by a change in *amplitude* or *frequency* of the continuous waves employed.

SPARC : Standards Planning and Requirements Committee.

SPARK : A visible disruptive discharge of electricity between two points of high potential difference, preceded by ionization of the path. A sharp crackling noise occurs because of the rapid heating of the air through which the spark passes. The distance travelled is determined by the shape of the electrodes and the potential difference between them.

SPARK FREQUENCY : The frequency with which the sparks occur in a radio-transmission apparatus on the spark system, i.e. the group frequency of the trains of waves and not of the waves themselves.

SPARK GAP : Under specified conditions, the distance between the electrodes is termed the spark gap. Specially designed electortdes are used to produce a spark over a given spark gap under particular conditions, as for ignition purposes in an internal combustion engine. The insulation is self-restoring when the potential across the spark gap falls below that required to produce the spark. When the electrodes are in the form of needle points the spark gap is termed a *needle gap*. A shpere gap is one that has spherical electrodes. A spark is of much shorter duration that an arc.

SPARK SYSTEM (of Radio Communication) : A system of radio-telegraphy in which successions of short traians of damped waves, obtained from oscillations provoked by a succession of spark discharges in a suitable circuit; e.g. the earlier forms of the Marconi System, and the systems employing the Quenched Spark. See SHOCK EXCITATION.

SPARK TRANSMITTER : A radio transmitting apparatus employing a spark discharge as the source of the oscillations. (See above RA) Cf. ARC TRANSMITTER and VALVE TRANSMITTER.

SPATIAL DATA : Locational data. Usually refers to distribution of a variable or the relationships between variables in a geographc region. Demographic features, marketing distributions, energy resource data, and topographic data are examples of information readily.

SPE : Serial Poll Enable; universal bus command

SPECTRAL COLOUR : Colour of a single waveleneth on the visible protion of the electromagnetic spectrum.

SPECTROGRAPH MASS : See MASS SPECTROGRAPH.

SPECTROMETER : See "X" RAY SPECTROMETER.

SPECTRUM : *Refraction* or *diffraction* of white light into spectral hues in order of their wavelengths, beginnging with violet (shortest wavelength), through blue, yellow, orange, and ending with red (longest wavelength). The spectrum does not include black, white, or colours which are mixures of wavelenghts, such as purple (a mixture of red and violet light).

SPECTRUM ANALYZER : A test instrument using a cathode-ray tube that measures and displays the intensities of the frequency components of a complex waveform throughout the frequency range of the waveform.

SPEECH AMPLIFIER : An audio-frequency amplifier specifically designed to amplify speech frequencies, which are generally in the 300 Hz-3 kHz range.

SPEECH CLIPPER : A clipper circuit that limits the peaks of speech frequency singanls for increasing the average percent modulation of a transmitted signal.

SPEECH COMPRESSOR : A device which eliminates certain speech frequency signals to reduce its bandwidth affecting intelligibility.

SPEECH FREQUENCY : See VOICE FREQUENCY and Cf. AUDIO FREQUECNY.

SPEECH INVERTER : Apparatus employed in Suppressed Carrier Wave Single Side Band Telephony which results in the high and low speech frequencies being interchanged, rendering the speech unintelligible if picked up without replacing the carrier wave. Cf. SCRAMBLER.

SPEECH MODULATED CONTINOUS WAVES : Continous waves, modulated in accordance with the vibrations produced by the sound waves of speech as employed in radio-telephony. (Sometimes called Type A3 Waves.)

SPEECH RECOGNITION/UNDERSTANDING : Is the perceptual problem of transducing some auditory input, represented as a two-dimensional graph of intensity over time, into a representation of the language it encodes, when the problem becomes are of natural language processing.

SPEECH RESTORER : Apparatus at the receiving end of a Suppressed Carrier Wave telephone system the inverted speech received to its original frequencies.

SPHAEROPHONE : An electrical musical instrument in which the frequency is controlled by a variable capacitor.

SPHERICAL COORDINATES : Spatial coordinates defined by two angles and a distance. Some robots are programmed in spherical coordinates.

SPHYGMOPHONE : A special form of microphone for attachment to the wrist to render the sound of the pulse clearly audible in a telephone receiver.

SPIKE : A sudden, high-intensity burst of electrical power that can damage electornic equipment including computers, stereos, and televisions. Also called power *spikes* or power *surges*. See PULSE.

SPINTHARISCOPE : An instrument for demonstrating radio-activity in which a fragment of radium is mounted in front of a fluorescent screen which continuously emits flashes owing to its bombardment by alpha rays.

SPLINE : In part programming, a special cubic interpolation routine for fairing of curves. The slopes of the curve as adjusted by an interative procedure until curvature is continuous over the length of the curve.

SPLIT ANODE MAGNETRON : A Magnetron used as a high frequency oscillatior in which the cylindrical anode surrounding the cathode filament is divided longitudinally into halves between which oscillations are produced.

SPOOLING : The simultaneous operation of peripherals using a disk to store output and/or input for multiple programs at the same time.

SPOT SIZE : Diameter of spot produced by beam on CRT surface, expressed in mills or thousands of an inch: gives stroke display line width.

SPREAD SHEET : A program used for calculating numeric results. Common spreadshcets include Lotos 1-2-3, Microsoft Excel, and Quattro Pro.

SPPS : Statistical Package for the Social Sciences.

SPURIOUS RADIATION : Electromagnetic waves effect at any frequency which can effect a radio-communication system outside the band assigned to the transmission in question.

SQUARE LAW CONDENSER : A form of variable air condenser, used for wave meters and for tunings radio circuits generally in which the plates are so shaped that angular displacement is proportional to the square of the capacitance and a straight line scale of wave length is obtained. Cf. STRAIGHT LINE FREQUENCY CONDENSER.

SQUARE LAW DETECTION : Detection in radio-receivers in which the output (direct) current is proportional to the square of the peak value of the applied oscillations.

SQUARE WAVE : A fundamental waveform that alternately equals two fixed values for equal lengths of time. A pulse train that consists of rectangular pulse the markspace ratio of which is unity. See PULSE Cf. SAWTOOTH WAVE.

SQUEAL : In a radio receiver or audio amplifier, a high-pitched tone heard together with the desired signal.

SQUEGE : To oscillate and cut off for alternate periods of time.

SQUEGGER OSCILLATOR : A thermionic oscillator constructed to give short successive pulses of oscillations by the action of a grid leak which permits of a grid bias building up during the oscillating periods to a prohibitive value and dying down to a restarting value during the quiescent periods.

SQUELCH : To automatically quiet a receiver by reducing its gain in response to the absence of an input signal.

SRAM : Static RAM. (See "Random Access Memory").

STABILISER : (in long distance telephony) : A device for preventing instability, i.e. avoiding self-oscillation, in a repeater circuit; sometimes consisting of a voice-operated relay producing an increase in amplitude in the direction of speech.

STABILIVOT : An apparatus consisting of a gas-filled tube, with a number of concectric coated iron electrodes, ionised by a D.C. voltage between the extreme inner and outer ones. The voltage between the intermediate electrodes is found to remain remarkably constant in spite of a considerable range of variation of the terminal voltage, so that the apparatus can be used as a source of practically constant e.m.f. with apparatus consuming very small currents.

STACK ARCHITECTURE : An architecture wherein any portion of the external memory can be used as a last-in, first-out stack to store/retrieve the contents of the accumulator, the flages, or any of the data registers. Many units contains a 16-bit stack pointer to control the addressing of this external stack. One of the major advantages of the stack is that multiple-level interrputs can be handled easily, since complete system status can be saved when an interrupt occurs and then be restored after the interrupt. Another major advantage is that almost unlimited subroutine nesting is possible.

STANDARD CAPACITOR or CONDENSER : A capacitor carefully made so that its capacitance is not likely to vary, and adjusted accurately to a value such as half a micro-farad: usually provided with arrangements where by it can be kept short-circuited to avoid possibility of effect of Residual charges.

STANDARD CELL : A Primary Cell which, when made according to a given specification, can be relied upon to give with considerable accuracy a known E.M.F. under known conditions of temperature; used for adjustment of potentiometers, etc.

STANDARD OHM : The actual value of a resistance standard such as the International Ohm, or a Standard Resistance, made to have a value as nearly as possible equal to it use in resistance comparisons.

STANDING WAVE : A ore wave in which the ratio of an instantaneous value at one points to that eat another point does not vary with time, which is produceing buy one two waveser of the same frequency travelling in oposite directions. Stationary wave. A wave that reamins staionary, i.e. the displacement at any given point is always the same and a given displacement, such as that of node, is not propagated along the wave. Standing waves results from the superimposition of two or more waves of three same period and usually occur when a wave is reflected totally or partially from a given barrier. Compare travelling two wave.

STAR : Self-Test and Repair. (Computer)

STAR CONNECTION : The method of connecting up poly-phase circuits in which one end of each phase is connected to a common or natural point, which may be earthed, insulated or connected to a wire to which all the others neutral points on the system are connected.

TRARISTEPPING : Jagged raster representation of diagonals or curves; corrected by anti-aliasing.

STAR-NETWORK : A system in which all stations radiate from a common controller.

START BIT : A bit or group of bits that identifies the beginning of a data word.

START TIME : The time interval required to commence some physical motion, as the movement of magnetic tape past the rearwrite head.

STAT : A prefix to a unit, indicating its use in the obsolete CGS electrostatic system of units. 1 statampere = 3.336×10^{-10} ampere, 1 statvolt= 2.998×10^2 volts. 1 statohm = 8.988×10^{11} ohms.

STATE : An condition or characteristic of something, for example, a relay. Defines the immediate condition of the interface, excluding transitions, as indicated by the control signals.

STATE DIAGRAM : A diagram which show what binary numbers a register goes through.

STATEMENT : An instruction to the computer and a high-level language instruction to the computer to perform some sequence of operations.

STATE SPACE : A representation of a problem domain as a set of states, or configurations of the problems elements, and operators, or manipulations that convert one state into another, that can be drawn as a directed graph in which nodes are states and arcs are operators.

STATE VARIABLE FILTER : And active filter using operational amplifiers as summing amplifiers and integrators to produce simultaneous low-pass, highpass, bandpass, and notch filter responses. Also called a universal filter.

STATICAL ELECTRICITY : Electricity mainfesting itself in charges at high potential, e.g. that produced by frictional or influence machines, formerly thought to be distinct from Galvanic or Voltaic Electricity, which manifests itself as a current.

STATIC CHARACTERISTICS (of a Thermionic valve) : Characteristic curves taken under non-oscillatory conditions. See ANODE-CURRENT CHARACTERISTIC and GRID CURRENT CHARACTERISTIC.

STATIC DIAGRAM : A memory whose contents remain indefinitely, requiring no refresh.

STATIC DISPLAY : Display circuitry where the presentation of each decimal digit takes place simultaneously (see BCD output, parallel mode).

STATIC ELECTRICITY : The transfer of static charge from one object to another either by direct contact or by a spark that bridges an air gap between the objects. See STATICAL ELECTRICITY.

STATIC FREQUENCY CHANGER : Any form of frequency changer not containing moving parts, such as those sometimes used in radio-communication for obtaining multiples of there original fundamental frequency by reinforcement of harmonics by resesonance. Also called Doublers and Triplers in the field. The cases of twice and three times the original frequency, or, in general, Frequency Multipliers. See also MERCURY VAPOUR FREQUENCY CHANGER.

STATIC GENERATOR : See ELECTORSTATIC GENERATOR.

STATIC HARMONIC ABSORBER : An Harmonic Absorber consisting of a transformer connected to capacitors so as to form a three-phase resonating circuit to absorb one particular harmonic such as the fifth. Cf. ROTARY HARMONIC ABSORBER.

STATIC RAM : See SRAM.

STATIC WAVE CURRENT : A term used in electormedical practice for the sudden discharge of an insulated patient who had been raised to a high potential by an electrostatic generator.

STATUS DATA : A device-dependent message indicating the actual status of a device.

STDM : Synchronous Time-Division Multiplexing.

STEDY STATE : A state reached by a system under steady operating conditions after any transient effects resulting from a change in the operating conditions have died away. A steady state occurs, for example, with forced oscillations.

STEATITE : A mineral, akin to tale, also known as soapstone, composed some-times used as an insulator where heat resisting property is an advantage.

STELJES PRINTINGS TELEGRAPH SYSTEM : A system emloying the A.B.C. junction with a receiver in which a type wheel, instead of an indicating pointer, is stopped at a point corresponding to a particular letter after the appropriate number of current impulses have been sent over the line, and then impressed upon and fa paper strip by the action of a second electromagnet. Cf MURRY, CREED, HuGHES SYSTEMS)

STENODE CIRCUIT : A form of superhetreodyne radio-receiving circuits of great selectivity in which a piezo-electric oscillator is employed to limit the intermediate frequency band. A special correcting circuit employed to avoid distrotion.

STEP BY STEP AUTOMATIC TELEPHONE SYSTEM : An Automatic Telephone System, such as the Strowger system in which the basic principle is the feeding forward of the selectors one step at a time by a succession of current impulses, as distinguished from Rotary Systems.

STEPPING MOTOR : An electric motor whose windings are arranged in such a way that the armature can be made to step in discrete rotational increments (typically 1/200 of a revolution) when a digital pulse is applied to an accompanying "driver" circuit. The armature displacement will stay locked in this angular position independent of applied torque, up to a limit.

STEPPING RELAY : A relay with a contact arm that rotates through 360 in two or more discrete steps.

STEPS IN ALGORITHM : The major steps involved in the alogrithm can be categorised as : (i) Partitioning, (ii) Computation, and (iii) Communication.

STEP UP TRANSFORMER : See TRANSFORMER.

STEREOPHONIC RECEPTION : A system of radio reception producting a sense of direction in components of received sound by the employment of two receivers having a phase difference in the sound emitted. Analogous to *stereoscopic vision*.

STIPPLE : Pattern of pixel illumination chosen to produce variations in raster colour or intensity.

STOP BAND : The frequency range in which a filter has high attenuation. Also called rejection band.

STOP BIT : A bit or group of bits that identifies the end of a data word and defines the space between data words.

STOP TIME : The time interval required to bring to a halt some physical motion, as the movement of magnetic tape past the read-write head.

STORAGE : Memory the retention through time of symbols representing information.

STORAGE BATTERY : A battery that is formed from secondary cells.

STORAGE BUFFER REGISTER : A register that holds data to be moved to or form main memory.

STORAGE CATHODE RAY TUBES : Storage cathode ray tubes produce a visual display of controllable duration. The tube has two electron guns—the *writing gun* and the *flooding gun;* it also has a phosphor viewing screen and two fine mesh metal screens. One of the metal screens, the storage screen, is coated with a thin dielectric material to from the target and the other serves as an electron collector Fig. (a). A positive charge image is produced on the stroage screen by scanning with a high resolution intensity-modulated writing beam from the writing gun. It remains until it decays or is erased. Information is extracted by *flooding* the storage screen with an electron beam from the flooding gun.

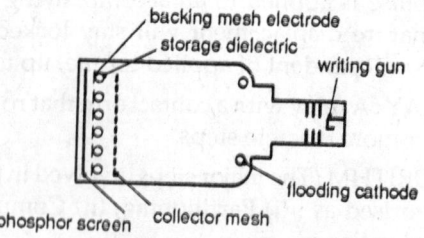

a Storage cathode-ray tube

STORAGE DUMP : A readout or printout of the contensts of a storage device. Usually only the contents of internal storage are dumped, and usually the readout is onto magnetic tape or punched cards, and the printout onto paper via a line printer or typewriter.

STORAGE PROGRAM : A set of directions representing a program read into storage from input and expressed in the same type of symbols as used to express the operand data.

STORAGE REGISTER : A collection of electronic circuits which allows data (usually one or more computer words) to be stored until needed.

STORE/COMPUTE TRADE OFF : See prepare/deliberate trade off.

STORED PROGRAM COMPUTER : A computer controlled by internally stored instructions that can *synthesize, store,* and sometimes *alter* instructions as if they were data, and that can then execute these instructions.

STORED PROGRAM DIGITAL COMPUTER : A device, essentially equivalent in computational power to a Turing Machine, which gets its instructions from its own writable memory rather than a separate control store; therefore, its programs can modify themselves and be replaced without building a whole new machine.

STRAIGHT FORWARD JUNCTION TELEPHONE SYSTEM : A system in which a junction line operator is automatically connected to the subscriber's operator at the called exchange.

STRAIGHT LINE FREQUENCY CONDENSER : A variable condenser with plates so shaped that the readings of the uniformly graduated dial are proportional to the frequency of the oscillation to which it is tuned to repsond.

STRAIGHT LINE WAVE-LENGTH CONDENSER : See SQUARE LAW CONDENSER.

STRAIN GAUGE : An instrument that measures strain at the surface of a solid body means of changes in the electrical properties of associated circuits.

STRATEGIC PLANNING DECISIONS : Decisions of a long-term nature that deal with setting the strategy and objectives of the firm.

STRAY CAPACITANCE : Capacitance which exists incidentally between portions of a circuit at different potentials as opposed to capacitance intentionally placed in the circuit.

STREAKING : A display condition in which objects appear to extend horizontally beyond their normal boundaries.

STREAMING (DATA) : The process of transferring information in a non-interlocked manner to achieve faster transfer rates.

STREES : Structural Engineering Systems Solver.

STRETCH : A CAD design/editing aid that enables the designer to automatically expand a displayed entity beyond its original site.

STRING : A set of letters, numbers, or other characters that are related to each other.

STROBE PULSE : A pulse that samples binary information.

STROBODYNE RECEPTION : A system of reception in radio-telephony similar to Superhetero-dyne Reception, in which the intermediate frequency is obtained by a somewhat different form of interaction between the frequencies of the auxiliary oscillator and the received waves, on an Interference principle analogous to Stroboscopic Methods.

STROBOGLOW : A form of Neon-electric Stroboscope employing a thermionic oscillator to effect the intermittent illumination of the lamp.

STROBOSCOPIC DIRECTION FINDER : A Direction Finder with a rotating frame aerial system connected through amplifiers to a pair of neon tubes or a disc rotating at the same speed which light up momentarily, the position corresponding to the bearing of the received signal.

STROBOSCOPIC METHOD : A method of measurement of speed, frequency, slip, etc., depending upon the observation of a moving object at regular intervals only; e.g. by viewing it by the light of a series of sparks of known frequency, of by the light of an alternating current lamp or through regularly placed slits in a revolving disc.

STROBOSTRON : A caesium activated cathode gas discharge tube particularly adopted for stroboscopic purpose.

STROKE CHARACTER GENERATOR : Electronic processor that forms alphanumeric characters from the line segments.

STRONG AI : Is the name given by John Searle to a research program whose goal is the creation of a complete artificial mind that will actually understand and experience cognitive states in the same way the human mind does. (See also robotics, weak AI).

STROWGER AUTOMATIC TELEPHONE SYSTEM : A system of automatic telephone exchange working characterised by the use of successive step-by-step Selector switches, actuated by current impulses produced by the rotation of a dial on the subscriber's instrument. The selector switches are entirely electormagnetic and contain a number of tiers of fixed contacts, each row arranged in a semicircle, and a moving contact arm which first rises to the height of the tier required, and then swings round and stops over the required contact.

STRUCTURED DESIGN : An approach to design that attempts to provide discipline for the designers and to clarify the desing itself.

STRUCTURED PROGRAMMING : A modular approach to program development that emphasizies stepwise refinement, simple control structures, and short non-entry point one-exit-point modules.

STRUDL : Structrual Design Language.

STYLE SHEET : A collection of specifications for formatting text. Stylesheets may include information for font, size, style margins and spacing. Applying a stylesheet to text automatically formats the text according to the stylesheet's specifications.

STYLUS : Device analogous to a pencil, used with a data tablet to input coordinate data.

SUB-AUDIO FREQUENCY TELEGRAPH SYSTEM : A system in which telegraphic signal currents at frequency below audibility are superposed on an ordinary telephone circuit by selected by suitable filter circuits.

SUB-CARRIER : A carrier signal that is applied as a modulating signal to modulate another carrier.

SUB-HARMONIC : A sinewave whose frequency is an integral submultiple of a fundamental frequency. A fifth subharmonic is one-fifth the fundamental frequency. See SUB-HARMONICS also.

SUB-HARMONICS : Oscillations or alternating currents of a frequency which is a submultiple of the fundamental frequency. Cf. HARMONIC.

SUB-MARINE CABLE : A cable laid under water on the bed of the sea, river, lake, etc., for telegraph, telephone, or power purposes. Usually, in the case of telegraph and telephone cables, insulated with gutta-percha, and in some cases carrying steel armouring. See ATLANTIC DEEP SEA, INTERMEDIATE, and SHORE END CABLES. Submarine telephone cables in many cases are provided with loading coils or distributed inductance. Submarine power cables are only used for comparatively short distances.

SUB-MARINE TELEGRAPHY : Telegraphy through Submarine Cables; differing from land telegraphy mainly in the long distances dealt with and in the lower speed of signalling possible on account of the effect of the capacitance of the cable, which hinders both the growth and extinction of each current impulse.

SUB-MODULATOR : An amplifying valve used between the microphone circuit and the modulating valve in a radio transmitter.

SUB-ROUTINE : A part of a routine, or a routine. A program that does a specific set of operations and is part of a larger program. A series of computer instructions to perform a specific task for many other routines. It is distinguishable from a main routine in that it requires as one of its parameters a location specifying where to return to the main program after its function has been accomplished.

SUBSCRIBER'S LINE : A telephone line connecting a subscriber's apparatus with a telephone exchange.

SUBSCRIBER'S LOOP : The part of a telephone circuit, including the Subsrciber's Line and apparatus.

SUBSTANDARD : Denoting a measuring instrument that is used as a standard instrument but is not quite so accurate as the primary standard. It may be used to check or calibrate a device but in turn needs itself to be checked against the primary standard.

SUBSTATION : A complete assemblage of plant, equipment, and the necessary buildings at a place where electrical energy is received and where it may be either converted from alternating current to direct current, *stepped up* or *stepped down* by means of transformers, or used for control purposes. The substation usually receives power from one or more power stations. An establishment, either in a separate building or forming part of a consumer's premises, where electrical energy is transformed. converted or controlled.

SUBTRACTIVE COLOUR MIXTURE : Mixture in which light interacts with a colourant (pigments, dyes, filters, etc.) which filters out or subtracts some of the colour from it by absorption. All colours (or the subtractice primaries) added equally together make *black. Yellow* absorbs *blue* light, *magenta* absorbs *green* light, and *cyan* absorbs *red* light. A *white* colour results from all colouts being reflected. Complementary pairs combine equally to form neutral grays.

SUCCESSIVE APPROXIMATION CONVERTER : A system of a A/D conversion whereby the D/A voltage is generated by successive approximation.

SUCCESSIVE MULTIPLICATION : A process used for converting a decimal fraction to another *radix*.

SUMMER : An analog device used for addiing currents or voltage.

SUMMIING AMPLIFIER : Also called a summer, an operational amplifier circuit which takes the instantaneous algebraic sum of two or more input signals.

SUM OF PRODUCTS : A Boolean expression formed by ORing ANDed terms.

SUPER ADUIO FREQUENCY : A frequency higher than that of audible sound.

SUPER-AUDIO TELEGRAPH SYSTEMS : System employing current at superaudio frequencies superposed on circuits carrying other traffic to provide additional telegraph channels. Frequencies up to more than 4000 cycles are employed for this purpose.

SUPER COMPUTERS : Very large and fast computers designed for scientific computations. Computer capable of handling data at more than 500 m-flops and with internal RAM of 256 mb.

SUPER CONDUCTIVITY : A phenomenon that occurs in certain metals and a large number of compounds and alloys when cooled to a temperature close to the absolute zero of thermodynamic temperature. At temperatures below a critical transition temperature. The electrical resistance of the material becomes vanishingly small and the material behaves as a perfect conductor. Currents induced in superconducting material have persisted for several years without significant decay.

SUPER-EMITRON CAMERA : A modification of the Emitron television Camera in which greater sensitivity is obtained by separating the functions of photo emission and charge stroage. An optical image is thrown upon a continuous photoelectric screen the emission from the back of which is focused electromagnetically on to a *mosale* screen scanned by an electron beam as in the earlier apparatus.

SUPER FREQUENCY WAVES : A term sometimes used for electromagmetic waves pf frequency between 3×10^6 and 30×10^6 k.c. per sec.

SUPER HET : Abbreviation or contraction for superheterodyne. Popular term for a Superheterodyne Receiving Set. See SUPER-HETERODYNE

SUPER HETERODYNE : A type of radio receiver in which modulated input carrier frequencies are converted or mixed to a fixed, or intermediate, frequency by a local oscillator, after which amplification takes place. Receiver tuning is done by varying the frequency of the local oscillator. Also called *superhet*.

SUPER HETERODYNE RECEPTION : The most widely used type of radio reception in which the incoming signal is fed into a mixer and mixed with a locally generated signal from a local oscillator. The output consists of a signal of carrier frequency equal to the difference between the locally generated signal and the carrier frequencies but containing all the original modulation. This intermediate frequency (or IF) signal, is amplified and detected in an intermediate-frequency amplifier and passed on to the audio frequency amplifier. Abbreviation for Supersonic Heterodyne Reception. A system of radio reception in which local oscillations (or a harmonic therof) slightly different frequency from the received oscillations, obtained from an auxiliary valve or from the main rectifying valve are superposed upon the received oscillations so as to produce beats of a frequency intermediate between radio and audio-frequencies at which amplification can be conveniently carried out. This system is suited for long distance reception, with a number of amplifying stages, and can be made of great selectivity. Cf. STROBODYNE and ULTRADYNE RECEPTION, and see also HYPERDYNE RECEPTION and STENODE CIRCUIT.

SUPER HIGH FREQUENCY : Abbreviated SHF, any frequency in the region from 3 to 30 GHz. See frequency band.

SUPERMALLOY : Trademane . An alloy of iron, nickel, and molybdenum that is similar to permalloy but that has a higher magnetic permeability.

SUPERMEDUR : Tradename. A square-loop material used as a magnetic core.

SUPER MINI COMPUTERS : Minicomputers that are very fast and that overlap with small mainframe computers.

SUPER PHANTOM CIRCUIT : A telegraph circuit with two separate loops for sending and receiving respectively, each superposed on a telephone Phantom circuit.

SUPERPOSED OR SUPER-IMPOSED CIRCUIT : A circuit for telegraph or telephone purposes, formed by superposing current upon a circuit, or parts of more than one circuit, being used at the same time for other messages. See RETARDATION METHOD, PHANTOM CIRCUIT, CARRIER CURRENT TELEGRAPHY.

SUPER-POWER VALVE : A Power Valve of large size, capable of dealing with grid sweeps of 30-40 volts, and in its later forms with a Screened Grid giving an amplification factor up to 100. See also PENTODE.

SUPER REGENERATIVE RECEPTION : A method of reception used for ultrahigh frequency radiowaves in which the detector is a squegging oscillator. The frequency at which the oscillations are quenched, the quench frequency is a function of the frequency of the received radio waves. Very large values of amplification can be obtained using this method of reception as a result of the positive feedback employed in the detector. Compared however to superheterodyne reception, the selectivity is realtively poor.

SUPER RETROACTIVE RECEPTION : A system of radio reception in which Retroaction is applied to an extent that can produce strong self-oscillation, while an auxiliary oscillator of a much lower frequency is employed to check these oscillations periodically, after which they build up again to an extent proportional to the strength of the received oscillations, so that the resulting effect, duly amplified, will represent the received signal. The arrangement is also called Superretroaction and Super-regeneration.

SUPERSONIC AMPLIFICATION : See SUPER-HETERODYNE RE-.CEPTION.

SUPERSONIC FREQUENCY : A frequency higher than that of sound waves within the limits of audition. Cf. AUDIO-FRQUENCY. See SUPERSONIC FREQUENCY.

SUPERSONIC HETERODYNE RECEPTION & SUPERSONIC RECEPTION : See SUPER-RETERODYNE RECEPTION.

SUPERVISOR : The control program that schedules and manages the computer's resources.

SUPERVISORY LAMP : A signal lamp on a telephone exchange switch-board controlled by relays in the operator's Cord Circuit, which lights up when the operator inserts a calling plug into the jack of the wanted line, and remains alight until the called subscriber takes his instrument off its hook, remaining extinguished until he hangs it up again. When the supervisory lamp and the calling subscriber's Line Lamp have both lighted up again the connection can be cleared.

SUPPLEMENTARY ANODE : A small extra anode used locally in

electro deposition to improve the uniformity of the deposit.

SUPPLY MAIN : The wires, cables, etc., belonging to a supply authority, extending into a consumer's premises as far as the terminals protected by double pole fuses, where his own installation commences; thus including both the outside Distributing Main and the Service Lines, etc.

SUPPRESSED AUTOMATIC VOLUME CONTROL : See DELAYED AUTOMATIC VOLUME CONTROL.

SUPPRESSED CARRIER WAVE TELEPHONY : A System of carrier wave (line or radio) Telephony in which the excess of unmodulated carrier wave is filtered out and not transmitted, but is reintroduced in the receiving apparatus in sufficient quantity to prevent distortion. See also SINGLE SIDE BAND TELEPHONY.

SUPPRESSED ZERO INSTRUMENT : Syn. : Set-up scale instrument. A measuring or recording instruments in which the zero position falls outside the dynamic range of the instrument; the moving part is not deflected until a predetermined value of the measured signal is reached.

SUPPRESSOR GRID : A vacuum tube element positioned between the screen grid and the plate, which is kept at a fixed potential less than either the screen grid or plate to prevent the transfer of secondary electrons from the screen grid to the plate. Also called the third grid G3.

SUPARCONDUCTIVITY : See SUPERCONDUCTIVITY.

SURFACE OF REVOLUTION : Surface produced when a line is rotated about an axis.

SURFACE RESISTIVITY : The resistance between two opposite sides of a unit square of the surface of a material. The measured value can vary greatly depending on the method of measurement.

SURFACE WAVE : A radiowave that travels along the surface separating the transmitting and receiving aerials. The surface wave is affected by the properties of the ground along which it travels. See GROUND WAVE.

SURGE : An abnormal transient electrical disturbance in a conductor. Surges are produced from many sources, such as a lightning storke, sudden faults in electrical equipment or transmission lines, or switching operations.

SURGE DIVERTER : A device that is connected between a conductor and earth and diverts the major part of any excessively large voltage surge. A lightning arrester is a particular example of a surge diverter. The most common arrangement used consists of one or more spark gaps connected in series with a material, such as silicon carbide, that exhibits a decrease in electrical resistance with increasing voltage. This material assists the spark gaps to return quickly to their normal condition following the passage of a surge.

SUSCEPTANCE : The reactance of an A.C. circuit or part thereof divided by the square of its impedance, or the wattless component divided by the voltage.

SUSCEPTIBILITY : The ratio of the Intensity of Magnetisation (J) to the Magnetising Force (H) producing it. Positive in Ferromagentic and Para-magnetic materials, zero for air, etc., and of slight negative value for Diamagnetic materials.

SUSCEPTIBILITY DIFFERENTIAL : See DIFFERENTIAL.

SUSCEPTOR PHASE ADVANCER : A Phase Advance controlling the magnetising current of the induction motor in conjunction with which it works, by the effect of the component of the voltage generated at right angles to the secondary voltage, e.g. the Kapp Vibrator and the Leblanc Recuperator.

SWAC : Standards Western Automatic Computer.

SWAMPING RESISTOR : An unbypassed resistor placed in the emitter lead of a bipolar transistor circuit to minimize the effect of temperature on the AC base-emitter junction resistance.

SWAP : Society for Wang Application and Programs.

SWG : See BRITISH STANDARD WIRE GAUGE.

SWING : The total variation in the frequency or amplitude of a quantity, such as the voltage swing.

SWING OSCILLATIONS : Oscillations in a Thermionic Valve circuit independent of the natural frequency of the circuit, caused by sudden change of stability conditions.

S-WIRE : The wire in a telephone exchange connected to the Sleeve of the plug. (Also called the "C" Wire. Testing Wire, Holdinig Wire, and Third Wire).

SWITCH : An instruction, added to a command, that designates a course of action, other than default, for the command process to follow : (1) A physical device having two or more position sensible by the computer. (2) An instrument in a program which is modified by the program to cause the computer to take alternative courses of action. The instruction usually is an unconditional transfer or a SWITCH NO OPERATION: the alternative condition (symbol value) of which is usually determined by one or more prior conditional transfer. For example, as a result of a conditional transfer, the computer might change at a later point in the control sequence a NO SWITCH OPERATION to an UNCONDITIONAL TRANSFER, thus changing the course subsequent processing (3) A device that opens or closes a circuit (4) A device that causes the operating conditions of a circuit to change between discrete specified levels. (5) A device that selects from two or more components, parts, or circuits in the desired element for a particular mode of operation. (6) An appliance for opening or closing a circuit or making some alteration of connections at will (7) An old-fashioned term for a (telegraph or) Telephone Switchboard.

SWITCH ROOM TELEPHONE : A room in a telephone exchange in which the actual switching operation are carried out either manually or automatically.

SWITCH TYPE VOLTAGE REGULATOR : An A.C. voltage regulator consisting of a variable ratio transformer in which the number of effective turns is varied by a multiple contact switch. Cf. MAGNETO and INDUCTION VOLTAGE REGULATORS.

SYLLOGISM : A statement in formal logic for arriving at a valid conclusion.

SYMBOL : An elementary object in a traditional information processing AI theory, in which intelligent behaviour is believed to arise from the manipulation of discrete symbols that refer to real-world concepts or combinations of them; in natural language SYMBOL processing, a symbol is any element in the alphabet of the language under condsideration.

SYMBOLIC LANGUAGE : "Human-oriented" programming language. Any programming language prepared in coding other than the specific machine language, and thus must be "translated" by compiling, assembly, or other means.

SYMBOLIC LOGIC : Exact reasoning about non-numerical relations using symbols that are difficient in calculation. One type of symbolic logic is Boolean algebra.

SYMBOLIC PROGRAMMING : The use of arbitrary symbols (often nomonic) to represent addresses and command in order to facilitate programming work.

SYNCHRONISE : Two currents, voltages, etc., are said to synchronise when they are of like *phase* and *frequency*. The verb is also used for the causing of two currents, etc., to remain in like phase and frequency, and for making connection between two circuits in which this is the case, e.g. connecting an alternator in parallel with others at a moment when its voltage synchronises with that of the others. The relationship between two periodically varying quantities when they are in phase. The state of affairs when two pieces of apparatus are running in step at equal speed, or currents in two circuits agree in frequency and phase.

SYNCHRONOMETER : A device that counts the number of cycles of a periodically varying quantity that occar during a predetermined time interval.

SYNCHRONOUS : Refers to the operation of the interface in a timing dependent manner that's device- dependent causing waiting for clock signals although all other signals at a paritcular logic gate were available. See asynchronous. A type of computer operation in which the execution of each instruction or each event is controlled by a clock signal: evenly spaced pulses that enable the logic gates for execution of each logic step. A synchronous operation can cause time delays by Events that are coordinated and controlled.

SYNCHRONOUS CLOCK : A mains-operated electric clock in which the speed of the driving motor is a function of the mains frequency and therefore the time-keeping is controlled by the mains supply. A clock driven by a synchronous motor for use on a.c. systems where the frequency is accurately controlled.

SYNCHRONOUS COMPUTER : A digital computer in which the performance of commands is initiated by equally spaced signals from a master clock.

SYNCHRONOUS LOGIC : A logic system whereby all elements are synchronized to a master clock.

SYNCHRONOUS OR A SYNCHRONOUS CONDENSER OR CAPACITOR : A *synchronous* or *asynchornous* machine used to provide leading current for the improvement of the power-factor of a circuit, and thus acting in for providing lagging or leading current for voltage regulation in a transmission system.

SYNCHRONOUS TELEGRAPH SYSTEMS : Telegraph systems depending upon the rotation of contact makers or other apparatus at the same speed and in step at the sending and receiving stations. e.g. Multiplex, Hughes, and Baudot systems.

SYNCHRONOUS TRANSMISSION : Data communications wherein a synchronizing, or clicking, signal between sendinig and receiving devices is used. (Contrast with "Asynchronous Transmission".)

SYNTAX : A level of analysis in language processing that decides whether a sentence is in a language and how its structural features make it a member of the language; it interacts strongly with the levels of semantics and pragmatics. Rules of statement structure in a programming language. The physical structure of a programming language or statement. (1) A set of rules describing the structure of statement allowed in computer language. To make grammatical sense, commands and routines must be written in conformity to these rules. (2) The structure of a computer command language.

SYNTHETIC PROGRAM : A specially constructed program (but one that is not used for production) used to measures and evaluate the performance of a computer system.

SYNTONY : Two Oscillating Circuits are said to be in "syntony" when their natural periods of oscillation are the same, i.e. when one is capable of Resonating to the waves produced by the oscillations of the other.

SYSTEM : (In computeronics) A collection of *hardware, software,* and *firmware* that is interconnected to operate as a unit. (1) In systems analysis, a combination of machine service, material service/and for example, the method of processing orders for merchandise. (2) Hardware, such as an automatic computer system; or some item or software, as a compiling system. (3) Any organized entityand labour service to accomplish information handling operations.

SYSTEMS PROGRAMMER : A programmer who works on the software associated with an operating or supervisory system.

TABLE : A collection of data, each item being uniquely identified either by some label or by its relative position.

TABLE LOOK-UP : The preparation of listings, as with a punched card line printer. (2) A procedure for obtaining the function value corresponding to an argument from a table of function values.

TABLET : (1) In computer graphics, an input device with a writing surface having direct correspondence between positions on the tablet and addressable points on the display surface of a display device. (2) An input device which digitizes coordinate data designed by stylus position.

TABULAR REDUCTION : The Quine-McCluskey method of reducing a Boolean expression.

TACHYGRAPH : A recording Tachometer.

T-AERIAL : An aerial consisting of one or more horizontal wire with the down lead taken from the centre thereof.

TALK ADDRESS : An address which selects on device as data source (talker) and disables all other potential talkers.

TALKER : A device that is able to transmit data on the IEC bus after it is addressed with its talk address by the controller.

TALKING RADIO BEACON : A radio beacon the impulses from which are synchronised with audible signals so that distance can be estimated from the time lag between their reception.

TALLY : A count of or number of somthing, for example, the number of records being processed or the number of records yet to be processed, often done with the aid of a B-register.

TANDEM : A method of connceting two machines or quadrepole networks so that the two output terminals of one machine or network are connected to the input terminals of the other. See also CASCADE.

TANDEM CONNECTION : See TANDEM,Cf. CASCADE CONNECTION.

TANDEM TELEPHONE SYSTEM : A method employed in dense

areas in which calls are passed through intermediate or
"Tandem" exchanges by automatic or manual means.

TANK : An acoustic delay line.

TANTALUM : A metal, atomic number 73, that has an extremely
high resistance to corrosion and is used for applications where
this property is desirable.

TANTALUM DETECTOR : A detector of electrical oscillations of the
self-restoring coherer type, in which the variable contact is
formed by the tip of a very fine tantalum wire just touching
the surface of a pool of mercury in a glass bulb.

TAPPING: A conductor, usually a wire, that makes an electrical connection with a point between the ends of winding or coil. The
number of turns included in the active portion of the coil can
then be selected. More than one tapping may be made to a
particular winding, such as that of a transformer.

TCXO : Temperature compenstaed crystal (xtal) oscillator; comprises
a crystal oscillator and a thermally controlled circuit that compensates for frequency changes over the specified temperature range.

TDM : Time Division Multiplexing.

TEACH : To program an manipulator arm by guiding it through a
series of points or in motion pattern which is recorded for
subsequent automatic action by the manipulator.

TEACH BOX : A hand-held control with which a robot can be programmed.

TEARING : A display condition in which groups of horizontal lines
are displayed in an irregular manner.

TEASER TRANSFORMER : See THREE-PHASE TO TWO-PHASE
TRANSFORMER and SCREENED VALVE.

TEFLON : Tradename Polytetrafuluorethlene (PTFE). An insulator
that has an extemely hgih resistivity and is very resistant to
moisture and temperature.

TELAUTOGRAPH : (1) The original form of the writing telegraph instrument of Elisha Gray, the modern commercial form of which
is known as the Telewriter. The movement of a pencil in the
transmitting apparatus in an up and down direction. regulates the current in one circuit. The two components of the
movement of the pencil are recombined at the received instrument by means of two moving coil galvanometers which

control the vertical and transverse movement of a pen, according to the strengths of the currents in the two circuits thus reproducing the movement of the transmitting pencil. (2) The Korn apparatus for transmission of photographs, etc., telegraphically was also sometimes called a "Telautograph."

TELAUTOGRAPHY : The transmission of images by telegraphic means, particularly applied to radio methods.

TELECOMMUNICATION : The transmission of images by telegraphic means, particulary applied to radio methods. The transmission of information by line or radio-telegraphy, telephony, television or any other method of remote control. The transmission of signals over a long distance, through either *private* or *public* carriers. The study and practice of the transfer of information by any electromagnetic means, such as wire or radiowaves. Communication from one computer terminal or system to another via telephone lines.

TELECOMMUNICATION SYSTEM : The complete assembly of apparatus and circuits required to effect a desired transfer of information. Systems include television radio and telephony.

TELECORD : A form of recording *phonograph* which can be connected to a telephone.

TELECTAL : An aluminium alloy containing lithium, of high tensile strength, suitable for transmission lines.

TELECTROGRAPH : A system of Phototelegraphy in which the original is composed of lines on a metal base prepared by photographing through a single-line screen on a bichromated film. A metal stylus passes over this and makes contact with the metal base in the parts corresponding to the dark portions of the picture, so that a current sent over the line. The synchronously moving paper in the receiving apparatus is chemically prepared, so that a coloured mark is produced electrolytically whenever a current passes through the paper from the stylus to the metal drum on which it is mounted.

TELECTROSCOPE : An early experimental Television apparatus due to Senlecq (1877).

TELEFUNKEN SYSTEM OF RADIO TELEGRAPH : The name given to various successive systems developed in Germany.

TELEGRAPH : Any system of conveying messages by signals over a di ance. Formerly applied to semaphore and other visual sys ms, but now almost entirely to electrical systems. See

TELEGRAPHY. The term is, however, will be used for apparatus of a non-electrical as well as of an electrical nature for conveying orders from a ship's bridge to the engine room and other parts of the vessel and similar purposes. See references under.

TELEGRAPH CABLE : A single or multicore cable for telegraph purposes, laid under-ground or under water. See SUBMARINE CABLE.

TELEGRAPH CIRCUIT : The circuit over which signal currents are sent, in a telegraph system, between the transmitting apparatus and the receiving apparatus; sometimes consisting of an overhead line or cable and a return path through the earth. Various methods exist by means of which more than one message can be sent at a time through a single telegraph circuit. See DIPLEX, DUPLEX, MULTIPLEX, CARRIERWAVE, SUPERPOSED CIRCUIT, etc.

TELEGRAPHIC TYPE SETTING : The remote control over a telegraph line of type-setting and typecasting machinery for column printing, the signals being usually made through the intermediary of a perforated tape.

TELEGRAPH INSTRUMENT : The actual apparatus used in telegraphy for modifying the current so as to give the required series of impulses corresponding to the message, or for enabling the signals to be observed at the receiving station.

TELEGRAPH LINE : The insulated portion of the telegraph circuit which connects the transmitting apparatus with the receiving apparatus usually applied to an overhead line as distinct from a cable.

TELEGRAPH POLE : A pole, usually of wood treated with a preservative such as creosote, for the support of one or more overhead telegraph lines. See "H" POILE.

TELEGRAPH RECEIVER : An insturment for the detection, indication, or recording of telegraph messages. Cf. TELEGRAPH TRANSMITTER.

TELEGRAPH RELAY : A Relay used for detecting weak signal current impulses, and causing corresponding stronger impulses to be sent in a local circuit containing the receiving apparatus. See POLARISED RELAY and other references under RELAY.

TELEGRAPH REPEATER : A form of relay used at an intermediate point in a long telegraph line to detect faint or distorted messages and to retransmit them automatically in a clearer form over a further section of the line. See REPEATING STATION and REPEATER.

TELEGRAPH TRANSMITTER : An instrument used to modify the current in a telegraph circuit so as to give the required series of impulses corresponding to the message; consisting principally of a key or other contact or contacts, opened and closed rapidly in accordance with the required signals, either by hand or by special mechanism.Cf. TELEGRAPH RECEIVER.

TELEGRAPH WIRE : (1) A wire supported as insulators on poles, or otherwise, for the transmission of telegraph currents. (2) Wire of the quality used for this purpose, of *iron, copper* or *bronze.*

TELEGRAPHONE : An electrical apparatus by means of which sound may be recorded and subsequently reproduced in audible form. In the Poulsen Telegraphone, a band of hard steel travels through the field of an electormagnet through the coils of which the received telephone current flows. and magnetised patches are produced in it corresponding ot the sound waves. To reproduce the message, the magnetised patches are produced in it corresponding to the sound waves. To reproduce the message, the magnetised band is passed through a coil connected to a telephone receiver, when it induces (owing to the changes of flux produced) varying currents corresponding to the original sound waves of the original message. The name has also been appropriated for an apparatus which automatically records the number of a caller at the called station when no answer is made.

TELEGRAPHY : Communication by means of a telecommunication system that transmits documentary matter, such as written or printed matter or fixed images, and reproduces it at a distance. The matter is transmitted as a suitable signal code, such as international Morse code, either by means of wire or by radio (radio telegraphy). Picture are transmitted using facsimile transmission. A telegraph network is a complete system of stations, installations, and communication channels that provides a telegraph service. The sending of messages by signals consisting of a series of current impulses in a circuit connecting two places at some distance apart, or by trains of electric waves or otherwise, according to a code. The sending of messages by signals consisting of a series of current impulses in a circuit connecting two places at some distance apart, or

by trains of electric waves or otherwise, according to a code.
Cf. TELEPHONY.

TELEHOR : An experimental Television apparatus for small pictures
designed by Mihaly in 1922.

TELEMETER : To transmit coded information (usually obtained by
anglo-to-digital conversation) to a place at which it is used.
For example, dated on the flow of fluids through a pipe can
be telemetered to a central point where the data are used to
control the flow or to make a record of the flow. Although
usually meaning an optical or other insturment for measuring
distance, this term is sometimes used for an instrument for
indicating current, voltage, etc., at a remote point, or for an
electrical apparatus for distant reading of mechanical instruments.

TELEMETRY : Measurement at a distance. Data is transmitted over
a particular telecommunication channel from the measuring
point to the recording appparatus. A measuring instrument
that measures a quantity and transmits the measured data as
an electrical signal to a distant recording point is known as
a telemeter. Space exploration and physiological monitoring
in hospitals both require the use of telemetry.

TELEMIXTE SYSTEM : A system for working Teletype apparatus on
ordinary telephone exchange lines.

TELEPHONE : An apparatus by means of which sound waves such
as those of the human voice can be made to produce variations in an electric current which, when received in an
instrument some distance away, can produce sound waves
reproducing the original sounds with sufficient accuracy for
spoken words to be clearly distinguished. The original instrument of Reis (1869) bearing the name telephone was limited
to the reproduction of musical sounds. Intelligible speech was
not transmitted till Bell's instrument was produced in 1876.
See TELEPHONE RECEIVER.

TELEPHONE CABLE : A cable containing conductors for use as telephone circuits. Such cables for underground circuits are usually
of the dry-core or air-space type, with the two wires forming
each circuit twisted together, to minimise capacitance and
liability to interference between the different circuits in the
same cable. Submarine telephone cables are insulated with
gutta percha, and in many cases are provided with loading

coils or continuous loading coils or continuous loading to minimise distortion. The distance over which a telephone cable is practicable has also been greatly extended by the use of Repeaters, and when used for broadcast relays, where quality is important, by corrector circuits

TELEPHONE CIRCUIT : The circuit over which speaking and signal currents are sent in a telephone installation between the two stations in communication, usually consisting of two insulated conductors, as earth returns are now, rarely employed in telephony. The circuit can consist of *overhead* lines or *underground* or *submarine* cables. A large number of such circuits can be included in one cable. See also PHYSICAL CIRCUIT, PHANTOM CIRCUIT, SIDE CIRCUIT, CARRIER-CURRENT, SUPERPOSED CIRCUIT, etc., for methods by which the equivalent of more than one circuit per pair of wires can be obtained.

TELEPHONE CONDENSER : A small condenser placed in parallel with a telephone receiver with the object of providing a by pass to the higher frequencies.

TELEPHONE EXCHANGE : An establishment to which all the lines from the subscriber's apparatus in one district are connected; and where on receipt of calls, they can be connected to any other subscriber's line that may be required, or through Junction, Toll or Trunk Lines to other exchanges, where connection is made to the required Subscriber's Line. Such connections are either made by hand through the medium of suitable Switch boards or by automatic apparatus put in action by signals made by the calling subscriber. see TELEPHONE EXCHANGE.

TELEPHONE INTERFERENCE FACTOR OF A POWER CIRCUIT : The ratio of the Equivalent Disturbing voltage to the fundamental voltage of the circuit.

TELEPHONE JACK : See JACK.

TELEPHONE LINE : One or more telephone circuits running along a particular route, usally confined to an overhead line supported on poles, as distinct from cables.

TELEPHONE METER : A step by step counting instrument actuated electromagnatically through a relay by means of a key, for registering the number of calls carried by a telephone circuit; also called *Call Meter and Register*

TELEPHONE PAIR : An expression often used for a single pair of wires in a multiple cable or otherwise forming one metallic telephone circuit.

TELEPHONE RECEIVER : The actual instrument in which the variations of the current received over a telephone line are caused to reproduce the sound waves corresponding to the words spoken into the Transmitter.

TELEPHONE RELAY : An apparatus for producing variations in a local current corresponding to, but much more powerful than, those of a received telephone current. See BROWN TELEPHONE RELAY Cf. TELEPHONE REPEATER.

TELEPHONE RELEASE : A system requiring the intervention of a telephonist (operator) to clear the lines. Cf. CALLING PARTY RELEASE.

TELEPHONE REPEATER : An apparatus used at an intermediate point in a long telephone line to produce a current in the further portion of the line having variations corresponding to, but much stronger than, those of the current, received. The telephone repeaters now in use are chiefly of the Thermionic Valve type acting as Amplifiers. The use of such apparatus enables economy in copper to be obtained by the employment of lighter lines, and the inductive Loading which would otherwise be necessary to be reduced or to be dispensed with.

TELEPHONE SWITCHBOARD : A switchboard at a telephone exchange, or at an installation where a number of lines are dealt with, provided with the necessary apparatus and connections for receiving incoming calls from subscriber's lines, other exhanges, etc., and required lines.

TELEPHONE SWITCHING : The whole subject of interconnection of telephone circuits and telephone exchange working.

TELEPHONE TERMINAL : See BINDING POST.

TELEPHONE TRANSFORMER : See OUTPUT TRANSFORMER.

TELEPHONE TRANSMITTER : The actual instrument wherrby the sound waves of the words spoken are caused to produce modulations of the current sent over the line which cause the Receiver to reproduce the original sounds. See MICROPHONE.

TELEPHONOGRAPH : An instrument for recording messages received by telephone on a phonograph.

TELEPHONOMETRY : Practical quantitative testing and measurement realting to telephone circuits and apparatus.

TELEPHONY : Communication by means of a telecommunication system that is designed to transmit speech or sometimes other sounds. A complete telephone system contains all the circuits, switching apparatus, and other equipment necessary to establish a communication channel between any two users connected to the main system. Communication between two points takes place along suitable cables (telephone lines) execpt where this is inappropriate; a particular access point may then be connected to the main system by means of a radio link (radio telephone), as in ship-to-shore telephony. The reproduction at a distant point of sponken words or other sounds by electrical or other means, with or without a connecting wire. Cf. TELEGRAPHY.

TELEPHONY BI BAND, LINE, OPTICAL, QUIESCENT CARRIER, RADIO, SIDE BAND, SPACE, SUBMARINE SUPPRESSED CARRIER WAVE TWO BAND WAVE and WIRELESS: See BI-BADN TELEPHONY, LINE TELEPHONY, etc.

TELEPOTE : A name applied many years ago to proposed Television apparatus (see also PHEROPE), and later applied to a particular form (due to Dauvillier) employing an exploring beam actuated by vibrating mirrors controlled by tuning forks in the transmitter and a form of Cathode Ray Oscillograph as the receiver. See also RADIO-PHOTE.

TELEPRINTER : A form of start-stop typewriter that comprises a keyboard transmitter, which converts keyboard information into electrical signals and a printing receiver, which reverses the process. Teleprinters are used in telex systems and in some computing systems. A simplified form of start-stop printing telegraph instrument with a typewriter keyboard for transmission, and a motor-driven tape printing receiver; suitable for working on telephone lines now largely used in the British Post Office Telegraphs.

TELEPROCESS : See TRANSCEIVER.

TELESCOPE ELECTRON : See ELECTRON TELESCOPE.

TELE-STEREOGRAPH : A name sometimes given to the Belin System of Photo-telegraphy.

TELE-TEXT : An information service in which information can be displayed as pages of text and pictorial material on the screen of a commercial television receiver, transmitted as part of the commercial television broadcast signal.

TELE TYPE : A type Printing Telegraph instrument for a single conductor system similar to the Baudot system, but using a keyboard transmitter, and for a type of Start-Stop Printer.

TELETYPESETTER : A complete apparatus for type-setting or type-casting controlled by received telegraph signals.

TELETYPEWRITER : See TELEPRINTER.

TELEVISION : The reproduction of visible images of moving objects at a distant station by electrical means. In an elementary form of the apparatus a beam of light, moving rapidly in two dimensions is made to "scan" the object, and the variation in light reflected causes variations in the current through photo-electric cells. These variations of current, suitably amplified, modulate the waves sentout and in the receiving apparatus, control the strenght of a synchronously moving beam of light so that a similar image is produced upon a screen. See also TELEVISION TRANSMITTER, TELEVISION RECEIVER, CATHODE RAY TELEVISION, ICONOSCOPE, SCANING.

TELEVISION BROADCAST BAND : The frequencies extending from 54 to 890 MHz which are assignable to television broadcast stations.

TELEVISION CAMERA : A vision pick-up used in television trans-mission depending upon the scanning in some way of an optical image formed by a lens.

TELEVISION CHANNELS (U.S.) : A band of frequencies 6 MHz wide in the television broadcast and designated either by number or by the limiting frequencies.

TELEVISION RECEIVER : A device that receives a television broad-cast signal and effects the sound and vision reproduction of the original scene. An apparatus for reproducting visual images by television. In small screen high defintion apparatus, usu-ally employing the cathode ray system. Other methods employ a moving beam of light controlled in intensity by a Light Relay and in position by moving slots, lenses, or mirrors.

TELEVISION TRANSMITTER : The whole equipment for transmit-ting radio signals for television usually employing very short wavelengths.

TELEVISOR : A name given to an early class of Television apparatus.

TELEVOLT SYSTEM : A system of activating signals at various points on a supply system by the application of a momentary excess voltage sufficient to cause a neon lamp in the signal circuit to discharge and to permit the flow of current.

TELEWRITER : A device used in telegraphy that converts manually controlled movements of a pen over a plane surface into two currents whose instantaneous values are a function of the position of the pen. These are used to cause automatically corresponding movements of a similar pen at the receiver. See TELAUTOGRAPH.

TELEX : A telegraphy system that enables users to communicate directly and temporarily among themselves using teleprinters and the public telegraph system. A voice frequency Teleprinter system, used in the British Post Office for special subscribers in connection with the telephone.

TEMPERATURE COEFFICIENT OF RESISTANCE : The incremental change in the resistance of any material as a result of a change in thermodynamic temperature. In general conductors exhibit a positive coefficient of resistance; semiconductors and insulators have a negative coefficient of resistance.

TEMPERATURE COMPENSATED CRYSTAL OSCILLATOR : Abbreviated TCXO, a crystal-controlled oscillator in which temperature-compensation networks have been adjusted to match the crystal characteristics so that frequency stability is maintained over a wide temperature range.

TERA : A prifix to a unit, denoting a multiple of 10^{12} of that unit: one terahertz is 10^{12} hertz. Ten to the twelfth power 10^{12} 1,000,000,000,000 in decimal notation. When referring to storage capacity, two to the fortieth power 2^{40} =1,099,511,627,776 in decimal notation.

TERMINAL : (1) A device that provides input/output facilities to a computer often from a remote location. It may be used interactively and usually contains a keyboard and/or visual display unit. An intelligent terminal contains some local storage and processing ability and can perform simple tasks independently of the main computer. (2) Any of the points at which interconnecting leads may be attached to an electronic circuit or device and at which signals may be input or output. (3) A device used to communicate with a central computer from a remote location, usually featuring a typewriter like keyboard. A set of equipment that permits people to communicate with any microprocessor based circuit. Also an input/output device that usually contains a keyboard and a CRT.

TERMINAL BOARD : A plugboard.

TERMINAL IMPEDANCE : The complex impedance at a pair of terminals of a transmission line or other device under normal
operating conditions but undernoload conditions.

TERMINAL SYMBOL : A character in the alphabet for a particular
language which can appear in strings that are sentences of
that language, such as "a" or "b" for English.

TERMINATE-AND-STAY RESIDENT (TSR) : Describes a program
that resides in memory and can be used while running another
program.

TERTIARY WINDING : An additional secondary winding on a transformer. It can be used to supply a load when a different
voltage is required from that of the main secondary or when
a load must be kept electrically insulated from that of the
normal secondary. It may also be used to interconnect supply
sytems that operate at different voltages.

TESLA COIL : An induction coil
used to generate very
high frequency high-
voltage oscillatory cur-
rent (see disgram).

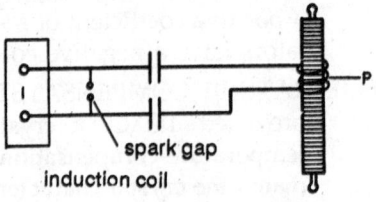

spark gap

induction coil

Tesla coil

TEST BOARD : (1) A board
arranged with suitable
instruments, switchgear,
and connections so that apparatus may be readily connected
thereto for testing purposes. (2) A board containing terminals
or jacks to which connections are brought from instruments,
circuits, etc. for convenience in testing.

TEST JACK : A Jack for making connections for routine telephone
testing.

TETRODE : Any electronic device that has four electrodes. The term
is most commonly applied to a thermionic valve that contains
a second grid. This auxiliary grid is usually a screen grid
designed to decrease the anode-grid capacitance and hence to
increase the resistance to high-frequency currents. It may also
be used either to decrease the anode-cathode resistance or to
modulate the main electron stream by injecting an alternating
voltage. A vacuum tube that has four elements: cathode, control
grid, screen grid, and plate. See FOUR ELECTRODE VALVE.

TETRODE MULTIPLEX. See MORSE MULTIPLEX SYSTEM.

TEXT STRING : Series of alphanumeric characters.

"T" FLIP-FLOP : A flip-flop element that toggles upon receipt of a pulse on the T input. It has only one input. Application of a pulse to this input causes the device to change state.

THE NORTON UTLITIES (IBM) : A group name of professional computer program's that are designed for use of computer operators and helpful in day to day working of personal computers of IBM or compatible design. The Norton Utilites has set the standard for utility program. DOS Easy to use, yet powerful for beginners and experienced computer owners. The Norton Utilites can protect your files from almost any disaster.

THEMACE UTILITES (IBM) : The Mace Utilities is more difficult to use than either. The Norton utilies or PC Tools Deluxe but is unrivaled in its desk recovery abilites. You can format a disk and The Mace utilitels can still recover all your data. If you need the best the file recovery, then The Mace Utilites is the best program for the job.

THEMATIC MAP : A map specifically designed to communicate geographic concepts such as the distribution of densities, relative magnitude, gradients, spatial relationships, movement, and all the required interrelationships and aspects among the distributional characteristics of the earth's phenomena.

THERMAL DETECTOR : A detector of electrical oscillations depending upon their heating effect, e.g. the *Barretter.*

THERMAL DRIFT : A departure of a value, such as the frequency of an oscillator, caused by internal heating of equipment during normal operation or by changes in environmental temperature.

THERMAL INSTRUMENTS : Measuring instruments depending on the heating effect of the current including Thermo-Couple Instruments and Thermal Expansion Instruments.

THERMAL MICROPHONE : A microphone depending for its action upon the variation of resistance of an electrically heated conductor according to its variation of temperature.

THERMAL RECEIVER : A telephone receiver in which the sound waves are reproduced either by the movements of the diaphragm, controlled by the variations of expansion of a wire heated by the telephone current, or by direct expansion and contraction of the air, due to the heating and cooling of the air, entirely without a diapphrag. Also called *Hot Wire Telephone* and *Thermophone.*

THERMAL RUNAWAY : A condition in transistors in which heating of the collector-emitter junction causes a rise in collector current, which in turn causes more heating. At its limit, the *junction temperature* rapidly approaches its maximum rating, beyond which the transistor will be destroyed. Also called *runaway*.

THERMAL TELEPHONE : See THERMAL RECEIVER.

THERMAUTOSTAT : An automatic heat-regulator in which the same set of resistors serve as thermometer resistance and heating elements, controlled by variation of the phase of an alternating current by a Thermionic Relay, the grid voltage of which is controlled by a light sensitive cell receiving the beam from a bridge galvanometer.

THERMIONIC : Pertaining to the emission of electrons as a result of heat.

THERMIONIC AMPLIFIER. : A Thermionic Valve used as an Amplifier or magnifier of the scale of the modulations of a telephone current, or trains of oscillations in radio-communication, by taking advantage of the fact that, under suitable conditions, small variations of e.m.f. applied between the Cathode and the Grid produce large variations in the current in the Anode Circuit, Several such valves may be used in cascade, i.e. with the Anode Circuit of one coupled to the Grid Circuit of the next, so that several successive stages of amplification are produced.

THERMONIC CURRENT : The current carried by electrons or ions through a Therminoic Valve.

THERMIONIC DETECTOR : A Thermionic Valve used as a Detector of electrical oscillations in radio-communication, either in its original two-electrode from (See OSCILLATION VALVE), acting simply as Rectifier in series with the receiving telephone, and thus enabling the oscillations to affect the diaphragm, or in its later and more usual *three-electrode form* by the Anode Bend or by the Cumulative (grid leak) methods of Rectification.

THERMIONIC EMISSION : Electron emission from the surface of a solid that is a result of the temperature of the material. An electron can escape from the surface with zero kinetic energy if it has thermal energy just equal to the work function of the material (compare photo emission). The numbers of electrons emitted increases sharply with temperature (see Richardson's equation). See also *Schottky effect*. The emission of a stream of negative electrons or ions from a heated cathode in vacuum tube. See THERMIONIC VALVE.

THERMIONIC GENERATOR : See THERMIONIC OSCILLATOR.

THERMIONIC INSTRUMENTS : Measuring instruments in which the amplifying effect of one or more thermionic valves is used.

THERMIONIC MAGNIFIER : An apparatus for increasing the strength of signals received over submarine cables in which the moving coil carries a second winding in which minute e.m.f.'s are induced and act on the grid of a thermionic amplifier valve. In another form, movement of a vane attached to a moving coil varies the capacitance coupling to the grid of two thermionic valves and thus affects their output.

THERMIONIC OSCILLATOR. : A Thermionic Valve, used as a generator of Continuous Oscillations, by suitable coupling of the grid circuit to the tuned Plate Circuit.

THERMIONIC RECTIFIER : A Thermionic Valve used as a *rectifier*.

THERMIONIC RELAY : A Thermionic Valve used to perform the function of a Relay. Relays in which the heating effect of a current is used to operate contacts or the effect of a heatinig coil on a bimetallic strip is employed. Gas filled relays, such as the thyratron, have also been used but these are being superseded by solid-state relays using SCRs.

THERMIONIC TUBE : See THERMIONIC VALVE.

THERMIONIC VALVE : A Discharge Tube with a Cathode formed by a heated filament, usually of tungsten, with or without the addition of other material or independently heated, and a metal Aode, sometimes called the Plate, upon which a stream of negative electrons in projected from the hot cathode when a sufficient D.C. voltage is applied across the tube. In its original form with two electordes, such a tube can be employed as a retifier, and consequently as a detector. See OSCILLATION VALVE. In most cases a third or Control Electrode, in the form of a Grid of wire, is employed as a screen between the cathode and the anode. Variations in the potential of this grid have a controlling effect upon the current through the tube of such a character that, within a certain range, a small change in the grid voltage produces a large change in the plate current. Additional grids are sometimes provided to compensate for various disturbing effects.

Such tubes can be employed, not only as Detectors but also as Amplifiers and as Oscillator, and since their introduction, have revolutionised all methods of radio-communication, See THERMIONIC AMPLIFIER, THERMIONIC DETECTOR, DIODE, TRIODE, RETROACTION.

THERMIONIC VOLTMETER : An instrument for measuring small voltages, by observing the change in anode current produced in a Thermionic Valve, when applied between the cathode and the Grid.

THERMIONIC WATTMETER : An apparatus for measuring power in a circuit, in the simplest form of which the sum of and the difference between the potential drops across a resistance in the main circuit, and a portion of a shunt across it, are applied to the grids of two thermionic valves, so connected that the power in the main circuit is proportional to difference between their anode currents which is read on a differential galvanometer.

THERMIONS : Ions emitted by the heated cathode of a Thermionic Valve tube.

THERMISTOR : Clipped thermal resistor, a device that makes use of the change in the resitivity of a semiconductor with a change in temperature. A thermistor has a high negative temperature coefficient of resistance so that its resistance decreases as the temperature increases. It is used chiefly in critical circuits to compensate for temperature variations in other components. A resistor, usually fabricated from semiconductor material, that has a large nonlinear negative temperature coefficient of resistance. An electrolyte, such as a viscous solution of water glass, is sometimes used as the temperature sensitive element. The thermistor is usually shaped as a *rod, bead,* or *disc* and named accordingly. Applications include compensation for temperature variations in other components, use as a nonlinear circuit element, and for temperature and power measurements. A thermistor bridge is an arrangement of thermistors for measuring power. Compare *ballast resistor, barretter.*

THERMO-AMMETER : An ammeter that measures a current by means of the heating effect produced by the current. The two main types of *thermoammeter* are the thermo-couple ammeter and the hot wire ammeter. Both these instruments may be used for either direct current or alternating current measurements since the heating effect is proportional to the square of the current. They must be calibrated empirically against known values of current.

THERMO-CONVERTER : A self contained combination of heating resistances and thermo-couple enabling an ordinary milli-volt-meter to be used as a thermo-couple by the current to be measured, e.g. Thermo-Galvanometer, Thermo-Ammeter, etc.

THERMO-COUPLE : A two-terminal device, based on the Seeback effect, which is composed of two disimilar metals that pro-duce a voltage across this junction that is linerly proportional to the temperature of the junction. Two dissimilar metals joined at each end to form an electrical circuit. If the two junctions are maintained at different temperatures an electromotive force (e.m.f.) is developed between them as a result of the Seebeck effect (see thermoelectric effects).

THERM-ODETECTOOR. See THERMAL DETECTOR.

THERMO-DYNAMIC TEMPERATURE : Syn. absolute temperature,. Temperature that is measured as a function of the energy possessed by matter and as such is a physical quantity that can be expressed in units, termed kelvin. In the thermody-namic temperature scale changes of temperature are inde-pendent of the working substance used in themometer. The zero of the scale is absolute zero. The triple point of water is defined as 273.16 KELVIN.

THERMO-ELECTRIC COUPLE : A pair of metals forming a Thermo-Electric Junction.

THERMO-ELECTRIC EFFECT : The e.m.f. arising from the heating of disimilar metals. (Also called *Seebeck Effect*) Phenomena that occur as a result of temperature differences in an electrical circuit. See See-beck effect, *pelder effect* and *kelvin effect*.

THERMO-ELECTRIC ELEMENT : One of the two pieces of different metals forming a Thermo-Electric Junction.

THERMO-ELECTRIC GENERATOR : An apparatus for generating electric currents by heating the junctions of dissimilar metals. See THERMOPILE.

THERMO-ELECTRIC INVERSION : The reversal of the direction of a thermo-electric current between a particular pair of metals above the temperature know as the *Neutral Temperature* for the two metals in question.

THERMO-ELECTRIC JUNCTION : A contact surface between two

different metals forming part of an electric circuit, which, when maintained at a different temperature from the contacts of the other ends of the pieces in question (with each other or with other parts of materials completing the circuit), will produce an e.m.f. which can cause a current to flow in the circuit deriving its energy from the source of heat maintaining the difference of temperature.

THERMO-ELECTRIC SERIES : An ordering of the metal elements so that if a thermocouple is formed from two of them, the direction of current flow in the hotter junction is from the metal appearing earlier in the series to the other.

THERMO-ELECTRICITY : The phenomena of thermo-electric currents and the differences of potential producting them. See SEEBACK EFFECT, PELTIER EFFECT. THOMSOM EFFECT, THERMO-ELECTRIC INVERSION, NEUTRAL TEMPERATURE, etc.

THERMOSTAT : An apparatus for maintaining constancy of temperature, e.g. an electrical thermometer in which the indicatinig instrument is fitted with contacts which control the relays, etc. There are also numerous forms of non-electrical thermostat. It is an automatic temperature-control switch that is used in conjunction with heating systems, such as an immersion heater, to maintain the temperature of a given medium within predetermined limits. It contains a temperature-sensing device, such as a bimetallic strip, that is used to operate a relay so that the source of heat is interruped when the temperature falls to a lower value. A two-terminal switch that opens or closes a circuit when the temperature changes from a preset value or range of values. Also called a *thermorelay*.

THINK C : *Think C* and *Think Pascal* are the most popular language compilers for the Macintosh. Most popular, including FOXBase+/Mac and Pagemaker, are written using Think C or Think Pascal.

THOMOSON EFFECT : The production of an e.m.f. between portions of the same metal in a circuit at different temperatures, or, conversely, the evolution or abstraction of heat when a current flows from one portion of the same metal to another at a different temperature. Cf. THERMO-ELECTRICITY..

THREE-ADDRESS : See *multi-address*.

THREE-IN-ONE VALVE : A Thermionic Valve with additional grids arranged so that it can be used simultaneously as a *detector,* *high frequency amplifier* and *low frequency amplifier.*

THREE-PHASE : Pertaining to an alternating current system where the circuit is divided into three branches or "Phases" the currents in which are displaced one from another in phase by 120. See STAR CONNECTION and MESH CONNECTION.

THREE-PHASE TRANSFORMER : A transformer that has three independent sets of windings, each usually with the same turns ratio, and is suitable for use with a three-phase (see polyphase) mains input.

THREE-QUARTER BRIDGE : A bridge circuit in which one of the diode rectifiers is replaced by a resistor. The average output voltage is 0.84 times the input RMS voltage.

THRESHOLD FREQUENCY : The frequency at which a particular phenomenon such as the photoelectric effect or photoconductivity, just occurs or the frequency below which an electronic device, such as a high pass filter, does not operate. In the latter case the term cut-off frequency is also used.

THRESHOLDING : The process of quantizing pixel brightness to a small number of different levels (usually two levels, resulting in a binary image). A threshold is a level of brightness at which the quantized image brightness changes.

THRESHOLD VOLTAGE : The input voltage level at which a binary logic circuit changes from one to the other.

THROUGHPUT : (1) The amount of processing done by a system in a given unit of time. (2) Transmission and processing from the instant data *input* into a computer until it becomes *output.*

THUMB-WHEEL : Graphic input device that controls the horizontal or vertical movement of a line across the display surface.

THYRATRON : Syn. Gas filled relay. A trade name for a particular type of Gas-Filled Relay. A gas-filled tube with three electrodes in which the voltage applied to one of the electrodes the grid (or control) electrode is used to initiate the discharge but does not limit it.

THYRATRON STROBOSCOPE : A special form of thyratron in which a sudden illumination is produced periodically by discharging a condenser across the tube when the grid voltage is allowed to attain a sufficient value. This is done at regular intervals by an arragement similar to that employed in a Neon electric Stroboscope.

THYRISTOR : A semiconductor device that contains three or more junctions and that has current-voltage characteristics similar to those of the thyratron. The term has most often been applied to silicon-controlled rectifiers but also applies to other *pn* devices.

TICCIT : Time-Shared Interactive Computer-Controlled Information Television.

TICKLER : A small coil connected in series with either the plate circuit of a vacuum tube or the collector circuit of a transistor, which inductively couples the grid or base circuit coil to provide positive feedback in regenerative dector and oscillator circuits.

TIES : Total Intergrated Engineering System.

TIME BASE : Generally a chain of decade dividers which together with the crystal oscillator, forms the time reference for the counter/time. A voltage that is a predetermined function of time and that is used to deflect the electron beam of a cathode-ray tube so that the luminous spot traverses the screen in a desired manner. One complete traverse of the screen, usually in a horizontal direction, is termed a sweep (or time base).

TIME CONSTANT : The time required for a unidirectional electrical quantity, such voltage or current, to decrease to $1/e$ (approximately 0.368) of its initial value or to increase to $(1-1/e)$ (approximately 0.632) of its final value in response to a change in the electrical conditions in an electrinic circuit or device.

TIME DIVISION MULTIPLEXING : A system of transmitting multiple channels over a single facility whereby each channel is assigned a recurring period of time.

TIME INTERVAL AVERAGE MODE : The function where the counter/timer measures a selected number of successive time intervals.

TIME INTERVAL MODE : The function where the counter/timer measures the time interval between, for example, two *levels* of the same signal (rise time) or between two *pulses* on different lines (delay.)

TIMER : (1) A class of integrated circuit devices which can be wired to function as a monostable multivibrator. (2) A special clock mechanism or motor-operated device used to perform switching operations at predetermined time intervals.

TIME SHARING : An on line system that provides computers services (including computational capacity) to a number of users at geographically dispersed terminals. The sharing of a main computer facility by many users, each of whom has a remote terminal. Processing time is "shared" so the users are unaware of each other.

TIMEX : Texas Instruments Minicomputer Information Exchange.

TIMS : The Institute of Management Science.

TINSEL CORD : A very flexible cord for telephone instruments, in which the *strands* are strips of thin meatl foil or *tinsel* instead of wire. Less liable to break when *kinked* than ordinary flexible cord.

TINT : In a subtractive system, a *hue* plus *white*.

TIP : Terminal IMP (Interface Massage Processor)

TIP WIRE : (of a Telephone plug) The rounded point of a telephone plug forming the projecting end of the central portion. Cf. SLEEVE.

TIP WIRE (in telephony) : See "T" WIRE.

TMR : Triple Modular Redundancy.

T-NETWORK : See QUADRHOLE.

T-NETWORK, BRIDGED : See QUADRIPOLE.

TOGGLE : (1) Alternation of function between two stable states (2) A flip-flop switch.

TOGGLE TO : Change a flip-flop to the opposite state.

TOKEN : A group of bits, such as eight 1s, used in some bus networks to signal network access by a particular station.

TOKEN SHARING NETWORK : A system by which all stations are attached to a common bus and an access token is passed from station to station.

TOLERANCE : A term used in connection with the generation of a non-linear tool path comprised of discrete points. The tolerance controls the number of these points. In general, the term denotes the allowed variance from a given standard, i.e. the acceptable range of data.

TOLL LINE : A term use in America for a long distance interexchange telephone line for converstaion over which an extra charge is made. Such lines in England are more commonly called *Trunk Lines;* The term "toll line" being limited to shorter lines of this nature.

TONE : In a subtrative system, a hue plus gray.

TONE COMPENSATED VOLUME CONTROL : Volume Control combined with Tone Control to prevent undue weakening of the bass at low volume levels.

TONE CONTROL : The adjustment of the properties of loud speaker circuit to alter the degree of response over different parts of the frequency range ot emphasise the proportion of *treble* and *bass* at will.

TONE FREQUENCY TELEGRAPH SYSTEM : See VOICE-FREQUENCY TELEGRAPH SYSTEM.

TONE GENERATOR : An apparatus for obtaining an audio-frequency intermittent or alternating current, for producting an audible not in a telephone receiver for signalling purposes or for testing audio-frequency apparatus generally.

TONE SOURCE : See TONE GENERATOR.

TONE WHEEL : An apparatus for the reception of continuous waves in radio-telegraphy, consisting of a wheel with fine teeth, separated by insulating material, against which a contact brush presses. When revoled, the apparatus serves as a contact brush presses. When revolved, the apparatus serves as a high speed interrupter. Can be used in a receiving telephone and run at such a speed that the frequency of the interruptions is slightly different from that of the incoming waves, resulting in Beats which produce an audible result in the receiver in a similar manner to Heterodyne Reception, or as a rectifier.

TONIC TRAIN : Modulated Keyed Continuous Waves in which the modulation is approximately singusoidal.

TOOL : A term used loosely to define something mounted on the end of the robot arm; for example, a hand, a simple gripper, or an arc welding torch.

TOP-DOWN : (Parsing/Processing) is the characteristic approach of an analytic system that starts with a high-level description of the input to be recognized, such as an expentation of what it might be, and attempts to fit the data to that hypothesis (if possible); if it fails, it can try different hypotheses, possibly derived from preliminary bottom-up processes.

TOP-DOWN DESIGN : Planning of a system by looking first at the major function, then atits subfunctions, and so on, until the *scope* and *details* of the system are fully understood.

TOP SPEED C : See Top Speed PASCAL (IBM).

TOP-SPEED MODULA (IBM) See TOPSPEED PASCAL (IBM)

TOP-SPEED PASCAL : (IBM) Top Speed C, Top Speed C++, Top Speed Modula-2, and Top Speed Pascal were written by former employees of Borland International who worked on the original Turbo Pascal. All the Top Speed compliers work together so you can write program combining C, Pascal and Modula-2.

TORNADO : Tornado works as memory-resident program, available to use from within any other programs. It lets you store random information including letters, database records, or spreadsheets stored as text or ASCII characters.

TORQUE : The moment exerted by a force acting on a body and tending to cause rotation about an axis. It is given by the product of the perpendicular distance from the axis to the point of application of the force, and the component of the force in the plane perpendicular to the axis.

TOTAL EMISSION (in a thermionic valve) : The maximum thermionic current that can be obtained through a valve.

TOTALIZING MODE : The fuction where the counter counts incoming pulses until a manual stop signal is given.

TOUCH SENSITIVE DISPLAY : Display surface that receives data through physical contact.

TOUCH-TONE : a registered trademark of AT&T, which is a telephone pushbutton dialing system using two of seven possible nonharmonically related, audio tone frequencies for the numbers 0 through 9 and the symbols # (pound) and * (star). The audio tones are divided into two frequency range groups so that any number or symbol uses only one frequency form each group in the matrix arrangement at the bottom of the page.

TRAC : Text Reckoning and Complier.

TRACE : (1) To follow the control sequence instruction by instruction, usually producing a print-out reporting the consequences of each instruction; a rountine for doing tracing. (2) Scanning path of the beam in a raster display.

TRACE INTERNAL : The time during which a visible raster line is scanned.

TRACK : A specific area on a moving-storage medium, such as a diskette, disk, or tape cartridge, that can be accessed by the drive heads. One of the conncentic circles formed by recorded data on a disk. In a serial device, a channel; in a parallel device, a group of channels.

TRACK BALL : A device that work like an upside-down mouse. Instead of moving the mouse to move the cursor on the screen a trackball lets you roll a ball around to move the cursor. An electronic aidon to home computers designed for the children to play games on computer. A ball is rolled in its socket to move an object on computer's Television's screen Compare : JOY STICK and MOUSE.

TRACTOR : A belt with pins that push or pull computer paper through printer.

TRAILING EDGE (of a pulse) : See PULSE Cf. LEADING EDGE.

TRADE OFF : The pros and cons of different alternatives; for example, one often is forced to *trade* cost savings for performance.

TRANSACTIONS PROCESSING SYSTEMS : Basic systems that process routine transactions in an organization, such as the entry of customer orders.

TRANSCEIVER : Communication equipment produced and marketed by IBM.

TRANSDUCER : Syn. *sensor.* Any device that converts a non-electrical parameter, e.g. sound, pressure, or light, into electrical signals or vice versa. The variations in the electrical signal parameter are a function of the input parameter. Transducers are used in the electroacoustic field. Gramophone pick-ups, microphones, and loud-speakers are all *electroacoustic* transducers. The term is also applied to a divice in which both the input and output are electrical signals. Such a device is known as an *electric transducer.*

TRANSDUCTOR : Syn. *Saturable reactor.* Acronym from *transfer inductor.* A device that consists of a magnetic core carrying several windings. The state of magnetic flux density in the core is controlled by a fixed alternating current in one of the windings. This current is sufficiently large to cause saturation of the core. Small variations in the current of one of the other windings the signal winding them cause large variations in the power in another circuit coupled by another winding the power winding. The device thus operates by magnetic modulation and is used in control circuits, such as lighting circuits, and, particularly, in aircraft. A general term used for any device receiving input power from one system and supplying output power corresponding to the input in certain characteristics (e.g. wave from) to another system, which may be *electrical, mechaninal,* or *acoustic,* and thus including transformer, amplifiers, filters, microphones, loud speakers, etc.

TRANSFER CIRCUIT : A circuit common to two operator's position in the same exchange.

TRANSFER COMMAND : A signal which conditionally or unconditionally specifies the location of the next instruction and directs and automatic computer to that instruction.

TRANSFER OSCILLATOR : A type of frequency converter in which the input signal is mixed with a harmonic of a VFO, the fundamental of which is measured by the counter.

TRANSFER PULSE : The transfer pulse enables data counted in the counter decades to be transferred to the memory.

TRASFORMATION : Performance of mathematical calculations such as matrix algebra to rotate, scale, or otherwise manipulated a graphic image whose coordinates are stored in the computer.

TRASFORMATIONAL GRAMMER : A syntactic theory that extends context- free grammer by allowing various surface-level transformation that maintain the "deep structure" of a parse tree for a sentence while changing aspects like interrogative or declarative case; it therefore represents different forms of the same sentence essentially identically.

TRANSFORMER : An apparatus that has no moving parts and that transforms electrical energy at one alternating voltage into electrical energy at another (usually different) alternating voltage without change of frequency. It depends for its action upon mutual induction (see electromagnetic induction) and consists essentially of two electric circuits coupled together magnetically. The usual construction is of two coils (or windings) with a magnetic core suitably arranged between them. One of these circuits, called the primary, receives energy from an a.c. supply at one voltage; the other circuit, called the sceondary, delivers energy to the load, usually at a different voltage.

TRANSIENT : A phenomenon, such as damped ocillations or a voltage or current surge, that occurs in an electrical system following a sudden change in the dynamic conditions of the system and that is usually relatively short-lived. A transient may be caused by the applicaion of an impulse voltage or current to the system or by the application or removal of a driving force. The nature of their transient is a function of the system itself but the magnitude depends on the magnitude of the impulse or the driving force. The transient response of an electronic device, such as an amplifier is the change in output that occurs as a result of a specific sudden change in the input.

TRANSIENT CURRENTS : Rapidly changing currents in a circuit, due to temporary causes such as *Surges*.

TRANSIENT DISTORTION : The alternation in form of a telgraphed signal impulse in transmission due to the influence of the characteristics of the circuit on the building up conditions.

TRANSIENT EFFECT (in telephony) : A form of phase distortion occurring in very long lines in which the high frequency components arrive later than those of low frequency.

TRANSIENT RESPONSE : See transient.

TRANSISTANCE : Acronym from *transfer* resistance. See *transfer parameter*.

TRANSISTOR : Contraction of transfer resistor and designated by the letter Q on schematic diagrams, a solidstate, current-gain device having a *collector, base* and *emitter* terminals made using both and type materials transistor are available in either PNP or NPN configurations. Also called a bipolar transistor. See also NPN transistor and PNP transistor. A multielectrode semiconductor device in which the current flowing between tow specified electrodes is modulated by the voltage or current applied to one or more specified electrodes. The semiconductor material is usually silicon. Transistors have now replaced thermionic valves as the general purpose active electornic device except for some very specialized uses. See UNIJUNCTION TRANSISTOR, BIPOLAR TRANSISTOR.

TRANSISTOR PARAMETERS : Parameters deviced for writing up relationships between input and output current or voltages based on equivalent circuit of transistors. Hybrid pararmeters are most commonly used for bipolar junction transistor, they are so called because the dimensions are mixed and a true matrix representation is not possible. The y patameters are frequently used for circuits containing field effect transistor z parameters are impedance parameters and are the inverse of y parameters, and g parameters are the inverse of h parameter.

TRANSITION : The edge of a pulse as it changes from one state to the other.

TRANSLATING REALAY : A telegraph Relay which also acts as *Translator*.

TRANSLATOR : A program that accepts a *source* language and produces an output,or *target*, language that differs in some respects from the *source* language. (1) A telegraph Repeater. (2) An instrument which receives messages according to one method of working and passes them on to another line according to another method of working e.g. from Wheatstone. Automatic to mirror or syphon recorder (see CREED TRANSLATOR) or from single Current to Double Current. (3) A four winding auto-transformer forming a connecting link between a four wire three-phase system and single phase lighting distribution.

TRANSMISSION : The act of conveying information in the form of electrical signals from one designated location to another by means of a wire, waveguide, transmission line, or radio channel and using any circuits, devices, or other equipment that may be necessary. A term employed in addition to its use in connection with transmission by mechanical power, with or without the intervention of electrical methods, to signify the dispatch of singals, etc., electrically, in connection with line or radio telegraphy, telephony, etc. and in connection with heavier currents to denote their conveyance over ling distances from one area to another ascompared to distribution in one area.

TRANSMISSION LEVEL (In a Telephone Circuit) : The ratio of the power at any point to the transmitted power in a standard reference circuit expressed in decibles or nepers.

TRASMISSION LINE (1) Syn. *power line.* An electric line, such as an open wire that is used to convey electrical power from a power station or substation (2) An electric cable or waveguide that conveys electrical signals from one point to another in a telecommunication system and that forms a continuous path between to two points.

TRANSMISSION UNIT : (T.U.) : A logarithmic based unit introduced in America to supercede the standard cable system of telphone transmission constants such that the two amounts of power transmitted by two circuits under the conditions differ by n units when they are in ratio of $1:10^{n(0.1)}$.

TRANSMITTED WAVE : That part of a wave propagated along a circuit which is transmitted beyond a Transition Point as compared with the Reflected Wave.

TRANSMITTER : An apparatus for the sending out of telephone or other electically transimtted meassages, signals, etc. Used in line telephony for the actual apparatus, usually of the Microphone type, which receives the sound waves of the voice and produces modulations in the amplitude of the current; but in radio telegraphy and telephony, usually including the whole apparatus for generating the waves.

TRANSMITTING AERIAL : An aerial used in the radiation of electromagnetic waves sutiably modulated to form signals.

TRANSMITTING CIRCUIT : The apparatus and connections of the equipment of a radio-telegraph or telephone station, used exclusively for transmission of message. Cf. RECEIVING CIRCUIT.

TRANSMITTING STATION : A station from which radio or other telegraph messages are sent out. Cf. RECEIVING STATION.

TRANSMITTING VALVE : A Thermionic Valve for use as an oscillator for the production of electromagnetic waves.

TRANSPARENT : Describes a computer operation that does not require user intervention.

TRANSPONDER : A radio device that receives an incoming signal and automatically retransmits it on the same or on a different frequency.

TRANSVERSE ELECTRIC WAVE : An electromagnetic wave in which the electric field vector is everywhere perpendicular to the direction of propagation.

TRANSVERSE ELECTROMAGNETIC WAVE : An eletromagnetic wave in which both the electric and magnetic field vectors are everywhere perpendicular to the direction propagation.

TRANSVERSE VOLTAGE : A term used in connection with Interference in communication circuits for voltages induced between two conductors of the same circuit.

TRAPEZIUM DISTORTION : A distortion of a trapezoidal pattern on the screen of a cathode ray tube instead of a square one and occurs when the deflecting voltage applied to the plates is unbalanced with respect to the anode.

TRAVELLING WAVE : An eletromagnetic wave that is propagated

along and is guided by a transmission line. In the case of a hypothetical lossless line of uniform cross section and infinite length, a sinusoidal a.c. supply at one end of the line (the sending end) causes electrical energy to be transmitted along the line with instantaneous values of current and voltage at any given point varying sinusoidally.

TREBLE : (1) High audio frequencies such as those handled by a tweeter in a sound system. (2) Audio frequencies above middle C (256 Hz)

TREE : A directed acyclic graph with unique parents, a sturcture consisting of nodes and arcs between them arranged in a tree shape with the single root node at the top. (See also node, arc, children, root, leaf.)

TREE STRUCTURED DIRECTORY : A file organization structure, consisting of *directories* and *subdirectories* that, when diagrammed, resembles a tree.

TRIAC : Acronym for *triode*. AC semiconductor switch, a five-layer semiconductor device that is equivalent to two silicon controlled rectifiers in anti-parallel, having a common gate and two anode terminals. It provides switching action for either polarity of applied voltage and can be controlled in either polarity by the gate terminal.

TRIAD : (1) Three electron guns grouped in a traingle and used with a shodow mask; each gun excits a red, green, or blue colour phosphor. (2) Colour at the vertices of an equilateral or isosceles triangle superimposed on the color wheel. These are considered to be a harmonious grouping, as are diads (complementary pairs), tetrads (square, rectangle, or trapezoid), hexads, and so on.

TRIANGLE WAVE : A wave form that is a repeating ramp function with equal positive and negative slopes. See PULSE.

TRIANGULAR PULSE TRAIN : See PULSE.

TRICHROMATIC : Three-coloured. In computer graphics, trichromatic generally refers to the three primary colour combined to create all other red, green, and blue.

TRICKLE CHARGE : A small continuous charge applied to a secondary battery in order to maintain it in fully charged condition during storage. The charging current is maintained at a value that just compensates for internal dissipation due to local action within the battery.

TRICKLE CHARGER : An arrangement by which an accumulator is kept connected through a high resistance to a source of supply so that when not in use it is continually receiving a gradual charge at a very low rate.

TRIGGER : Any stimulus that initiates operation of an electronic circuit or device. Also the act of initiating operation in the circuit or device. In general the response to a trigger continues after the cessation of the stimulus. A circuit, such as a filp-flop or multivibrator, that is used to trigger other circuits is known as a trigger circuit.

TRIGGER WINDOW : See *hysteresis.*

TRIMMER : A small auxiliary condenser for fine adjustment of one or two ormore ganged condensers.

TRIOD PENTODE : A thermionic valve consisting of a triode for use as an oscillatior and a variable mumixing valve in one envelope with a common cathode but separate anodes to each system.

TRIODE : (1) A vacuum tube that has three elements *cathode, control* grid and plate. A Thermionic Valve with three electrodes, viz., the *Cathode,* the *Anode,* and the *Grid;* and consequently four terminals Cf. DIODE. Any electronic device with three electrodes, such as a bipolar junction transistor field-effect transistor, or thyratron. The term is particularly applied to a thermionic valve that has three electrodes.

TRIPLE FREQUENCY HARMONIC : A superposed sine wave, three times the frequency of the fundamental, on the wave form of an alternating current or voltage. Owing to the variation of permeability at different inductions, and to hysteresis effects, there is a somewhat pronouned triple-frequency harmonic component in the magnetising current of a transformer. This is liable to produce circulating currents in the closed circuits of Delta connected windings of three-phase transformers, or in Star connected windings through the neutral fourth wire (if any). Similarly, triple frequency currents are liable to circulate through the neutrals of a number of star connected alternators running in parallel with all their neutral points earthed. For this reason, it is usual either only to earth the neutral of one machine at a time, or the earth them all through separate resistances. See THIRD HARMONIC.

TRIPLES TELEGRAPH SYSTEM : A system in which two messages

in one direction and one in the other can be sent simultaneously over a single circuit.

TRIPPING DEVICE : A device that normally constraints a circuit-breaker in the 'on' position until actuated, when the circuit-breaker is allowed to break the circuit. Mannual operation is common in many types of tripping device. Other types are operated electromagnetically. A typical example is the trip coil, which consists of a coil that controls a movable plunger or armature; the plunger controls the action of the circuit-breaker.

TRIPPLE-CLICK : To press and release the mouse button three times rapidly in quick succession.

TRISTALE LOGIC : A type of gate output that can be either totem pole or high impedance, depending on a control lead.

TRISTIMULUS VALUES : Relative amounts of three primary colour that are combined to create a colour.

TROJAN HORSE : A program designed to attract your attention on the screen by displaying graphics or a fake program such as work, processor. While you watch the screen, the Trojan horse secretly erases or damages the hard disk files. See also *Virus.*

TRUNCATION : To end a computation according to a specified rule; for example, to drop numbers at the end of a line instead of rounding them off, or to drop characters at the end of a line when a file is copied. The losses of digits or bits (usually the least significant), as from overflow or from shiftinig, to produce a less precise representation.

TRUNK (1) : See Trunk Line. (2) In automatic telephone exchanges, a Link (3) Used in the U.S.A. for a Junction Line.

TRUNK CIRCUIT : (1) A circuit within an exchange, connected to a Trunk Line. (2) Such a circuit includinig the Trunk Line. A telephone circuit conneting two exchanges in different areas, called a Long Distance Line in the U.S.A.

TRUNK FEEDER : A Feeder inter-connecting two generating stations or two distributing networks.

TRUNK HUNTING : An expression used in automatic telephone working for the automatic movement of a contact arm in a selector until it comes to rest on a contact connected to an idle line or selector.

TRUNK MAIN : See TRUNK FEEDER.

TRUNK SWITCHBOARD : A telephone exchange switchboard for making connection to Trunk Lines.

TRUNK SWITCHING CIRCUIT : The part of a telephone circuit at a trunk exchange connecting a Subscriber's Line or an incoming Junction Line to an outgoing Trunk Line.

TRUTH MAINTENANCE SYSTEM (TMS) : A device for recording the justifications of conclusions so that if a once proved theorem should become invalid, those theorems whose proof relied on it can be retracted or rejustified without disturbing others that are still valid.

TRUTH TABLE : A table indicating the output of a combinational logic circuit for all input states. A table used in formal logic that lists the truth or falsity of the outcome when a logical operator, such as 'and' or 'or', is applied to combinations of logical statements. The truth table has been adapted to describe the operation of logic circuits by listing the outputs of a binary logic gate, such as a flip-flop, for all possible commbinations of inputs. The 'true' state corresponds to the voltage level representing a logical 1 and 'false' to logical 0.

T-SECTION : See QUADRIPOLE.

TTL : Transistor-transistor logic; a family of digital integrated circuit logic Transistor-transistor logic; a logic system similar to diode-transistor logic in which the logic diodes are replaced by a multiemitter transistor.

TUCC : Triangle Universities Computing Centre.

TUNED AMPLIFIER : An amplifier in which the load is a tuned circuit. Consequently the load impedance and amplifier gain vary with frequency.

TUNED ANODE CIRCUIT : An Anode Circuits of a thermionic valve with the inductance and capacity adjusted to give it a natural period of oscillation or *resonance.*

TUNED ANODE COUPLING : A method of coupling high-frequency amplifier valves in which an inductance and a variable condenser in parallel are used to tune one anode circuit and the e.m.f. across them (which is high owing to the high impedance to the frequency in question) is applied through a condenser to the grid of the next valve.

TUNED CIRCUIT : An LC circuit which can be adjusted for resonance at a desired frequency. A circuit with inductance and capacity so adjusted that it resonates to a particular frequency.

TUNED CIRCUIT AMPLIFIER : *See* RESONANCE AMPLIFIER.

TUNED RADIO-FREQUENCY RECEIVER : A radio receiver whose amplifier stages are tuned to resonate at the carrier frequency of the received signal by a ganged variable capacitor. The amplified signal at the carrier frequency is demodulated and is further amplified and fed to a speaker or headphone.

TUNED RELAY : A Relay employed in some protective systems provided with mechanical or other resonating arrangements so that it responds only to currents of the normal frequency of the system.

TUNER : Any apparatus for making an adjustment to an Oscillating Circuit to bring it into *resonance* with another.

TUNGAR RECTIFIER : A rectifier for obtaining unidirectional currents from alternating current circuits acting on the principle of the Thermionic Valve, with a heated filament as cathode and a graphite disc as anode in argon at a low pressure.

TUNGSTEN : A heavy metal, atomic number 74, that has an extremely high melting point and is extensively used to form lamp filaments. It has also been widely used to form thermionic cathodes.

TUNING : The adjustment of the inductance or capacitance (or both) of an oscillating circuit to get the maximum degree of Resonance with received waves of a particular wave length.See SYNTONY.

TUNING COIL : See TUNING INDUCTANCE.

TUNING CONDENSER : A variable condenser used for Tuning an oscillating circuit.

TUNING CURVE : See RESONANCE CURVE.

TUNING ERRORS IN DIRECTION FINDERS : See LOOP TUNING ERROR and PLAIN TUNING ERROR.

TUNING FORK DRIVE (in radio transmitters) : A method of independent drive for ensuring constancy of wave length in which a high harmonic of the oscillations controlled by the continuous vibration of a tuning fork, selected by filter circuits and strongly amplified determines the frequency of the Main oscillator. Cf. CRYSTAL CONTROL.

TUNING IN : Adjusting the tuning of a radio receiving circuit to give the maximum response to waves from the station which it is desired to hear.

TUNING OUT : Adjusting the tuning of a receiving circuit to give the minimum responses to waves from sources which it is not desired to hear.

TUNING SWITCH : A switch for making adjustments necessary for tuning a radio-receiving circuit.

TURBO ASSEMBLER (IBM) : Turbo C, Turbo Pascal and Turbo Assembler are direct competitors to Macrosoft's language compilers. Turbo Pascal has set the standard for Pascal compliers for the IBM. Turbo C runs faster and provides more help of novice C programers, Turbo Assembler assembles programs faster than Microsoft's Macro Assembler.

TURBO C : See TURBO ASSEMBLER (IBM)

TURBO PASCAL : See TURBO ASSEMBLER (IBM)

TURBO PASCAL (MACINTOSH) : See TURBO ASSEMBLER (IBM)

TURING MACHINE (TM) : A formal model of computation that is maximally general in that any possible computational process can be executed by an appropriately constructed TM.

TURING TEST (TT) : An empirical test that decrees a computer system intelligent if a human examiner connot distinguish (at better than chances level) between a computer and an human witness in anonymous conversation.

TURN-AROUND DOCUMENT : A computer-prepared document, usually a punched card or printed report, this is sent to a customer. When returned to the sender, the document frequently can be reentered into the computer without modification.

TURN AROUND TIME : The length of time elapsing between the submission of input and the receipt of output.

TURNKEY SYSTEM : A complete computer system with software installed for customer use. Pertaining to a computer system in which a supplier is totally responsible for building, installing, and testing the system, including hardware and software.

TURNS RATIO : Syn. *transformer ratio*. The ratio of the number of turns, active in the secondary circuit of a transformer to the number of turns, in the primary winding.

TWEAK : To make a small refining adjustment on a circuit after it has been installed. Also called trim.

TWEETER : A loudspeaker that handles only the higher audio frequencies, generally above 3 kHz.

TWIN-T NETWORK : See QUADRIPOLE

T-WIRE : The wire in a telephone circuit within the exchange, connected to the Tip of the plug and to the "A". Wire of the line Cf. "S" WIRE and "R" WIRE.

TWO ADDRESS : See MULTI-ADDRESS.

TWO AND A HALF DIMENSIONS : The effect produced by two dimensional viewing software when overlapping objects are given a display priority.

TWO-BAND TELEPHONY : See BI-BAND TELEPHONY.

TWO-ELECTRODE RECTIFICATION : Rectification by an ordinary three-electrode thermionic valve, using the unilateral conductivity between the cathode and the anode, while the grid takes no part in the rectification, although it is given a positive bias to decrease the resistance and space charge effects.

TWO'S COMPLEMENT : A method of representing negative binary numbers in one's complement +1 form.

TWO-WAVE RECTIFICATION : See FULL-WAVE RECTIFICATION.

TWO WIRE CIRCUIT : A circuit that consists of two conductors insulated from each other and that provides simultaneously a two way communication channel in the same frequency band between two points in a tele-communication system. The circuit may be a phantom circuit in which case it is formed from two groups of conductors. A circuit that operates in the same manner but that is not limited to only two conductors (or groups of conductors) is termed a two-wire type conductor type circuit. Compare FOUR-WIRE CIRCUIT.

TWO-WIRE TELEGRAPH OR TELEPHONE CIRCUIT : A metallic circuit consisting of two separate wires without earth return.

TWO-WIRE TELEPHONE CIRCUIT : A metallic circuit composed of two conductors alongside each other or twisted together.

TYPE-OUT : See PRINT-OUT.

U

UHF : Abbreviation for Ultra High Frequency.

ULTRA ACOUSTIC TELEPRINTER WORKING : The superposition on a telephone circuit of teleprinter signals of a frequency above the audio-frequency range, e.g. about 2900 cycles per sec.

ULTRADYNE RECEPTION : A system of radio reception, similar to the Superheterodyne system, in which the intermediate frequency is obtained from auxiliary oscillations superposed upon the anode circuit of the firstvalve.

ULTRA FREQUENCY WAVES : A term sometimes used for electromagnetic waves of frequency between 300,000 and 3,000,000 kc. per sec.

ULTRA-GAMMA RADIATION : See PENETRATING RADIATION.

ULTRA-HIGH FREQUENCY : Abbreviated UHF, any frequency in the region from 300 MHz to 3 GHz.

ULTRA-HIGH FREQUENCY (UHF) : See FREQUENCY BAND.

ULTRA-MAGNIFIER : See RETROACTION CIRCUIT.

ULTRA-PHOTIC RADIATION : A term sometimes used for ULTRA-VIOLET RAYS.

ULTRA-SHORT WAVES : Electric Waves of lengths from 100 cm. to 10m. Cf. MICRO WAVE. See also DECAMETRIC WAVES.

ULTRA-SONIC COMMUNICATION : Underwater communication at ultrasonic frequencies using suitably modified *sonar*.

ULTRA-SONIC FREQUENCY : Any frequency in the region above the audible range, roughly above 20 kHz.

ULTRA SONIC-LIGHT VALVE : A device that can be used to transmit video information. It consists of a piezoelectric quartz crystal immersed in a transparent liquid. The crystal is excited by ultrasonic-frequency alternating current and the resulting mechanical vibrations set up compressive waves in the liquid. If the crystal is fed a modulated video signal, corresponding changes in the compression result. The system acts as a liquid diffraction grating to a light beam that is shone through it. The video information may be recorded on photographic film or detected with a suitable photodetector. Also known as supersonics. The study and application of sound fequencies that lie above the limits of audibility of the human ear, i.e. frequencies above about 20 kilohertz.

ULTRA-UDION : A special type of thermionic valve receiving circuit employing *Retroaction.*

ULTRA-VIOLET : Abbreviated UV, pertaining to electromagnetic radiation at those wavelengths from 10 to 380 nanometers, which are beyond the visible (violet) end of the light spectrum.

ULTRA-VIOLET RADIATION : Electromagnetic radiation lying between light and X-rays on the electromagnetic spectrum. Radiation of frequency close to that of the range is far *ultraviolet.*

ULTRA-VIOLET RAYS : Radiation of a higher frequency than that which can affect the eye, i.e. beyond the visible violet end of the spectrum. Such rays have powerful actinic effect and have the power of causing Ionisation, and can produce phosphorescence and photoelectric effects. They are also employed extensively for curative purposes, and to stimulate plant growth. Their wave length ranges from about 136 to 4,000 Angstrom units, of which the range from 2,000 to 3,200 A° has the greatest physiological effect, and that below 3,200 A° the greatest photographic effect. See ULTRA-VIOLET LAMP.

ULTRA X RAYS : Very penetrating rays of even shorter wave length than Gamma Rays.

UMBRELLA AERIAL : A form of aerial for radio-transmitting stations in which the component wires radiate from a central pole or tower downwards to the ground.

UNABSORBED FIELD STRENGTH : The field strength of a radio-wave that would exist at the receiving point in the absence of any absorption between the transmitting and receiving aerials.

UNBUNDLING : The separation of prices for computer services and hardware.

UNCERTAINTY : Lack of knowledge about a state or event.

UNCOL : Universal Computer-Oriented Language

UNCONDITIONAL TRANSFER : A command which causes the following instruction to be taken from the address which is not the next one in the sequence.

UNDAMPED OSCILLATIONS AND WAVES : Oscillations or Waves which continue with undiminished amplitude as long as they last.

UNDERCURRENT RELEASE : A switch, circuit-breaker, or other tripping device that operates when the current in a circuit falls below a predetermined value. A current that causes the release to operate is termed an undercurrent. Compare *overcurrent release.*

UNDERFLOW : The situation that exists when the number is too small to be recorded as larger than zero and is taken to be zero by the computer.

UNDERSHOOT : An initial transient response to a change in an input signal that precedes the desired response and is in the opposite sense. Compare *overshoot.*

UNDERVOLTAGE RELEASE : A switch, circuit-breaker,or other tripping device that operates when the voltage in a circuit falls below a predetermined value. A voltage that causes the release to operate is an *undervoltage.* An undervoltage release is sometimes used in conjunction with circuit-breaker in order to prevent the circuit-breaker closing when the supply voltage is too low. This application is known as undervoltage no close release.

UNDISTORTED WAVE : A periodically varying quantity that consists of sinusoidal components of which none is present at one point that is not present at all points and in which both the attenuation and velocity of propagation are the same for all the sinusoidal components.

UNDO : A command that lets you take back the previous command. Some programs provide multiple undo levels, letting you take back commands you gave in the past.

UNDULATOR : (1) A telegraph receiving instrument for use on long lines, consisting of a delicate polarised relay, to the arm of which a syphon pen is attached, making a wavy line record similar to that of a Siphon Recorder on a paper strip. This instrument is used in conjunction with the Gulstad Relay for cable circuits. (2) An apparatus of the nature of an Ionic Valve used for converting D.C. into A.C.

UNIBUS : A bus that is a set of electrical conductors which carry specific signals to several other electrical circuits.

UNIDIRECTIONAL : Refers to the use of signal lines that can never be asserted by both the *host* and called or receiver unit, either concurrently or successively,

UNIDIRECTIONAL AERIAL : An aerial arranged to send out waves of maximum strength in a particular direction but nearly zero in the opposite direction. Cf. DIRECTIONAL AERIAL.

UNIDIRECTIONAL CURRENT : See CURRENT.

UNIDIRECTIONAL RECEIVER : A radio-receiver having maximum sensitivity in a particular direction but practically zero in the opposite direction. Cf. DIRECTIONAL RECEIVER.

UNIDIRECTIONAL TRANSDUCER : Syn. for *unilateral transducer*. See *TRANSDUCER*.

UNIDIRECTIONAL TRANSMITTER : A radio-transmitter employing a Uni-directional Aerial.

UNIDYNE RECEPTION : A name given to an arrangement of radio-receiving apparatus in which the same battery is used for the filament and anode circuit, and an additional electrode is employed to prevent an excessive space charge.

UNIFORM COST SEARCH : A state-space search algorithm which uses a cost function to compute the expense of paths in the search tree and always expands the partial path of minimum cost; it degrades to breadth-first search when all operators have equal cost.

UNIJUNCTION TRANSISTOR : Abbreviated UJT, a three-terminal semiconductor device with a single emitter lead and two base leads, used primarily as a switching device.

UNINTERRUPTABLE POWER SUPPLY (UPS) : A battery-powered device that protects against power *spikes* and power *outages*. If the power goes out, the UPS continues supplying power to the computer so you can continue working or safely turn off your computer without losing any files.

UNIT (1) : A functional division of an automatic computer. (2) A distinct part of any while. (3) The basis for measurement or expression. Slave on the interface that operates under control of the host and can, in turn, control one to sixteen devices.

UNIT RECORD : A physically separate record that is similar in form and content to other record in its group, for example, a summary of a particular employee's earnings to let in connectionist, parallel distributed processing networks, are highly interconnected, tiny processors that compute simple functions of their input signals from other units to produce a similar output result.

UNITY-GAIN BANDWIDTH : (1) The frequency at which the frequency response of a system decreases to unity voltage gain, or 0 dB. (2) A measure of the gain-frequency response of an amplifier.

UNIVAC : Universal Automatic Computer.

UNIVERSAL COMMAND : Command causing every instrument on the IEC bus equipped to do so to perform a specific interface operation.

UNIVERSAL LOGIC : Any logic hardware that can be used as an AND, OR, and invert element.

UNIVERSAL MAINS RECEIVER : A radio-receiver which can be worked on A.C. or D.C. mains of the same voltage, without alteration.

UNIVERSAL MOTOR : An electric motor that may be used with a direct current power supply or an alternating-current supply. It incorporates a commutator in order to achieve this operation. Small motors of this type are commonly used in domestic appliances, such as vacuum cleaners or portable drills.

UNIVERSAL SHUNT : Resistance with suitable tappings. A galvanometer shunt that is tapped so that it can pass designated fractions (0.1, 0.01, etc.) of the main current and can thus be used with galvanometers of widely varying internal resistance.An example is the Ayrton shunt, which is a relatively large.

UNIVERSAL VALVES : Thermionic Valves for use in Universal Mains Receivers. Usually of the indirectly heated type the heaters of which can be connected in series direct on the mains with or without a further resistance but without a transformer.

UNIX : A popular, multiuser operating system. IBM computers can use a special version called Xenix and Macintosh computers can use their own version called A/UX. Bell Lab's portable, multilanguage operating system that runs on many types of computers. It is a general-purpose, interactive, multiuser operating system written in the C language.

UNL : Unlisten; bus command.

UNLOADED AERIAL : An aerial with no added inductance or ca pacitance. See LOADED AERIAL.

UNLOADED WAVELENGTH : The wavelength corresponding to the frequency of the free oscillations in an aerial without the addition of inductance or capacitance.

UNMODULATED KEYED CONTINUOUS WAVE : Continuous waves broken up for telegraph signalling purposes by a definite change of amplitude or frequency. (Sometimes called Type All Waves,)

UNT : Untalk; bus command.

UNTUNED AERIAL : See APERIODIC AERIAL.

UP-COUNTER : A counter that increments by one binary number each time a clock pulse is received.

UP-DATE : (1) To bring up to date by altering; (2) To conform to changes since the last such correction was made, as to correct a Diwali card mailing list by recording changes in few addresses.

UP-DOWN COUNTER : A counter capable of operating as an up counter or down counter, depending on a control lead.

UP-GRADE : To expand a system by installing options or using revised software.

UP-LOAD : To send data to another computer through a modem and a telephone line.

UP-PACKS : To break into parts, as a packed word into its constituent fields.

UPPER SIDE BAND : A Side Band composed of frequencies of the sum of the Carrier and modulating Frequencies. Sometimes eliminated by wave filters,etc., together with the carrier frequency wave, so that only the Lower Side Band is transmitted in order to economise power to take up a narrower range of frequencies, and for other reasons. When this is done, it is necessary to re-introduce the carrier frequency in the receiving apparatus in order to obtain the correct resultant audio-frquencies for intelligible speech.

UPTIME : The time during which an automatic computer is operating from of component failures, plus the time it is energized and capable of such operation. (This definition is not universally accepted).

USART : Universal synchronous-asynchronous receive-transmit chip that controls serial data transmission.

USASCII : United States of America Standard Code for Information Interchange (see ASCII)

USER-FRIENDLY : Term that means that a computer or program is easy to use. Rarely used correctly.

UTILITY FUNCTION : Computer programs, dedicated to one particular task, that are helpful in using the computer. For example, FDISK, for setting up partitions on the fixed disk.

UTILITY PROGRAM : Standard programs prepared and generally used to assist in the operation of data-processing systems.

UTILITY ROUTINE : A routine to perform functions auxiliary to the running of other programs. Example are storage dump routines, bootstarp routines, tape label rountines.

UUA: Univac Uses Association

VACUUM JUNCTION : A thermo-couple attached to a fine heating wire in an exhaused bulb, for the measurement of alternating or high frequency current in the latter by a galvanometer connected to the former.

VACUUM SWITCH : A switch in which the contacts are contained in a substantiallyevacuated container in order to minimize spark formation.

VACUUM TUBE : An electron tube evacuated to a sufficiently low pressure that its electrical characteristics are independent of any residual gas. Compare *gas-filled tube*. A term formerly applied to all forms of Discharge Tubes, but properly limited to those in which the vacuum is high enough for the residual gas not to affect the discharge. See INVERTED VACUUM TUBE.

VACCUM TUBE LAMP : See DISCHARGE LAMP.

VACUUM TUBE VOLTMETER : Abbreviated VTVM, a high impedance electronic voltmeter in which the measured voltage is amplified using a vacuum tube prior to being read by a meter.

VACUUM VALVE : A Vacuum Tube having rectifying properties. See OSCILATION VALVE, LODGE VALVE, etc.

VALENCE ELECTRONS : Electrons that occupy the outermost (lowest energy)energy levels of an atom and are involved in chemical and physical changes.

VALENCY : The number of hydrogen (or equivalent) atoms with which an atom will combine or that it will displace. The two most important types of valency are *electrovalency* and *covalency*.

VALLEY CURRENT : The minimum current in a negative resistance device, such as a unijunction transistor or tunnel diode.

VALUATOR DEVICE : Graphic input device, such as a control dial, that input graduated values within a user-defined range.

VALUE : Comparison of a chromatic colour to an achromatic colour, situating it along a gray from white to black. Other words used synonymously are brightness, brilliance, intensity, lightness, luminosity, and luminance. Sometimes a distinction is made that value is the perceived nonblackness of a colour, whereas brightness is the measurable amount of energy in a colour. The brightness definition is used when colours must be chosen so as to remain distinguishable on a black-and-white monitor, whereas value can be used when fully saturated or pure hues are given equal weight. Lightness, the amount of energy present in a colour, refers to non-self-luminous objects, while brightness refers to self-luminous objects. Luminosity is a subjective term for the amount of light emitted, transmitted, or reflected. Luminance is an objective term for the amount of radiant energy per unit area. Syn. *electron tube.*

VALVE : An active device in which two or more electrodes are enclosed in an envelope, usually of glass, one of the electrodes acting as a primary source of electrons. The electrons are most often provided by thermionic emission (in a thermionic valve) and the device may be either evacuated (vacuum tube) or gas-filled (gas-filled tube). The name derives from the rectifying properties (see rectifier) of the devices, i.e. current flows in one direction only. The word valve is gradually becoming obsolete and is being replaced by electron tube. Valves are being replaced by semiconductor devices for everyday applications.

VALVE ADAPTOR : An accessory whereby a Thermionic Valve with one type of terminal can be used in a holder designed for another type.

VALVE AMPLIFIER : See THERMIONIC AMPLIFIER.

VALVE BASE OR CAP : The cap on a Thermionic Valve which carries the contact pins fitting into the sockets in the holder.

VALVE COUNTER : An apparatus for detecting single alpha or similar projected particles by the sudden ionisation produced in a special chamber connected through amplifying valves to an oscillograph.

VALVE DETECTOR : See THERMIONIC DETECTOR.

VALVE DIODE : See DIODE THERMIONIC VALVE.

VALVE DRIVE : *Independent Drive* in which the frequency of the master oscillator is determined by the natural period of a thermionic valve oscillator. *Cf. Mechanical Drive*

VALVE ELECTRIC : Originally any device which only allows the flow of current in one direction, i.e. a Rectifier, but now generally used for an evacuated vessel with electrodes through which a current can passby electronic or ionic flow, whether used as a rectifier or for any other purpose, e.g. as an amplifier.

VALVE HOLDER : An appliance for holding and making contact with the terminals of Thermionic Valves, usually arranged so that the valve can be inserted only the right way round.

VALVE NOISE : Interfering noise heard in radio-receivers, due to causes withinthe valves themselves.

VALVE OPERATED INDUCTION FURNACE : A high-frquency induction furnace in which the supply, at a suitable frequency, is derived from a Thermionic Oscillator.

VALVE OSCILLATOR : See THERMIONIC OSCILLATOR.

VALVE RECEIVER : A radio-receiving set employing Thermionic Valves, as detector, amplifiers, etc.

VALVE RECTIFIER : See VALVE AND RECTIFIER.

VALVE RELAY : A thermionic valve connected to an oscillator and arranged to act as a Relay (not an Amplifier) by producing sustained oscillation in the anode circuit as soon as the weak oscillations in the grid circuit reaches a sufficient value to react on it.

VALVE TRANSMITTER : A radio-transmitting apparatus employing *Thermionic Valves* as the source of the oscillations.

VALVE VOLTMETER : See THERMIONIC VOLTMETER.

VAN DE GRAAFF ACCLERATOR : An accelerator in which the high-voltage terminal of a Van de Graaff generator acts as as a charged-particle source. Ions derived from the high-voltage terminal are injected into an evacuated accelerating tube. An appropriate voltage gradient is maintained along the tube. Van-de Graaff accelerators can produce ion current of several hundred microlamps and electron current of one milliamps at voltage of about five mega volts. An electrostatic

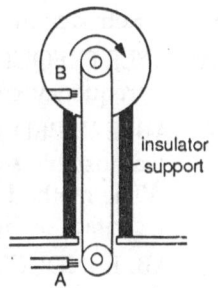

Van de Graaff generator

voltage generator that can produce potentials of millions ot volts. An electric charge from an external source is applied to a continuous in sulated belt at points A (see diagram). The belt travels vertically up into a large hollow metallic sphere and the charge is collected at points B. The charge resides on the exterior of thhe sphere and the possible voltage generated is limited only by leakage.

VAPORWARE : A program promised by a publisher but never released.

VAR : The unit of reactive power of an alternating current. It is identical to the watt, being the product of volts times amperes. Practical unit of Reactive Power, being one reactive volt-ampere. Abbreviation for Reactive Volt amperes.

VARACTOR : Acronym from *variable reactor*. A semiconductor diode operated with reverse bias so that it behaves as a voltage-dependent capacitor. The depletion layer at the junction acts as the dielectric and the *n*- and *p*-regions from the plates diode intended for use as a varactor generally has a particular impurity profile designed to give an unusually large variation of junction capacitance and to minimize the series resistance.

VARIABLE : A quantity that can assume any of a set of values as a result of processing data. A quantity that can take on different values.

VARIABLE CONDENSER : A condenser the capacitance of which can be adjusted by a relat-ive movement of the electrodes or otherwise,e.g. a *Tuning Condenser*.

VARIABLE COUPLING : Inductive coupling, the tightness of which can be varied by relative movement of the primary and secondary windings; used in radio-communication to adjust the sharpness of the tuning, and to control the amount of Retroaction in regenerative circuits.

VARIABLE FREQUENCY OSCILLATOR : An oscillator whose output frequency can be varied over a given range.

VARIABLE IMPEDANCES : Impedances, such as capacitors, that are adjustable so as to present variable values of the impedance. The method of adjustment is usually provided by a movable contact or by physical adjustment of the size of the device.

VARIABLE INDUCTANCE : An inductance, the value of which can be varied, e.g. a Tuning Inductance. See also VARIOMETER.

VARIABLE LENGTH RECORD : A record in which the number and/or length of fields may vary from those of other records accessed by the same program.

VARIABLE MAPPING : A technique of reduction in which variables are placed directly on the map.

VARIABLE MU VALVE : A type of Thermionic Valve in which the amplification factor and mutual conductance can be varied by adjusting the grid bias.

VARIABLE RESISTANCE : See RHEOSTAT.

VARIABLES : Input wires in a logic system.

VARIABLE SPEED SCANNING : A method of scanning in cathode-ray television in which the speed of deflection of the scanning beam is governed by the optical density of the object (in this case a film). The beam in the receiver is of constant intensity and copies the motion of the beam in the transmitter owing to inter-connection of the deflecting circuits and as the apparent illumination of the screen depends upon of the beam, the picture is reproduced.

VARIOMETER : A form of variable inductance composed of two or more coils, the angular position of which relative to each other can be wired.

V-BAND : A band of microwave frequencies ranging from 46.0 to 56.0 gigahertz. See *frequency band*. The radio-frequency band from 46 to 56 GHz.

VCO : Voltage-Controlled Oscillator; basically a VFO turned by the setting of a dc voltage.

VDI : (Virtual Device Interface): A device-level graphics interface that has become the de facto standard in North America.

VDU : Visual Display Unit

VECTOR : Line drawing or "calligraphic." A vector display device stores and displays data as line segments identified by the x, y coordinates of their endpoints. (contrast with "Raster".)

VECTOR GRAPHICS DISPLAY: A display system in which the electron beam 'paints" the desired image on the display screen; unlike raster-scan displays, vector-graphics displays do not scan horizontal lines to create image.

VELOCITY MODULATION : A process that introduces a radiofrequency (r.f.) component into an electron stream and thereby modulates the velocities of the electrons in the beam. Individual electrons will be either accelerated or retarded by the radio-frequency signal depending on the relative phase of the r.f. component at the point of interaction with the electrons.

VELOCITY OF PROPAGATION : The velocity with which an electrical disturbance is radiated as a waves through a medium, e.g. the velocity of light: equal, in the case of a dielectric, to $1/\sqrt{(xu)}$, where x and μ are the permittivity and permeability of the medium. In the case of waves propagated through wires, such as telephone current in a cable, depending also upon the frequency. Equal in free space to 3×10^{10} cm. per second, or about 186,000 miles per second. See DISTORTION.

VENN DIAGRAM : A diagram that represents ANDs and ORs by intersecting circles.

VENT PLUG IN ACCUMULATOR CELLS : A form of stopper for enclosed cells which permits of escape of the gases evolved during charging without allowing the acid to splash over.

VENTURA PUBLISHER : Ventura Publisher leads the desktop publishing field in the IBM market. Considerd to be one of the best desktop publishing programs of all. Ventura Publisher is best for creating lengthy documents such as books or magazines, but not for creating short documents like brochures or letters. Compare PAGEMAKER.

VERIFY : To attempt to ascertion the accuracy of something, for example, of the manual transcription of information from source document to a punched card.

VERSION SPACE : A structure used by the candidate-elimination algorithm for single concept learning; using feature-vector representations, it evolves separate necessary and sufficient conditions for category membership.

VERTEX : A point at the junction of two edges in a visual scene; when scenes are constrained to come from the blocks world, there are a limited number of legal vertex types, which can be identified using Waltz's Algorithm. (See also node.)

VERTICAL RESOLUTION : In raster-scan graphics systems, the number of visible raster lines displayed by a monitor, or the number of display-memory addresses representing pixels along the vertical axis of the display. In video systems, the number of horizontal test-pattern lines which can be reproduced by a camera and monitor (typically equal to 70% of the displayed raster lines). See RESOLUTION.

VERY HIGH FREQUENCY : Abbreviated VHF, any frequency in the region from 30 to 300 MHz. (VHF). See *FREQUENCY BAND*.

VERY HIGH FREQUENCY WAVES. : A term sometimes used for

electromagnetic waves of frequency between 30,000 and 300,000 k.c. per sec.

VERY LARGE SCALE INTEGRATION (VLSI) : The production of computer chips with hundreds of thousands or even millions of components on each chip.

VERY LOW FREQUENCY : Abbreviated VLF, any frequency in the region from 10 to 30 khz.

VERY LOW FREQUENCY WAVES : A term sometimes used for electromagnetic waves of a frequency below 30 k/hz

VERY SHORT WAVES : An expression sometimes used for wavelengths below 10 metres (above 30,000 kc.). Also called Ultra-Short Waves. Cf. LONG, MEDIUM, and SHORT WAVES, and MICRO-RAYS. See also METRIC, DECAMETRIC and CENTIMETRIC WAVES.

VFO : Variable-frequency oscillator; an oscillator whose frequency can be varied (manually or electrically) with a certain range.

VHF : Abbrev. for Very High Frequency. See FREQUENCY BAND.

VIBRATING REED INSTRUMENTS : Instruments for measuring frequency, etc., consisting of a row of spring a different and known period of mechanical vibration, in the field of an electromagnet excited by an alternating current. The reed whose natural frequency corresponds most nearly with that of the current vibrates in resonance to the field, and can be clearly distinguished from the others, which remain at rest, and the frequency can be read off on a scale along side the row of reeds.

VIBRATING REED RECTIFIER : An apparatus for rectifying an alternating current by periodical reversal by the movement of a vibrating member of magnetic material in a control field produced by a permanent magnet acted upon by a superposed field produced by the alternating current.

VIBRATION : See VIBRATOR. KAPP VIBRATOR.

VIBRATOR : (1) A vibrating contact-breaker. See TREMBLER. (2) A portable apparatus driven by a small motor for producing mechanical vibrations by a rapidly reciproacting part, for massage, etc. (3) Themoving system of an Advancer with a reciprocating armature.

VIBRATORY PHASE ADVANCER : A Phase Advancer with reciproacting moving parts. e.g. the Kapp Vibrator and the Recuperator. Cf. ROTARY PHASE ADVANCER.

VIBROGRAPH : An apparatus for recording vibrations of machinery, etc. In one electromagnetic form relative movement between fixed and spring supported portions of the apparatus alters the relative length of the air gaps of two high-frquency transformers in opposition and modulates the resultant output which is amplified and recorded.

VIDEO AMPLIFIER : A wideband amplifier capable of amplifying video signals in television receivers, radar, or computer displays.

VIDEO DISPLAY : Television-type display (raster format), which uses an analog signal. A digital-to-analog converter transforms the digital information to a video signal that is used for display.

VIDEO FREQUENCY : Picture-carrying signals in the 15 Hz-5 MHz range. The frequency of any component of the output signal from a television camera. Video frequencies are within the range 10 hertz to two megahertz. An amplifier that is designed to operate with video-frequency signals is termed a *video amplifier*. See VISION FREQUENCY.

VIDICON : A low-electron-velocity photoconductive *camera tube that is widely used in closed-circuit television and as an outside braodcast camera since it is smaller, simpler, and cheaper than the image orthicon.

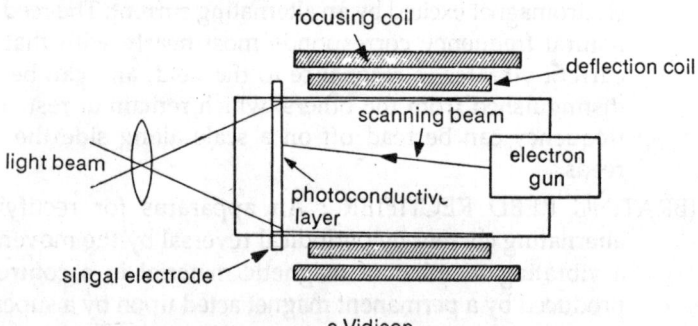

a Vidicon

VIE SURFACE : 2-D display surface mapped to represent 3-D normalized device coordinate space.

VIEWING OPERATION : An operation that maps positions in world coordinates to positions in normalized device coordinates. It also specifies the portion of the world coordinate space that is to be displayed.

VIEW PLANE : Logical surface for 2.D image projection; can be extended to create a 3-D viewing volume.

VIEW POINT : Specified window on display surface that marks limits of a display.

VIEW SITE : Coordinate point on an object through which the view vector passes; used to determine position of object in relation to hypothetical viewer.

VIG FILTER : An yttrium-iron-garnet microwave filter based on the properties of small YIG spheres.

VILLARI REVERSAL : The reversal, above a particular field strength, of the effect of stress on the permeability of iron, which tends to increase with stress at low fields and to decrease at high fields.

VIM : Name of Control Data Corporation 6000 series users organization.

VINT : See VOLT

VIRTUAL DEVICE METAFILE (VDM) : Stored device-independent display that can be moved from one system to another.

VIRTUAL GRAPHICS ARRAY (VGA) : Graphics standard for the IBM that can display 16 colours 640 x 480 pixel resolution.

VIRTUAL MACHINE : The computer system as it appears to the user. The term was first used to refer to the extension of main memory to almost infinite capacity by the automatic use of secondary storage. The operation system automatically moves portions of a program that are too large for primary memory to and from secondary memory.

VIRTUAL MEMORY : Addressable space beyond physical memory that appears to the user as real; it is provided through a combination of hardware and software techniques.

VIRTUAL STORAGE : Addressable space that appears to the user to be real storage, from which instructions and data are mapped into real storage locations. Virtual-storage size is limited by the addressing scheme of the computing system (or virtual machine) and by the amount of available auxiliary storage, rather than by the actual number of real storage locations.

VIRUS : A program that attaches itself to other files on a floppy or hard disk and duplicates itself without the user's knowledge. After this virus attacks the computer by erasing the hard disk. Viruses are written by miscreants with malafide intensions to harm the user, out of completion. Most of the virues are generated when pirated softwares are used by the computer owners/users. Some famous viruses are : stone virus, Brain,

Happy Birthday Mr. Joshi, Rain Drop. See also TROJAN HORSE.

VISIBLE RADIATION : Radiation between the limits of frequency corresponding to the limits of the visible spectrum, i.e. odinary light. to 8,000 Angstrom Units. Cf. ULTRA-VOILET RAYS and INFRA-RED RAYS.

VISION : is the process of receiving, transducing, and understanding images represented initially as light and finally as an appropraite and useful conceptual structure. (See also computer vision.)

VISION FREQUENCY GENERATOR : The apparatus in a television transmitter which produces currentmodulated in accordance with the scanning of the object or·its image; e.g. a Television Camera.

VISION FREQUENCY (In Television) : The frequency of any component in the modulation produced by the scanning device in the transmitter. Also called Video-Frequency.

VISION PICK UP : See VISION FREQUENCY GENERATOR.

VISUAL DISPLAY UNIT (VDU) : A display device used with a computer that displays information in the form of characters and line drawings on the screen of a cathode-ray tube. A VDU is most often assoicated with a keyboard and/or light pen, which allow the information in the displine.

VISUAL TUNING : Arrangements in a Radio Receiver whereby an indicating strip of light glows when the circuit is accurately tuned to an incoming carrier.

VITA RAYS : A term sometimes used for the range of ultra-violet rays which have a physiological effect. A term sometimes used for the range of ultra-violet rays which have a physiological effect.

VITREOUS ELECTRICITY. An old name for Positive Electricity because vitreous bodies, such as glass, become positively charged by friction. Originated by Dufay in 1735.

VLSI CIRCUIT (Very-Large-Scale Integrated Circuit): High-density ICs characterized by relatively large size (perhaps) 1/4 inch on a side) and high complexity (10,000 to 100,000 gates.) VLSI design, because of its complexity, makes CAD a virtual necessity.

VM : Virtual Memory

VOGAD : Abbreviation for Voice Operated Gain Ajusting Device. A method of mainatining constant speech output level in long distance speech output level in long distance radio-telephony, one a similar principle to that of Automatic Volume Control.

VOICE FREQUENCY : The frequency range of ordinary speech and singing, i.e. from about 100 to 2,000 cycles per second. Cf. AUDIO-FREQUENCY.

VOICE FREQUENCY TELEGRAPH SYSTEM : A system in which several channels are obtainable in one pair of conductors by using a different audio-frequency for each message, derived from tuning fork controlled thermionic oscillators, superposed on the line and selected by suitable filter circuits after amplification at the receiving end; extensively used in the British Post Office up 18 channels in one circuit.

VOICE INPUT DEVICE : Graphics input device that receives and interprets spoken data.

VOLATIC TROUGH : An early form of multi-cell zinc-copper single fluid battery due to volta.

VOLATILE MEMORY: A memory system in a computer or control system which requires a continual source of electric current to maintain the data it is storing intact. Removal of power from a volatile memory system results in the loss of the data being stored.

VOLT : The practical unit of Electromotive Force, Potential Difference,etc., being 10^8 C.G.S. units and that e.m.f. which produces a current of one ampere through a resistance of one ohm. The SI unit of electric potential difference, and electromotive force. It is defined as the potential difference between two points on a conductor when the *current* flowing is one ampere and the *power* dissipated between the points is one watt. See INTERNATIONAL VOLT.

VOLTA EFFECT : The E.M.F. between dissimilar metals in contact.

VOLTAGE : The potential difference between two points in a circuit or device. Electromotive Force or Difference of Potential expressed in Volts.

VOLTAGE AMPLIFICATION FACTOR : See AMPLIFICATION FACTOR.

VOLTAGE CIRCUIT : The circuit in a meter, etc., carrying a current proportional to the voltage and connected, sometimes in series with a resistor directly across the mains.

VOLTAGE DIVIDER : A network consisting of series-connected impedances connected across a voltage, from which one or more voltages can be obtained across any portion of the network.

VOLTAGE DROP : The voltage between any two specified points of an electrical conductor, such as the terminals of a circuit element or component,due to the flow of current between them.

VOLTAGE DOUBLER : An arrangement of two rectifiers that produces an output voltage amplitude twice that of a single rectifier. In a typical circuit using diode rectifiers (see Diagram) each rectifier separately rectifies alternate half cycles of the input alternating voltage and the two outputs are than summed.

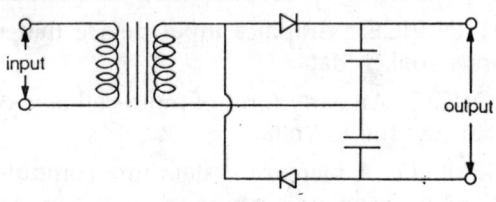

. Voltage doubler

VOLTAGE REGULATOR : An apparatus, such as a Shunt Rheostat, for regulating the voltage of a circuit.

VOLTAGE STABILIZER : A circuit or device that produces an output voltage that is subs-tantially constant and independent of variations either in the input voltage or in the load current,i.e. acts as a constant voltage source. Such a device is most often used as a voltage regulator.

VOLTAGE TRANSFORMER : Syn. *potential trnsformer.*

VOLTAIC PILE : A source of e.m.f. consisting of a pile of pairs of discs of dissimilar metals, with a moistened pad between those composing each pair, thus forming a number of elementary cells in series; produced by Volta in 1796.

VOLTAMPERE : The SI unit of apparent power, defined as the product of the root-means-square values of voltage and current in an alternating-current circuit. See *POWER;ACTIVE VOLT-AMPERES; REACTIVE VOLT-AMPERES.*

VOLTAMPERE HOUR METER : A meter integrating volt-amperes.

VOLTAMPERE METER : An instrument for indicating volt-amperes. irrespective of power-factor.

VOLTAMPERES (VA) : The product of the voltage and current in a circuit; otherwise known, in the case of alternating currents, as Apparent Power. See also KILOVOLTAMPERES.

VOLTMETER : A device that measures voltage. Voltmeters in common use include d.c. instruments (such as permanent-magnet moving-coil instruments), digital voltmeters, and cathode-ray oscilloscopes. In order to provide the minimum disturbance in the circuit contain a large series resitance however is required in the case of the moving-coil voltmeter and the electrostatic voltmeter in order to increase their input impedances. The electrostatic voltmeter is a voltmeter based upon the principle of operation of a quadrant electrometer or other type of electrometering the voltage to be measured, voltmeters are required to pass very little current and therefore require a very high input impe-dance. Digital voltmeters, cathode-ray oscilloscopes, and the now little used valve voltmeter comply with this requirement.

VOLUME CONTROL : Adjustment by variating retroaction, grid bias, etc., in a radio-receiver or in sound repropduction generally to control the loudness of the output.

VOLUME EXPANSION : An arrangement similar to Negative Automatic Volume Control employed in sound film production to extend the volume range and to reduce the proportion of background noise.

VOLUME LABEL : The name for the contents of a diskette or a partition on a fixed disk.

VOLUME RESISTIVITY : Resistivity defined as resistance per unit length of unit area at a given temperaturs, i.e. Specific Resistance as usually understood. Symbol: Cf. MASS RESISTIVITY

VON NEUMANN BOTTLENECK : A problem that arises in serial, stored-program digital computers when the channel between memory and central processors is not "wide enough" to accommodate the enormous traffic and competition for resources; it has led to the development of "non-Von Neumann" parallel architectures, as in the Connection Machine.

VP-PLANNER 3D VP-Planner : 3D provides a five dimensional spreadsheet with DBASE III Plus file certain and loading abilities. For extensive condsolidation needs, VP-Planner 3D proves superior to most other spreadsheets on the market.

VS : Virtual Storage.

VSAM : Virtual Sequential Access Method

WAFER : A large single crystal of semiconductor material, usually silicon, that is used as the substrate during the manufacture of a number of chips. The number of viable chips that can be produced from a single wafer, typically up to 10 cm in diameter, depends on the size and complexity of the circuits or components on the chips.

WAGNER EARTH CONNECTION : A method of connection used with an alternatin-current bridge circuit that minimizes the admittance to earth to the bridge.

WAIT STATES : A condition that occurs when a processor runs faster than its meory chips can retrieve data, thereby forcing the processor to wait periodically. Fast computers have no wait states, slower computers have one or two wait states.

WALKER PHASE ADVANCER : A Phase Advancer of the Expedor type, consisting of a generator with a three-phase star-connected armature with a commutator andthree field poles. Each of the three brushes is in series with one field coil, the other end of which is connected to one of the motor slip rings and the armature is driven directly by the main motor.

WALKIE-TALKIE : A light portable radio set designed to be carried and capable of allowing two-way communication by the user while in motion.

WALK-THROUGH : Programming by giving the robot instructions one by one, with the robot executing each before receiving the next. The speed of the robot is increased when rogramming is satisfactory. A teach box is usually used.

WALTZ'S ALGORITHM : An extension and fundamental improvement on work by Huffman and Clowes on determining the orientation of vertices in three-dimensional blocks world imagers; it introduced the techniques of realworld constraints and their propagation through networks to solve problems.

WAN (Wide Area Network) : Connecting of computers over longer distance than a single building, involving distance of miles rather than feet. cf: LAN

WANDERING : Change in apparent direction of received radio signals irrespective of the adjustment of the transmitter or receiver.

WANDER PLUG : A plug on a flexible cord for making connections to any one of several sockets such as those forming tappings on dry batteries for the anode circuits of thermionic valves.

WARM COLOURS : The *yellow, orange, red* portion of the spectrum.

WARM-UP : The period after a particular electronic device, circuit, or apparatus has been switched on during which it reaches a state of thermal equilibrium with its surroundings. A circuit may not be fully operational until after the warm-up period, particularly if it contains a thermionic cathode.

WATCHMAN'S RECORDER (Electrical) : An instrument arranged to make a record of the times at which a watchman sends signals from points on his rounds.

WATCH-RECEIVER : A compact watch shaped Telephone Receiver.

WATER COOLED VALVE : A large Thermionic Valve with water-cooled Anode.

WATFIC : Successor to WATFOR.

WATFOR : University of Waterloo FORTRAN.

WATT : The SI unit of power. It is defined as the power resulting when one *joule* of energy is dissipated in one second. In an electric circuit one watt is given by the product of one ampere and one volt. The practical unit of electric power (equal to 10^7 ergs per second, 1 joule per second or 1/746 h.p.). The power in a circuit in watts is equal to the number of volts multiplied by the number of amperes and, in the case of alternating current, by the power factor. See INTERNATIONAL WATT HOUR.

WATTHOUR (WH) : The practical unit of electrical energy, equal to 3600 joules, being that expended by 1 watt, flowing for 1 hour. One thosandth of a kilowatt hour. See also KILO.

WATTHOUR EFFICIENCY OF AN ACCUMULATOR : The ratio of the energy obtainable during discharge to that put in during charging. Cf AMPERE-HOUR EFFICIENCY.

WATTHOUR METER : Syns. *intergrating wattmeter; recording wattmeter.* An integrator that measures and records electrical energy in watt hours or more usually in kilowatt-hours. An integrating meter measuring energy in watt-hours or kilowatt-hours. Cf. AMPERE-HOUR.

WATTLESS COMPONENT : (1) (of current) See REACTIVE CURRENT (2) (of volt-amperes) See REACTIVE VOLT AMPERES.(3) (of voltage) See REACTIVE VOLTAGE.

WATTMETER : An instrument that measures electrical power and is calibrated in watts, multiples of a watt, or submultiples of a watt. The most commonly used type of wattmeter is the electrodynamic wattmeter (see electrodynamometer). In circuits that have substantially constant currents and voltages an induction wattmeter (see induction instrument) may be used. For standardization and calibration purposes an electrostatic wattmeter is used. This consists of a quadrant electrometer that is arranged in a suitable circuit containing noninductive resistors so that it measures power directly. The National Physical Laboratories, London, use this type to calibrate other wattmeters. An instrument containing one circuit in series with the current and another connected across the voltage, which react on one an-other in such a way that a deflection is obtained from which the power in the circuit in Watts can be read directly. See DYNAMOMETER TYPE INSTRUMENTS etc.

WAVE : A periodic disturbance, either continuous or transient, that is propagated through a medium or through space and in which the displacement from a mean value is a function of *time* or *position* or both. Sound waves, water waves, and mechanical waves involves displacements of particles in the medium; these small displacements return to zero after the disturbance has passed. With electromagnetic waves (see electromagnetic radiation) it is changes in the intensities of the associated magnetic and electric fields that represent the disturbance and a medium is not required for propogation of the wave.

WAVE AERIAL : An aerial, usually supported only a few feet from the ground, of a length equal to a multiple of the wave length to be received, in which wave reflection from the ends is prevented by resistances; giving high efficiency of reception and considerable direction.

WAVE BAND : Waves of any wavelength between certain limits, e.g.the range of wavelengths set apart for broadcasting services.

WAVE CHANGER : A switching arrangement enabling connections to be altered rapidly in a radio-transmitting apparatus to cause waves of a different wave length to be transmitted.

WAVE DETECTOR : See DETECTOR.

WAVE DISTORTION : See DISTORTION.

WAVE ELECTRIC : (1) A regular undulatory electromagnetic disturbance radiated through a medium and therefore transmitting energy through it as the result of electrical oscillation in a system. Propagated at the same speed as light (3×10^8 cm., or 186,000 miles per secon), and essentially of the same nature but of greater wave length. The electric waves employed in radio-communication range up to several thousand metres in length. Electric waves can also be propagated along conductors as well as across a di-electric. See RADIATION, PROPAGATION, (2) The term "wave" is also employed to express the periodic variation of an alternating current.See WAVE FORM, SINE WAVE,

WAVE FILTER : See FREQUENCY FILTER.

WAVE FLATTENER : An apparatus consisting of surge absorbers, inductances, capacitors, etc., used to protect transformers, etc., against surges, by modifying the steepness of their wave fronts.

WAVE FORM : The actual shape of the curve representing the wave of an alternating current, etc. See PULSE, SAWTOOTH WAVEFORM, SQUARE WAVEFORM. FLAT TOPPED PEAKED AND SINE: See FLAT TOPPED WAVEFORM, PEAKED WAVEFORM, etc.

WAVE FRONT : The advance side of a current wave, i.e. the portion where the current is increasing. A term used principally with reference to the propagation of the start of a heavy current or surge, the effect of which largely depends on the "steepness" of the wave front, i.e. the rate of increase of the current. See also POTENTIAL FRONT.

WAVE LENGTH : in radio-communication, etc., range from a few cm, to thousands of metres. The distance in the direction of propagation at any moment between successive zero points or successive maxima of electric waves that are being transmitted: obtained by dividing the velocity of Propagation by the Frequency.

WAVE LENGTH FUNDAMENTAL NATURAL AND UNLOADED: See FUNDAMENTAL WAVELENGTH, NATURAL WAVE-LENGTH, etc.

WAVE METER : An instrument for measuring the wave length or frequency of electric waves by adjusting the inductance and capacitance(or both)of an oscillating circuit until the maximum degree of resonance is obtained, or otherwise.

WAVE SHAPE : See WAVE FORM.

WAVE TELEPHONY : See CARRIER WAVE TELEPHONY.

WAVE TILT : The forward inclination of the wave form of electric wave arriving along the ground, depending upon the electrical constants of the soil.

WAVE TRAIN : A short series of waves, such as that producing from the disturbance set up by each spark in the early spark systems of radio-communication.

WAVE TRAP : An arrangement to enable one station to be rendered inaudible in a receiving set in order that a further station of fairly near wavelength can be heard; consisting either of a Reject or Circuit for preventing oscillations of the unwanted frequency from passing, or an Acceptor Circuit where by they may be diverted.

WAVE VELOCITY (of Sinusoidal Waves in a Non Dispersive Medium): The product of the wavelength and the frequency. In the case of a dispersive medium in which waves of different lengths are transmitted at different velocities the corresponding quantity is termed Phase Velocity.

WEAK AI : is the name given by John Searle to a research program that views AI as a tool for exploring human behavior and cognitive processes and for producing useful computers applications. See also strong AI.

WEAK CURRENT ENGINEERING : See LIGHT CURRENT ENGINEERING.

WEAK METHODS : are techniques for problem solving in AI that are largely independent of any special knowledge or specific problem; heuristic search algorithms are the most common weak methods.

WEBER'S LAW : A name proposed at the Paris Congress in 1889 for a practical unit of magnetic flux equal to 10^8 C.G.S. units, but adopted in America in 1894 for the C.G.S. unit itself (named after W.E.Weber, 1804-1891). This was, however, renamed the "Maxwell" by Weber of a practical unit of flux equal to 10^8 Maxwells which had been provisionally called the pramaxwell the Paris Congress of 1900. (The term had been previously used as the unit of current but was replaced by the ampere in 1881, and earlier still as the unit of quantity now called the coulomb.) It is now agreed that the term weber be used as the name.

WEBER'S UNIT OF RESISTANCE : A proposed unit of resistance based on the millimetre instead ofthe centimetre in the C.G.S. electromagnetic system, and therefore equal to 10^{-10} ohms.

WEIGHTED CODE : A binary code in which each bit position is assigned as specified numerical value word: a binary number.

WELDING : The process of joining together two pieces of the same metal using either no extra metal or extra metal of the same material. Welding is performed electrically using either resistance welding, in which no extra metal is required, or arc welding, in which a filler rod of the same metal is used.

WENNER WINDING : A method of forming a wire-wound resistor in order to reduce the reactance. The resistor is wound on a former and alternate truns of wire are looped along the former. Compare bifilar winding.

WESTERN ELECTRIC AUTOMATIC TELEPHONE SYSTEM : A system employing power-driven rotary multiple contact switches, characterised by the fact that dialling impulses are stored up in an apparatus called a Register, which, in conjunction with other apparatus, proceeds to operate the selectors which make connection with the line required.

WESTON NORMAL CELL. A standard Cadmium Cell, constructed according to a certain specification, and having an e.m.f. of 1.01830 International Volt at 20°C.

WESTON PORTABLE CELL : See WESTON NORMAL CELL.

WESTON STANDARD CELL : A cell that has a substantially contant termi-nal voltage and is used as a reference standard for electromotive force. This cell is constructed in an H-shaped glass vessel. The positive electrode is mercury and the negative electrode is a *cadmium* and *mercury* amalgam with a saturated cadmium sulphate solution as the electrolyte. The e.m.f. developed by the cell at 20 C is 1.01858 volts. The cell has very low temperature-coefficient of e.m.f. Compare Clark cell. *Syn.* cadmium cell.

WET BATTERY : A collection of Wet Cells.

WET CELL.: A primary Cell with a liquid electrolyte. Cf. DRY CELL.

WFF : Well-Formed Formula

WHEATSTONE BRIDGE : A four-arm bridge used for measuring resistance. Each arm contains a resistance (see diagram), with the unknown and reference resistances, R_1 and R_2, being connected at a point. At balance, when a null response is obtained from the indicating instrument $R_1/R_2 = R_3/R_4$ The two arms R_3 and R_4 are known as the ratio arms and may take the form of a wire of uniform resistance, which is tapped by a sliding contact. R_3 and R_4 are proportional to the lenghts of wire, L_1 and L_2, on each side of the contact, so that $R_1/R_2 = L_1/L_2$.

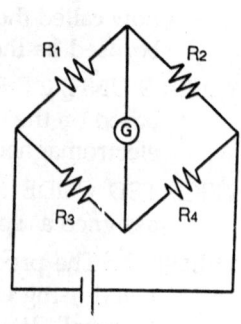

Wheatstone bridge

WHEATSTONE'S A.B.C. TELEGRAPH : See A.B.C. TELEGRAPH.

WHEATSTONE'S AUTOMATIC TELEGRAPH : See AUTOMATIC TRANSMISSION.

WHEATSTONE'S BRIDGE : An apparatus for measuring D.C. resistances in which the current from a battery divides into two parallel paths, each divided into two portions or "arms," which are adjusted so that there is no difference of potential between the dividing points on both sides, as shown by the absence of deflection of a galvanometer placed across them. The ratio of the resistances of the two pairs of arms is then the same, so that if the resistance of three of the arms is known, that of the fourth can be determined. See ORGANIZATION BRIDGE, SLIDE WIRE BRIDGE, POST OFFICE BRIDGE, etc.

WIDE AREA NETWORK : See WAN

WIEN BRIDGE : An A.C. bridge method of measuring dielectric losses in which the capacitance to be tested and a standard condenser form two arms, and variable resistances the other two. It is a four-arm bridge that is used for the measurement either of capacitance or frequency. A typical network is shown in the diagram. At balance, when a null response is obtained from the indicating instrument.

$$Cx/Cs = (Rb/Ra)- (Rs/Rx)$$

$$Cs\ Cx = P/w^2 RsRx$$

where w is the angular frequency. For the measurement of a

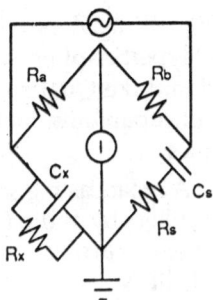

Wien bridge

frequency f, it is convenient to make $Cs = Cx$, $Rs = Rx$, and $Rb = 2Ra$ so that $f = (2\pi CR)^{-1}$

WIEN EFFECT : An increase in the conductivity of an electrolyte when subjected to a high voltage gradient (of the order of two megavolts per metre). At sufficiently high values of voltage gradient the ions in the solution move at such a rate that they pass very quickly.

WILLIAMS TUBE : A cathode-ray tube for electrostatic storage of information similar of the type originally designed by F.C. Williams.

WILSON EFFECT : Electrical polarization of an insulator when it is moved through a region containing a magnetic flux. The motion induces a potential difference across the material that results in its polarization since the insulating properties inhibit the creation of an electric current.

WIMSHURST MACHINE : An early electrostatic generator.

WINCHESTER DISK : A hermetically sealed electromechanical device for file storage that employs a rotating magnetic disk.

WINDING : A complete group of insulated conductors in an electrical machine, transformer, or other equipment that is designed either to produce a magnetic field or to be acted upon by a magnetic field. It may consist of a number of separate suitably shaped conductors, electrically connected, or a single conductor shaped to form a number of loops or turns. See RING WINDING.

WINDOW : In computer graphics, a bounded area within a display image which contains a scissored subset of the displayable data.

WING-Z (MACINTOSH) : Wing-Z provides its own programming language called HyperScript enabling you to create you own custom application. Like Quattro Pro, Wing-Z also offers an extensive range of capabilites, including colour and three-dimensional graphics.

WIRE BROADCASTING : A broadcasting service in which a porgramme is spplied to subscribers by audio frequency currents or modulated carrier frequency currents over permanent wire circuits. See also REDIFFUSION.

WIRE BROADCASTING, AUDIO FRQUENCY AND CARRIER FRQUENCY : See AUDIO-FREQUENCY WIRE BROADCASTING and CARRIER FREQUENCY WIRE BROADCASTING.

WIRED WIRELESS : A name sometimes used for Carrier Current Telephony and telegraphy, owing to the high frequency oscillatory currents employed being regarded as electric waves guided along wires.

WIRE-FRAME GRAPHICS : A CAD technique for displaying a three dimensional object on the CRT screen as a series of lines outlining its surface. (See "Finite-Element Model.")

WIRE GAUGE : B. & S., Birmingham British Standard, Brown & Sharpe, French, Legal Standard, Standard and Stubs. See B. & S. WIRE GAUGE, BIRMINGHAM WIRE GAUGE, etc.

WIRELESS RECEIVER : See RADIO RECEIVER.

WIRELESS TELEGRAPHY : See RADIO TELEGRAPHY.

WIRELESS TELEPHONY : See RADIO TELEPHONY.

WIRING DIAGRAM : A diagram of a piece of electronic equipment showing the interconnections between assemblies and subasseremblies. Such a diagram is particularly useful for the maintenance or repair or such equipment.

WOMMELSDORF MACHINE : A form of Influence Machine, with alternate fixed and moving plates and collectors in the form of steel wires dipping into grooves in the rims of the moving discs.

WOOFER : A speaker designed to reproduce bass frequencies, or the low end of the audio spectrum.

WORD : The set of bits comprising the largest until that the computer can handle in a single operation. A combination of bits that form a logical storage grouping. A word may be further sub-divided into bytes, which can be addressed by instructions. An ordered set of a admissible marks of a definite length which has at least one meaning and which is stored, handled, and transferred within the computer as a group of field. Basic unit stored in the computer's memory. The amount of information that can be handled by a computer at one time. For microcomputers, usually one or two bytes (8 to 16 bits). For minicomputers, two or four bytes (16 or 32 bits).

WORD FOR WINDOWS (IBM) : see MICROSOFT WORD (MACINTOSH).

WORD LENGTH : The number of admissible marks that constitute a field or word, often expressed in terms of the number of bits or decimal digits.

WORD PERFECT (IBM, MACINTOSH, ATARI ST, AMIGA) : WordPerfect has become the standard for IBM word processing. To make the program more accessible to beginners, the latest version of WorkPerfect includes pull-down menus and mouse support. WordPerfect offers one of the best telephone support policies in the industry with toll-free numbers you can call for help.

WORD PROCESSING : A system for creating and correcting text. Preparing text, letters, reports, and other documents, on a computer. This usually includes the ability to make changes in a body of text easily. A program designed for turning your computer into an electronic typewriter.

WORD TIME : The time required to move one word.

WORD WRAP : A feature that automatically moves a word to the next line if the word exceeds a predefined right margin.

WORK CELL : A manufacturing unit consisting of one or more work stations. (See "Workstation.")

WORK CENTRE : (1) An administrative or accounting subdivision of a department. (2) Production facility made up of one or more persons or machines.

WORKING MEMORY (WM) : In a production system holds transient data about the problem currently being solved, such as facts given or deduced about the problem; in psychology, this is called short-term memory, whereas longterm memory stores the system's actual production rules and the procedures they use.

WORK IN-PROCESS : Product in various stages of completion throughout the production cycle. This includes raw material released
for initial processing as well as completely processed material
awaiting final inspection and approval as finished product
for shipment to a customer.

WORK STATION : A powerful personal computer system to support
the tasks of an individual such as an *engineer* or a *stockbroker*.
Configuration of computer equipment designed to be used
by one person at a time. A workstation may be a terminal
connected to a large computer, or may be a "stand-alone"
with local processing capability. It generally consists of an
input device keyboard digitizer, etc.) a display device, memory,
and an output device such as a printer or plotter.

WORK VOLUME : A quantitative measure of the amount of work
done or to be done.

WORLD COORDINATE SYSTEM : A device-independent three-dimensional Cartesian coordinate system in which two- and
three-dimensional objects are described to a viewing system.

WRAP-AROUND : Effect of positioning a display item so it extends
beyond the device space boundary and a portion appears on
the opposite side of the display surface.

WRITE : Process of inserting data into memory. This is a destructive
process, in that any data already in a particular memory location is destroyed when new data is written into that location.

WRITE-INTO : To place words as fields in storage or some register
from some other register or some input medium such as magetic
tape.

WRITE NOW (MACINTOSH): WriteNow is one of the least expensive Macintosh work processor, but also one of the fastest
and easiest to use. In addition, WriteNow requires the least
amount of memory of any Macintosh word processor.

WRITE-OUT : To copy words of fields from storage or some register
onto some output medium, such as magetic tape.

WRITE-PROTECT : To prevent a computer from adding or modifying data stored on a disk.

WRITE PROTECT NOTCH : A cut-out opening in the sealed envelope of a diskette that, when covered, prevents writing or

adding text to the diskette, but allows information to be read from the diskette.

WRITE PROTECT TAB : A small plastic tab that slides over a hole in 3.5-inch floppy disks, or a sticker that covers a notch cut in the side of 5.25-inch floppy disks.

WRITING TELEGRAPH POLLAK VIRAG : See POLLAK-VIRAG WRITING TELEGRAPH.

WRITING TELEGRAPH SYSTEM : A system of telegraphy in which the message is written down automatically in characters resembling handwriting. See TELAUTOGRAPH.

X

X-BAND : A band of microwave frequencies ranging from 5.2 to 10.9 gugagertz. See FREQUENCY BAND.

XDS : Xerox Data Systems

XENIX : See UNIX

XEROGRAPHY : A photographic copying process that uses electrical effects to form the image. An image of the document to be copied is reproduced on a uniformly charged plate, usually coated with selenium:different areas of the plate are discharged to different extents by an intensity-modulated beam of ultra-violet radiation to leave a charge pattern corresponding to the brightness information of the original document.

X-GUIDE : A transmission line that is used for the propagation of surfacewaves and consists of a length of dielectric material with an X-shaped cross section.

X-MODEM : A method of transmitting data to another computer through modems. Transfers data in packages of 128K.

X-OGRAPH : See "X" RAY PHOTOGRAPH.

XOR : Exclusive OR XPUNCH : On IBM punched cards, a hole in the row above the zero-row, loosely; on *overpunch*.

X-RAY CRYSTAL ANALYSIS (or crystallography) The investigation of the structure of a crystal by the reflection or diffraction of "X" Rays by its facets.

X-RAY CRYSTALLOGRAPHY : See X-RAYS.

X-RAY EXAMINATION OF MATERIALS : (1) The examination of materials for hidden flaws, impurities, etc., by "X" Ray Photographs or by viewing with a fluorescent screen. (2) Examination of the "X" Rays Spectrum of the material.

X-RAY GENERATOR : Any apparatus for generating "X" Rays, including tubes of the ordinary form and powerful apparatus with revolving water cooled anti cathodes.

•X-RAY PHOTOGRAPH : A photographic record of the shadow of an object cast by "X"Rays; capable of revealing embedded articles and interio of structure in a way not possible with ordinary light, owing to the different range of penetrative power of "X" Rays. (Also called *Radiograph, Radiogram, Skiagraph, Skia-*

gram, and *Rontgenogram* and "X" ograph.

X-RAY SPECTROGRAPH SPECTROMETER AND SPECTROSCOPE: An apparatus for recording, measuring, and observing "X" Ray Spectra respectively.

X-RAY SPECTRUM : A diagram showing what are the wavelengths of the component in an "X" Ray beam. See also CHARAC-TERISTIC RADIATION

X-RAY TRANSFORMER : A special type of transformer for giving the high voltages necessary for the working of "X" Ray Tubes.

X-RAY TUBE : A highly exhausted Vacuum Tube in which a stream of Cathode Rays, given off from a cathode (sometimes heated to facilitate emission of electrons) falls upon an Anti-Cathode or Target within the tube, causing it to give off "X" Rays of wavelengths the lower limit of which depends on the applied voltage. See HARD, TUBE, SORT TUBE, GAS TUBE, ELEC-TRON TUBE, METAL "X" RAY TUBE, OSMO-REGULATOR, MERCURY AIR VALVE REGULATOR, PILON REGULATOR.

X-RAYS : Rays, also called, after their discoverer, W.K. Von Rontgen (in1895), *Rontgen Rays,* consisting of electric waves of much higher frequency than light (but not as high as that of Gamma Rays), produced when the velocities of rapidly moving elec-trons are altered, e.g. by the striking of *Cathode Rays* upon a soild substance. See "X" RAY TUBE. They possess the fol-lowing characteristic properties: They are not deflected by electric or magnetic fields. They can penetrate solid substances which cannot be penetrated by ordinary light, and cause considerable photographic and other chemical effect. See "X". They cause electrons to be emitted from solid bodies upon which they fall. They are not reflected or refracted under ordinary conditions, but are diffracted if reflected by crystals. (See "X" RAY SPECTRUM). They excite Fluorescence in certain substances, such as barium platinocyanide and they cause ionisation of gases through which they pass. Their wave lengths range from 0.06 to about 1,000 Angstrom units. See also ULTRA "X" RAYS.

X-Y PLOTTER : A graphical instrument that produces a chart show-ing the relationship between two varying signals. One of the signals causes the pen to move in the direction of the x-axis and the other independently causes it to move in the direc-tion of the y-axis.

XYZ SPACE : A three-dimensional coordinate system based on the 1931 CIE chromaticity diagram which plots X, Y, and Z as the tristimulus values of a colour.

Y : Symbol for Admittance.

YAGI AERIAL : A sharply directional aerial array from which most aerials used for television and radioastronomy have been developed. The active part of the aerial consists of one or two dipole aerials together with a parallel reflector aerial and a set of parallel directors. The directors are relatively closely spaced, being from 0.15 to 0.25 of a wavelength apart. When the aerial is used for transmission, the directors absorb energy from the back lobe of the dipole radiation pattern and rereflect it in the forward direction; the major lobe is thus reinforced at the expense of the back lobe.

YAW: (1) The angular displacement of a moving body around an axis which is perpendicular to the line of motion and to top side of the body, (2) In robotics, the rotation (especially of the "hand") in a horizontal plane when the "arm" is extended horizontally.

Y-AXIS : The vertical axis on the screen of a cathode-ray tube.

Y-CONNECTION : See STAR CONNECTION.

Y-DELTA STARTER : See STAR-DELTA STARTER.

YIELD : In IC manufacture, the number of good chips obtainable per wafer. It is inversely proportional to the area of the chip, that is, as the area of a chip increases, the number of good chips on that wafer will decrease proportionately.

YIG : Abbrev. for *yttrium-iron garnet*. A ferrite that is widely used for microwave applications. The magnetic properties are altered by the amount of trace elements in the material. The most common trace elements are calcium, vanadium, and bismuth.

Y-MODEM : A method of transmitting data to another computer through modems. Transfer data in packages of 1,024K, which makes it faster than the X-modem transmission protocol.

YOKE : A piece of ferromagnetic material that is used to connect permanently two or more magnetic cores and thus complete a magnetic circuit without surrounding it by a winding of any kind.

Y-PUNCH : On IBM punched cards, a hole in the row nearest the top of the card; also know a 12 punch.

Y-SIGNAL : The component of a colour-encoded display signal representing luminance information. The signal produces a black. and-white image on a standard monochrome monitor. It is made up by combining specified fractions of the red (0.30), green (0.59), and blue (0.11) COLOR SIGNALS.

Y-VOLTAGE : See VOLTAGE TO NEUTRAL.

Z

Z : (1) In handwritten work, the letter Z is distinguished from a number 2 by means of a short bar drawn across the letter symbol for Impedance.

Z-CLIPPING : Defining the front and rear limits of a 3-D display to set planes parallel to the view plane.

ZEEMAN EFFECT : The multiplication of the spectrum lines of light sources when in a strong magnetic field.

ZENER BREAKDOWN : The application of a small reverse voltage (of up to about six volts) is sufficient to cause electrons to tunnel directly from the valence band to the conduction band (see energy bands; tunnel effect). At the Zener breakdown voltage a sharp increase in the reverse current is obtained. Above the breakdown value the voltage across the diode remains substantially constant. A type of breakdown observed in a reverse-biased *p-n* junction that has very high doping concentration on both sides of the junction. The build-in-field (see *p-n* junction) is high and the depletion layer narrow as a result of the high level of doping.

ZENER DIODE : Syn. *voltage-regulator diode*. A *p-n* junction diode that has sufficiently high doping concentrations on each side of the junction for Zener breakdown to occur. The diode therefore has a well-defined *reverse breakdown voltage* (of the order of a few volts only) and can be used as a voltage regulator.

ZERO BEAT : An expression used in radio-telephony, etc., for the state of affairs when a receiving circuit is oscillating at the exact fre-quency of the incoming waves so that no beat tone is produced, although other objectionable freatures of re-ra-diation and distortion may still be present; also called *Dead Space*.

ZERO CROSSING : is a point or line in a filtered image between two pixel groups, one of which is all positive and the other all negative, that can be used to determine edges in the original image when the filtering procedure is appropriate.

ZERO GRAPH : An early form of start-stop printing telegraph apparatus, designed by L. Kamm.

ZERO LEVEL : An arbitrary reference level used in telecommunication systems to compare the relative intensities of transmitted signals or of noise.

ZERO METHOD : Any method of measurement in which adjustment are made until no current passes and therefore no deflection is produced in a galvanometer, etc., or in which no change in deflection is produced by some alteration of connections.

ZERO OFFSET : On an NC unit, this feature allows the zero point on an axis to be relocated anywhere within a specified range, thus temporarily redefining the coordinate frame of reference.

ZERO POINT : The origin of a coordinate system.

ZERO POINT ANODE : An additional Anode in a Grid Controlled Mercury Vapour Rectifier connected to the neutral point, which modifies the wave form in a way which permits of better regulation and increased output with better efficiency.

ZERO POTENTIAL : An expression often used for the potential of the earth which is taken as zero for convenience in comparison. See EARTH POTENTIAL.

ZERO PUNCH : See OVERPUNCH.

ZERO SUPPRESS : To replace leading or insignificant zeros by blank, as in editing before printing.

Z-FOR-1 : Of a coded system in which eight bits represent alphabetic characters, and only four bits represent numeric characters.

ZIG-ZAG CONNECTION : A modified star connection of polyphase circuits in which each branch contains portions of two consecutive phases.

ZINC : A metal, atomic number 30, that is used as an electrode in some electrolytic cells.

ZINC-LEAD ACCUMULATOR : A type of accumulator in which a zinc negative electrode is used in conjunction with a lead peroxide positive of the ordinary type in an electrolyte of dilute sulphuric acid.

ZINC VAPOUR LAMP : A Discharge Lamp with a certain amount of zinc vapour in the bulb having similar characteristics to a Cadmium Vapour Lamp.

ZIRCONIUM : A metal, atomic number 40, that can be used as a getter in hard vacuum electron tubes.

ZIRCONIUM OXIDE LAMP : An incandescent lamp with ziroconium oxide incorporated in the filament.

Z-MODULATION : (of a cathode-ray tube) : See intensity modulation.

ZONE COMMUTATING AND NEUTRAL : See COMMUTATING ZONE and NEUTRAL ZONE.

ZONE LEVELLING : Method of zone refining in order to distribute impurities evenly throughout the bulk of the material. Zone purification is the application of zone refining in order to reduce the concentration of an impurity in a material.

ZONE REFINING : A method of redistributing impurities within a solid material, such as a semiconductor, by melting parts of the material and causing the molten zones to move along the sample. Impurities travel along the sample in a direction determined by their effect.

ZOOM : In computer graphics, continuously scaling the elements of display image to more clearly perceive and manipulate details not readily perceived in the previous view.

ZOOM BOX : A box symbol that appears in the right corner of some program windows. Clicking in the zoom box causes the window to expand to fill the entire screen or to contract to a smaller size.

Z-PARAMETER : See TRANSISTER PARAMETERS.